**PROBLEMS
OF ECONOMIC GROWTH
IN LATIN AMERICA**

PROBLEMS OF ECONOMIC GROWTH IN LATIN AMERICA

by
BÉLA KÁDÁR

ST. MARTIN'S PRESS · NEW YORK

© Akadémiai Kiadó, Budapest 1980
Revised edition of *Latin-Amerika gazdasági dilemmái*,
published by Közgazdasági és Jogi Könyvkiadó, Budapest 1977

Translated from the Hungarian by Pál Félix
Translation revised by Dávid Biró

All rights reserved. For information write:
St. Martin's Press Inc., 175 Fifth Avenue, New York, N. Y. 10010
First published in the United States in 1980
ISBN 0-312-64758-1

Library of Congress Cataloging in Publication Data

Kádár, Béla, 1934–
  Problems of economic growth in Latin America.

  Rev. ed. of the author's Latin–Amerika gazdasági dilemmái published in 1977.
  Bibliography: p.
  Includes index.
  1. Latin America–Economic conditions. I. Title.
HC123. K32 1979 330.9'8'003 79-17824
ISBN 0-312-64758-1

Printed in Hungary

CONTENTS

Introduction 7

PART I. TRENDS OF DEVELOPMENT

Chapter 1. The Main Phases of Latin American Economic Development 13

   1.1. The Heritage of Pre-Columbian Society 14
   1.2. The Heritage of Colonization 18
   1.3. The Period of Economic Liberalism and Semi-Colonial "Outward-Looking" 31
   1.4. The Period of "Inward-Looking" Growth 36
   1.5. The Special Features of the Development of Brazil 40

Chapter 2. The Main Trends of Development of the Latin American Economy 46

   2.1. A Comparative Analysis of the Growth Performance of the Latin American Countries 46
   2.2. Problems of Income Distribution 54
   2.3. The Development of Agriculture 60
   2.4. The Characteristics of Industrialization 63
   2.5. The Service Sector 67
   2.6. The Development of Economic Power Relations in Latin America 68

Chapter 3. Problems of the External Economic Sector 74

   3.1. The Main Development Trends in Latin American Foreign Trade 74
   3.2. Trends in the Commodity Structure of Foreign Trade 79
   3.3. Problems of Geographical Dependence 83
   3.4. Trends in the Flow of Foreign Capital 88
   3.5. The Role of Foreign Capital in Economic Growth 95
   3.6. The Latin American Attitude towards Foreign Capital 99
   3.7. Latin America and the World Economy after 1973 107

Chapter 4. Regional Cooperation in Latin America 113

   4.1. Some Theoretical Problems of the Integration of the Developing Countries . 113
   4.2. The Social Problems of Integration 119
   4.3. The Development of Regional Cooperation in Latin America 125
   4.4. Main Characteristics and Underlying Factors of Regional Economic Cooperation 133
   4.5. The Role of Small Countries in Latin American Regional Cooperation 142

PART II. ROADS OF DEVELOPMENT

Chapter 1. The Nationalization of the Decision-Making Centres in Peru 151

   1.1. The Armed Forces and the Model of Economic Growth 152
   1.2. The Socio-Political Aims of Peruvian Development 159
   1.3. The Balance of Economic Growth 163

Chapter 2. The Faltering Economic Growth of Argentina 172

   2.1. Declining Rate of Economic Growth—A Case Study 172
   2.2. The Heritage of Protectionism 177
   2.3. The New Economic Strategy in the Early Seventies 184

Chapter 3. The Brazilian Way of Accelerating Growth 192

   3.1. Scope and Principal Methods of Brazil's Development Strategy 192
   3.2. The Achievements and Drawbacks of Brazilian Economic Growth 201
   3.3. The Impact of World Economic Changes on the Economic Growth of Brazil 208
   3.4. Recent Trends in the External Economic and Foreign Policy Orientation of Brazil 214
   3.5. The Character of the Economic Growth Process in Brazil 217

Chapter 4. Colombia: Case Study of a Switchover from Import-Substituting Growth to Export-Oriented Economic Development 221

   4.1. A Short Review of Colombia's Development 221
   4.2. The Means of Realizing an Export-Oriented Economic Policy 226
   4.3. The Growth Achieved by the Switchover 229

Chapter 5. The Mexican Model of Balanced Economic Growth 238

   5.1. The Institutional Characteristics of Growth 238
   5.2. Characteristics of Economic Growth 244
   5.3. The Main Factors of Mexican Industrialization 248
   5.4. The Role of *Ejidos* in Economic Growth 253
   5.5. The Problem of External Economic Dependence 258

Index 265

INTRODUCTION

It has become a truism that Latin America is the continent most interwoven with contradictions and most exposed to socio-economic changes. The capitalist development of Latin America has recently undoubtedly gathered momentum, the more developed countries of the region have gradually extricated themselves from traditional underdevelopment and entered the group of medium-developed countries. However, on account of its capitalist character, this accelerated development has been rather uneven: there is an ever greater gap between the different sectors of the economy and also between the individual regions. Consequently, socio-economic antagonisms have increased not only between the working class and the peasantry on the one hand, and great landowners and the capitalist class on the other, but contradictions have sharpened also within the ruling classes themselves, moreover between national and foreign capital, as well as between the forces interested in national development and those striving to preserve imperialist subordination.

The problems of Latin America are generally thought to be identical with those of the relatively backward, dependent countries. This, however, disguises quite a few over-simplifications and leaves unanswered a great number of questions relating to the development of Latin America. Rodney Arismendi is right when he observes that the position of the Latin American continent is identical with that of Asia and Africa inasmuch as it constitutes an integral part of the historic forces which have risen against imperialism but, at the same time, it differs from the other two ones on account of the degree of dependence, the level of capitalist development, and the social structure, i.e. by reason of the characterictics of its historic development.

Latin America is distinguished from other continents of the developing world by a number of peculiar developmental characteristics. Because of the multiple synthesis of different socio-economic organizations, the region has a highly individual historical heritage. The developmental level of the productive forces is generally much higher than the average standards of the Afro-Asian countries, and the structural indices of the economy are by no means characteristic of those of the Third World countries. As a result of the development of commodity production—which has taken place historically much earlier than in other developing regions—capitalist development and relations of production have reached a considerably higher level than in the countries of Africa and Asia. Owing to earlier achieved political independence, more indirect forms of subordination and exploitation have been established. Another interesting distinctive feature is the fact that, as a result of the contradictions of accelerated and uneven development, a number of the region's countries have entered the period of violent clashes between antagonistic ideas and forces. Presently, Latin America represents one of the biggest "experimental laboratories" of historical development. There are some countries which have long-standing traditions of armed struggle for the elimination of the obstacles to development. Over the last ten years the efforts of the national bourgeoisie aiming at the realization of structural reforms speeding up the rate of development, and the attempts to reduce dependence upon the developed capitalist powers, particularly on the United States have gained momentum. All this takes place naturally within the limits of capitalist development. In recent years even the government of the United States and the policy-making top-managements of the multinational corporations seem to have realized that owing to accumulating tension in the Latin American region, the defence of American government and business positions by traditional means has become not only untimely but unfeasible as well. Thus, by taking into account the new conditions, they have begun to change their strategies. The progressive forces of Latin America have also started to devise an up-to-date and flexible strategy, and—taking into consideration the trends both in international and national power relations—they do not rule out temporary arrangements with the national bourgeoisie, whereas the termination of dependence and the elimination of the impediments to development are planned to be effected gradually.

The specific and manifold problems of the capitalist way of development of the region, as well as the diverse attempts to solve them, make Latin America a veritable repository of experiences. The development of the region renders it possible to draw valuable conclusions regarding the tactics and strategy of the revolutionary movement, the conditions, possibilities and forms of the struggle waged against imperialism and for social transformation. The heritage of the past is still evident in present-day problems and the seeking of ways and means for their solution in the individual countries is highly instructive. As a modifying factor, it is of great help in surveying the mechanisms of effectiveness of economic policies in countries of identical institutions and development structures. The period of switchover from protectionist economic development to the exploitation of the advantages of the international division of labour illustrates economic mechanisms. The struggle of the Latin American countries for the abolishment of one-sided dependence, especially for ending the subordination to the United States, assumes a peculiar tint under the conditions of the switchover stage in economic growth. This struggle illustrates the possibilities as well as the limits of progress from one-sided dependence towards mutual interdependence, the dangers of traditional dependence becoming a new type subordination, and also endeavours to keep decision-making centres under international control. From the point of view of practical policy, as regards the methodology of external economy, the positive and negative lessons drawn from efforts aimed at both using and restricting foreign capital are of considerable significance. As a result of the changeover to a more export oriented economy and to a more energic stand against dependence, the position of Latin America in world economy and world politics assumes new aspects. Thus, more favourable conditions are brought about for the broadening of economic, political and cultural relations between the socialist countries and Latin America. The problems arising out of the results of regional cooperation efforts among the countries of this region, first and foremost the problems of integration of countries with different developmental levels and economic potential, are by no means of merely regional importance.

Introduction

The specialists of economic history—even if they recognize the inherent limitations of historic similarities—cannot but notice various characteristics originating from the belated modernization, medium development, and peripheral economic position of Latin America which may be traced to a greater or lesser extent and in a modified form also in the growth and/or in certain stages of development in the Central and East European countries as well. The Latin American continent possessing immense natural and human resources comes more and more into the focus of international economic and political interest. The searching of ways and means, the experiences gained in the course of abolishing backwardness offer great possibilities of research not only to students of power politics and growth processes but also for those interested in the progress of socio-economic development. The present book undertakes primarily the analysis of the process and experiences of economic growth, partly with the aim of enriching our knowledge of the continent, and partly because it might promote the struggle waged by progressive forces for the elimination of dependence, the termination of subordination, the bringing about of new-type international relations and the avoidance of uncalled-for historic detours. As the direction of future development is still uncertain, the mere definition of the dilemmas of growth may prove useful.

The first part of the book acquaints the reader with the peculiarities of Latin American development and also with the present-day economic situation. In the second part the development of five Latin American countries with different socio-economic backgrounds is illustrated. The book focuses mainly on the characteristic features of the growth process, and it is from this point of view that some aspects of the socio-political sphere are treated as well. This volume contains the revised and augmented lectures of the author given at various Latin American universities and at the Spanish Language Department of Budapest University (ELTE) between 1968–1973.

Béla Kádár

PART I

TRENDS OF DEVELOPMENT

CHAPTER 1

THE MAIN PHASES OF LATIN AMERICAN ECONOMIC DEVELOPMENT

As a preliminary, we should exactly define what is meant by the name Latin America. Latin America is an extensive geographical region on the American continent, south of the United States, of about 20 million square kilometers. In spite of the unity of the geographical notion and the numerous common geographical features of the countries in this region, the name itself also indicates that in the course of history, i.e. the colonization, the European Latin countries exerted a decisive influence on the development of the countries situated in the region. The specific way of economic development in the Latin American countries, differing both from that of the northern regions of the continent and that of Western Europe can only be understood in the context of history.

The examination of the development of the region throws light on the fact that in the case of Latin America it is not simply capitalist development which diverts the patterns of long-term development as for instance in Central Europe, but the formations which had taken shape in the periods prior to capitalism; socio-economic differences left strong marks on Latin American development as a whole. Thus, in Latin America we are facing not only a belated and deformed variety of West European development but also a radically differing model of that development trend. True, from the end of the 18th century on the impulses received from the world economic centres which have emerged first in Western Europe and then in the United States have considerably influenced the countries of the region, however, the capitalist-type development superimposed from outside on the historically established basis brought about specific and highly differing types of growth in the individual countries.

Let us now consider how the present-day specific features of Latin American development have taken form, distinguishing this continent from the other developing regions of the world, and how the phases of economic and historical development have shaped the problems of this continent. The concept illustrated by a definition by the Brazilian historian P. Cambon seems rather a simplification and thus not acceptable. According to him, the history of Latin America is the superposition of different periods. By reason of this school of thought, colonial feudalism was superimposed on traditional Indian society, and was later followed by a more backward form of capitalism. As a result of political compromises characteristic of this continent, every new superimposition left its own foundation virtually untouched. If this oversimplified view reflected reality the remedy would be rather simple: everything could be solved by a mere removal of the negative elements inherent in the individual superpositions, by going back to the early beginnings devoid of any deformations and/or by an undistorted reconstruction of the original structures following the example of the contemporary developed countries.

Therefore, it seems justified to demonstrate that on the one hand, the development of the Latin American countries has been a specific one, not characteristic of the contemporary developed nations, and on the other hand, this historical development should by no means be regarded as a mere superposition but as a sequence of long-lasting, irreversible, highly correlated changes. The specific character of this historical process supports the view that present-day problems can be interpreted and solved expediently only by taking into account the characteristics brought about in the course of historical progress.

1.1. THE HERITAGE OF PRE-COLUMBIAN SOCIETY

It would be a mistake to apply mechanically the viewpoints used for examining the evolutionary processes in pedagogy, psychology or medical sciences to the analysis of society. Researches made, however, as to the origins and past of society and economy may help us throw light on quite a few of our present problems. Latin America had joined in the universal history of mankind actually only with the beginning of the era of colonization. By the time of colonization the developmental level of Indian socio-economic organizations and the number of the Indian population had been an established fact. However, the influence of this society on contemporary Europe (new products, new knowledge, the monetization of the economy by the flow of precious metals, the acceleration of the unfolding capitalist development, etc.) logically leads to the assumption that the so-called Pre-Columbian socio-economic, organizational and behavioural forms which had taken shape prior to colonization influenced to a certain extent the further course of development on the continent. Let us review briefly the characteristic features of Pre-Columbian society, its heritage (the existence of which has been denied by generations of historians), taking naturally the risk of certain hypothetical statements following from our limited knowledge regarding Latin American evolutionary processes.

The original peopling of the Latin American continent, the appearance of organized society is presumably connected with the main direction of the migration of population and with the geographical conditions. Earlier assumptions regarding the external, i.e. Asian and Oceanian, origin of the American Indian population are today substantiated by both archeology and linguistics. Immigration across the Bering Strait and/or the Pacific had been of a north-south direction, and settlement had concentrated virtually on the Pacific coast. Under the circumstances of equatorial climate conditions, flora and fauna of the valleys of the high mountain ranges proved to be most advantageous for human conditions. Thus, pre-colonial Indian society reached its highest level of development in the valleys of the Central American Cordilleras and the Andean ranges (i.e. the territory of contemporary Mexico, Guatemala, Peru and Bolivia).[1]

Considerable amount of data is available as to the developmental level of the productive forces of Pre-Columbian society. The highly developed agrotechnics (even if compared to that of 15th-century Europe) which served as the economic mainstay and foundation of Indian societies, the high standards of architecture, ceramics, medical science, mathematics and arts have started

---

[1] F. Katz: *Vorkolumbische Kulturen*. München, 1969.

a heated debate in the last century about the definition of the developmental stage of Indian organizations, and this debate is still going on. The heart of the matter is the problem that, on the basis of the development of their productive forces, the highly cultured Aztec, Maya and Inca societies cannot be classified under any of the scientifically established evolutionary categories.

In his work on the origin of the family, private property and the state Engels indicated the place of the Indian societies—on the basis of Morgan's schema—at the middle stage of barbarism, i.e. on the level of tribal associations or at best of military democracy. However, owing to the well-known high level of the development of productive forces, the developmental level of Indian societies was much higher than indicated by the characteristics of barbarism, not to mention facts like the highly centralized strong state and the establishing of a power mechanism which could by no means be regarded as democratic. At the same time, Indian social formations did not fit the criteria of slavery inferring private land property and antagonistic classes. Neither could they be placed—similarly to Asian and African formations—within the framework of the Engels—Morgan system of evolutionary progress.

Mariategui[2] thinks to discern a settled social organization of agrarian character developing towards socialism, whereas Baudin sees an abstract kind of communism,[3] Anavitarte patriarchal slavery, while Choy and Lumbreras[4] discover imperialistic societies in the Inca and Aztec formations. According to recent views based on the Marxian development pattern (Métraux,[5] Roel,[6] Godelier[7]), the characteristics of the Asian mode of production are fundamentally determinative, and it is assumed that if no colonization had occurred, these societies, similarly to the Chinese, Japanese or Vietnamese models, would have reached feudalism without going through the stage of slavery. However, history cannot verify the truth of these assumptions: it is impossible to prove that these Indian formations would have developed according to a certain model, if there had been no external intervention. It cannot be ruled impossible either that a peculiar transition stage would have led to some kind of independent Aztec or Inca mode of production, or even that it reflected the characteristic features of such a mode of production.

The spread of the Asian mode of production is to be explained partly by the consequences of the peopling of the continent. The immigrants from Asia brought with them their traditional customs, resumed their ancient way of production, followed their rites, maintained their habitual social structure and political system. Thus, a cultural and social organizational form originating from another continent was established on the American continent and became the foundation of subsequent American societies. The contents and origin of this culture are verified by the similarity of the ethnological and anthropological elements of the ancient American and Asian cultures.[8]

It would not be superfluous, however, to take notice of some distinctive features as compared to the Asian societies; the effects of these made themselves felt on the later stage of development, too. The cultural progress and socio-

---

[2] J.C. Mariategui: *Siete ensayos*. Lima, 1927.
[3] L. Baudin: *L'empire socialista des Incas*. Paris, 1938.
[4] L.G. Lumbreras: *Evolución de la Sociedad Andina*. Lima, 1938.
[5] A. Métraux: *Les Incas*. Du Seuil, Paris, 1962.
[6] V. Roel P.: *Esquema de la evolucion económica*. Amauta, Lima, 1972.
[7] M. Godelier: *El Modo de producción asiatico*. Eudecor, Cordova, 1966.
[8] E.H. Gerol: *Dioses, Templos Y Ruinas*.

political transformation of the Indian social formations established in the western mountain ranges cannot be understood without taking into consideration their geographic as well as their economic positions. This socio-political organization and culture came into being always in close connection with their environment, i.e. the effects of soil, climate and altitude. The geographical environment left its marks on the characteristics of material and social life. Natural and geographical conditions account for the fact that adaptation to the requirements of farming in high mountains (strip cultivation; irrigation for lack of waterways; the building of an overland road network; production specialization due to terrain configurations and varied climatic conditions as well as the division of labour deriving from the latter) and the surmounting of nature's forces made common actions and the developing of the necessary organizations inevitable. The natural and geographical factors have certainly been instrumental in rendering *social cooperation more intensive, stable and lasting up to the present time* than in the case of Asian and African formations based essentially on big river systems and a homogeneous climatic and geographical environment.

In face of the natural forces in the Cordilleras and the Andean mountain ranges, individual small-holdings and private enterprise did not prove viable. The setting up of the Mexican *ejidos* or the Peruvian *ayllus*—economic units including a family society called into being by the merger of several families and serving as the framework for collective work—has been made necessary by adverse natural-geographical conditions. Here, the social working-system has necessarily taken the form of collective and compulsory communities. Since common efforts were of an equal nature and of equal value, the rights and shares of all members had also to be equal. Thus, the social contents of these communities had been an all-embracing egalitarianism, and subsequently this served as the main argument in favour of the newly discovered "Inca socialism".

The nature of the foundations of the Aztec and Inca states, as well as the increased centralism of the state superimposed on the communal organization—in connection with a cooperation of increased intensity—can be explained by a necessarily higher intensity of social cooperation.

The diversity of the ecological factors throws light upon the differences between the Afro-Asian and Indian social forms. Though of fundamental importance, the work organization of highland irrigation had been but one of the components of Indian societies. Marx already indicated that neither the Aztecs nor the Incas could be regarded as primary societies as their empires had come into being in the wake of the subjugation and assimilation of several former Indian tribes and cultures.[9] According to Marx, the Asian form does not necessitate conquest, whereas in the case of the Indian societies, according to Wittman,[10] conquest can hardly be separated from the origins and development of the state. The function of conquest is a further distinctive mark as compared to the Afro-Asian formations, and the peoples and cultures assimilated as a result of it have made valuable contributions to the Inca-Aztec-Maya models by introducing new characteristics into these cultures.

---

[9] K. Marx: *A tőkés termelés előtti tulajdonformák* (Pre-Capitalistic Ownership Relations). Budapest, 1963.
[10] T. Wittman: "A latin-amerikai 'feudalizmus' kialakulásának vitás kérdései" (Problems of the Rise of Latin American "Feudalism"). *Századok*, 1972/6.

It would be not to the purpose to start a debate on the quantitative definition of the importance of conquest. As a result of a more developed cooperation there undoubtedly accumulated enough surplus product within the Indian formations to create a relatively strong state organization. This state organization in turn broadened its own economic basis by conquest, expansion, and partly by organizing the division of labour between the conquered territories and the centre.

Gerol[11] refers to the fact that almost in no other civilization had the head of the state wielded such an almost unrestricted power and enjoyed worship due to divinity as the rulers of the Andes. In addition to their unlimited power of disposing with the repressive organizations and with the economies of their empires, the fact that they were the connecting links with the divine powers ensured the rulers absolute control over the minds of their subjects and established their ideological monopoly. All the key positions of the army, administration and clerical system were held by members of the royal families.

Within the framework of economic organization the state specified the minimum targets of production, took charge of the equipment necessary for the maintenance of productive capacity and also of irrigation and the supply of manure, calculated the number of labour force needed for public works, disposed of surpluses, prevented the accumulation of wealth regarded as pernicious, and held the monopoly power of settling simple exchange among the different geographical regions. The absolute personal power of the ruler, the full centralization of decisions, the lack of any alternative power-holding institutions and their inherently reduced capacity for rivalization, the decisive role of the state power in all economic and technological matters—these were the characteristic features of the organized Pre-Columbian society. It should be mentioned that contrary to levying produce rent as customary in the Asian model, the Indian system was based on labour rent (compulsory participation in public projects, tillage of the state-owned lands) which left a lasting mark on the socio-economic organization and from the very outset hindered the development of market relations.

In conformity with the above characteristics, there evolved behavioural norms, a social scale of values and national psychological traits moulded by the geographical environment and the social relations, the main characteristics of which were the following: a marked collective spirit, readiness for cooperation, strong discipline, obedience and submission to higher authorities, the entire lack of utilitarian views, accompanied by a very low degree of initiative and capacity for independent decision-making, as well as an introvert and rigid personality type.

Thus, collective farming, the powerful economic intervention of the state and the centralized socio-political structure ensured a relatively high level of the development of productive forces in the case of the Inca mode of production. Due to different geographical and other factors, this socio-political and economic system was rather *self-sufficient, "inward-looking" and static*. Its economic growth was primarily of an extensive nature. The above main characteristics serve also as a key to understanding the fall of these formations. The possibilities of extensive growth, further conquests and "rejuvenation" had been exhausted as early as the beginning of the era of colonization, whereas economic and technological development received no impetus what-

[11] E.H. Gerol: *op. cit.*

soever. The centralization of power rendered it possible that after having destroyed the peak of the power pyramid, the conquering Spanish forces were able to subdue the historic power organizations without any serious resistance.

From the point of view of further analysis, it is of great significance that only the power centres had been destroyed by colonization, while Indian social forms proved highly resistant, and succeeded in handing down numerous of their characteristics to subsequent stages of development.

The tenacious resistance power and transmitting capacity of Indian social systems can be explained by several factors. Tőkei[12] points out that formations "in which the individual does not become independent as against the community, put up much stronger resistance against disintegrating factors as for instance the Germanic form". Marx calls attention to the fact that it is much more difficult to split up formations in which production has preserved its self-sufficient nature and the unity of agriculture and craftsmanship. Finally, let us add that the relatively small number of the conquerors, the lack of the so-called 'demographic cement' has also contributed to the fact that in the course of ethnic merger beginning with the colonial period the Indian population of an estimated 15 million[13] was able to hand down to a large extent its original values and national psychological traits.

1.2. THE HERITAGE OF COLONIZATION

The recognition of the historic role played by Pre-Columbian social formations, particularly the highly topical problems of its vestiges in the mainly Indian populated countries, naturally does not explain either the development of the complex historical characteristics of Latin America or the specific problems deriving from it. The Mexican revolution, the Peruvian APRA, and even the present military government in Peru made immense efforts in order to return to the ancient sources of Pre-Columbian morale and ideology. History, however, is an irreversible process, and the results of four centuries cannot be regarded as a mere historic by-pass. One of the most specific features of Latin American development is that the social formation of the colonizing power has been adopted to such an extent that is unknown in any of the African or Asian countries. Thus, in the light of historic development, the efforts aiming at "the return to the ancient Pre-Columbian sources" are highly questionable. The origins of modern Latin American historical development are to be sought for most appropriately in the structures which have evolved under the impact of the actual conditions of colonization. These concrete conditions, in turn, were none other than the natural-geographical and social relationships of Latin America, as well as the peculiar socio-economic features of the colonial power.

Quite a few experts, setting out from the general patterns of the process of colonization, trace the specific features of Latin American development back to the backwardness of the colonizing powers (Africa begins at the Pyrenees) or to the fact that no industrial revolution had taken place in those

---

[12] F. Tőkei: *A társadalmi formák elméletéhez* (On the Theory of Social Formations). Budapest, 1971.
[13] A. Rosenblatt: *La poblacion indigena.* Buenos Aires, 1954.

countries. According to certain views, the present underdevelopment of Latin America is due to the base motivation of colonization.[14] The colonization of North America for instance, is to be explained by the religious and political non-conformity of the first colonist generations, that of Latin America is rooted in the conquistadors' greed for gold. To their mind the evolutionary peculiarities of Latin America and also the developmental differences between the northern and southern parts of the American continent are explained by the qualitative difference between the human stock of the two colonist strata. Some expert views refer to the burdens of Spanish feudal heritage, whereas others (Haring,[15] Arciniegas,[16] Ots Capdequi[17]) emphasize that the Spanish colonization of Latin America had been essentially the most outstanding individual venture of capitalism realized by enterprising personalities. It was their heritage which brought about that sort of anarchistic individualism which became the strongest characteristic of the Spanish conquistadors. The thesis of A.G. Frank[18] (expounded earlier by S. Bagu) has won great popularity in circles of the new left. According to him, original accumulation of capital, capitalism itself had been transplanted to Latin America by Spain in the 16th century, and thus, from the very beginning, the continent took the capitalist road of development. Underdevelopment, therefore, cannot be blamed on feudalism because the very regions which have made greatest progress on the road of mercantile-capitalistic export orientation are the most backward ones.

All these views paid little attention to the fact that in the 15th and 16th centuries Spain had shown a great number of characteristics divergent from the well-known West European pattern of historical development. We are not going to subject the Spanish model of development to a thorough analysis, nevertheless it would seem expedient to refer to some of its divergent features.

Historians generally share the view that the Iberian Peninsula had been one of the most profoundly "Romanized" provinces of the Roman Empire in cultural, legal, political as well as sociological respects.[19] However, the province was conquered and Romanized only in the late period of the Republic, at a time when the political and social organization of republican Rome was already on the downgrade. Concurrently with its becoming a world empire, there appeared more and more forms of control and government adopted from eastern societies, and thus, power became centralized increasingly in the person of the emperor, and the official civil bureaucracy (embracing ever wider fields of socio-economic life) strengthened its positions. In the last period of the Empire, after Christianity had become a state religion, and later, during the stage of the Byzantine model, a peculiar dualistic bureaucratic system came into being where the administrative bureaucracy was partly consummated, partly controlled by the church bureaucracy wielding an ideological monopoly. The late Roman or Byzantine model accomplished a highly integrated political and ideological power where the ruler exercised extreme

---

[14] A.W. Shanahan: *South America*. Methuen and Co., London, 1940.
[15] C.H. Haring: *The Spanish Empire in America*. Oxford University Press, New York, 1947.
[16] G. Arciniegas: *America, Tierra firme*. Ercilla, Santiago, 1937.
[17] J.M. Ots Capdequi: *El estado español en las Indias*. Fondo de Cultura Económica, Mexico, 1941.
[18] A.G. Frank: *Capitalism and Underdevelopment in Latin America*. Monthly Review Press, 1967.
[19] R. Altamira y Crevea: *Historia de España*. Liberia de Juan Gili, Barcelona, 1900–1902.

(cesaro-papal) rights of presentation. In the third century A.D. (i.e. under Emperor Diocletian) the Roman state, following the eastern pattern, played an ever-increasing role in the control of economic life, aiming not at developing production but rather at intervening in the regional distribution of incomes. Polányi[20] and other authors have pointed out that, under the conditions of this early interventionist economy, there could not develop any real markets, market institutions or market theories, because economic activity was decisively influenced by "macroeconomic" considerations.[21]

In Western Europe the Roman model was destroyed by the great migrations and was fused with the Germanic formation as described by Marx. Feudalism—based on contemporary production forces and reflecting the power balance between the relatively weak central power and the relatively strong local powers—gradually emerged and strengthened its positions.

On the Iberian Peninsula the Germanic invasion was delayed by the Pyrenees. After the fall of the West Roman Empire the southern part of the peninsula came under Byzantine rule. The Visigoths who conquered the central and northern parts of Iberia, arrived there already as the 'most Romanized' Germanic tribe.[22] They settled in rural districts and virtually left untouched the cultural and economic system of the town centres established under the rule of the Roman Empire.

Though the Iberian Peninsula was under Arab domination from the 8th to the end of the 15th century, the Arabs displayed far-reaching tolerance towards the historically developed social formation, but at the same time they consolidated the elements of the Oriental formation within the Spanish pattern. Under the Arab rule earlier economic relations with the more developed, i.e. Near Eastern part of the world economy were maintained and on the eve of Modern Times the social distribution of labour, as well as the standard of urbanization was higher on the Iberian Peninsula than in feudal Western Europe the economy of which was based on primitive agrarian economies.[23]

It is of no avail to argue whether the above developments would have made possible to reach a higher degree of social development by-passing the stage of pure feudalism. After endless wars lasting some seven centuries, the *reconquista* which was launched from the more backward northern parts of the peninsula liquidated the Arab rule. Due to the distribution of economic goods and servitudes of the regions reconquered from the Arabs, an ever growing part of the nobility's power position originated from the crown and not from its hereditary estates. An increasing part of the nobility became military nobility and civil servants. By playing off urban centres against the nobility, by stabilizing the positions of the professional bureaucracy, and by disposing of the goods of the reconquered regions, the central power secured the liquidation of all its institutional rivals.

The economic position of the power centre was further stabilized by the fact that, owing to the royal monopoly of wool exports, the main growth sector of the economy of medieval Spain, i.e. sheep-farming highly organized by the Mesta, was also connected with the interests of the power centre.

---

[20] K. Polányi–C.M. Arensberg–H.W. Pearson: *Trade and Market in the Early Empires.* Glencoe, Illinois, 1957.
[21] M. Weber: *General Economic History.* Glencoe, Illinois, 1950.
[22] J. Ortega y Gasset: *España invertebrada.* Revista de Occidente, Madrid, 1946.
[23] A.G. Frank: *Capitalism and Underdevelopment in Latin America.* Monthly Review Press, 1967.

It is true, however, that the trend of vigorous state intervention (*dirigismo*), much stronger than anywhere else in contemporary Western Europe, proved to be rather unfavourable from the point of view of the growth process. The central power strived to cream off the surplus value of the most dynamic economic activities, e.g. wool-production and trade, by methods little short of ruthless exploitation. Instead of stimulating the productive branches, it controlled trade turnover by an extensive system of customs and taxes (*alcabalás, quintos*) thus, to an ever greater extent hamstringing the branches of economic life which were primarily responsible for growth.

By the end of the 15th century the Spanish socio-economic system was characterized by the marks of the late Roman-Byzantine model, as well as by a historically almost unprecedented degree of centralization.

Thus, as early as the last stage of the Roman Empire, the development trends of Spanish history considerably diverge from those of contemporary Western Europe. This tendency, strengthened in a peculiar way by the great migration, the Islam, and also by the events of the Reconquista, preserved and reinforced the so-called Byzantine model. After the Iberian Peninsula had been reconquered in 1492, the Spanish monarchy established its rule without any military, political or economic countervailing forces.[24]

Though Tibor Wittman, an outstanding Hungarian expert on Iberia and Latin America derived the peculiarities of Spanish historical development from the characteristics of Hispanian feudalism differing from those of Western Europe,[25] he also pointed out the patrimonial character of the feudal Spanish state,[26] as well as the fact that the latifundia taking shape at that time were strongly linked to the Crown; he stressed the dominating role of state exploitation, and the subordination of the cities to the Crown. He also emphasized that the cities emerging in the era of Spanish feudalism did not come into being as a result of their struggle against the feudal liege lords and the breaking away of the towns from the agricultural system, but it was mostly the stimulating efforts of the state, the church and the foreign markets which helped their development. However, these incontestable processes described by Wittman should not be imputed entirely to the peculiarities of Spanish feudalism. Increased state exploitation, the parasitic character of the cities from the point of view of the economy were highly important attributes of the heritage of Oriental forms. It was not by chance that Marx[27] commented on the characteristics of the Spanish social system as follows: "...Absolute monarchy in Spain is rather on a level with Asian power forms than with those of the absolutistic states of Europe."

The Marxist model finds the source of the so-called West European course of development in the amalgamation of the antique and Germanic formations. The growth of West European feudalism, and later of capitalism, was based on the creation of privately owned land, the related social system of incentives, the extension of the social division of labour, as well as on the fusion of the antique and Germanic production, technical and legal forms. Such a process of amalgamation did not take place in Southern Europe. The Germanic formation

---

[24] M. Colmeiro: *Historia de la económica politica España*. Liberia de D. Angel Calleja, Madrid, 1863.
[25] T. Wittman: *op. cit.*
[26] J. Stanley–B.H. Stein: *The Colonial Heritage of Latin America*. New York, 1970.
[27] K. Marx–F. Engels: *Válogatott művek* (Selected Works), Vol. II. Budapest, 1963.

came in contact with the antique formation only in the northern regions of Portugal, whereas in Spain it hardly affected the Iberian version of the latter.

During the last centuries of the Roman Empire, simultaneously with the centres of gravity shifting towards the East, the Oriental model strengthened also in the Iberian provinces. If amalgamation is mentioned in connection with the development in South Europe, it is the meeting of the antique form and the Oriental model, which, in the case of the Iberian Peninsula, was promoted—in addition to the "late Roman" development—by the Arab invasion, and in the case of the Balkan Peninsula, by Slavic and later Turkish invasions.

Thus, the historic process of amalgamation in Southern Europe was rather different from that of Western Europe as early as in the migration period.

From among all the consequences of the different amalgamation processes only a few specific features of the long-term South European growth will be treated in the following passages, i.e. the significance of land-ownership, of the social division of labour, the role of the state and the effects of conquest.

The changing ownership relations of the antique model on the Iberian Peninsula reflect the weakness of the Germanic influence, and later, the effect of the Arab conquest. The Germanic community of land was not introduced, and the role of Roman villas, slaves and coloni was greater than in Western Europe. The Arab conquest extending over the major part of the Peninsula, and later the *reconquista* lasting seven centuries created a stratum of proprietors holding hereditary rights but depending heavily on the central power, as the distribution and entrance into possession of reconquered lands rested with the sovereign.

According to Marx, the development of Asian tyrannical rules was a natural course since peasants who had the means of production in their hands, paid land rent as subject to the state, thus realizing the commonly owned land. In Spain, land acquired through the *reconquista* was distributed by the King for services rendered. The army nobility and the coloni who were given land faced, unlike the West European feudal forms and similarly to the Oriental model, not private landowners but the central power. The formula *"soy el servidor del rey"* (I am the King's servant) often to be read in the literature of the Spanish golden age, reflected actual conditions of ownership and control.

Independence of the proprietor—the effect of which is still detectable in the national psychological character of Spaniards—had developed under interest relations which did not stimulate the proprietor to attain a maximum economic utilization of his property but to render maximum service to the central power, and thus, to acquire new property. Disinterest in economy and almost a thousand years of continual warfare gradually wore out the lower nobility and agriculture, so that concentration of landownership and low economic standards, as well as negligible interest of extensive big estates became general by the 18th century.

The social division of labour and the changing relations between towns and villages were closely related to the development of ownership relations. West European development was stimulated by the fact that part of the population was turning its back on agricultural production: the developing towns were becoming centres of trade and industry, and supplied an expanding market for agricultural products; social division of labour was growing incessantly. In South Europe, however, both a mechanism for stimulating the expansion of production and the social division of labour were missing or were but rudimentary. In the South European towns therefore it was not the economic but the military and administrative functions that came to the fore. The number of townspeople was inflated disproportionately by the precariousness of existence, the almost permanent state of war, and pauperism in the villages. Spanish and Portuguese historians often make reference to a higher degree of urbanization in the 15th and 16th centuries than in Western Europe, but these towns—similarly to East European urban development—were not the products and did not become the stimulants of an increased social division of labour. The most important economic functions of towns were related to foreign trade or to a comprador-type domestic trade. However, those engaged in trade were generally foreign or minority groups alien to the central power as well as to the direct socio-economic environment, constituting characteristic social, ethnical and cultural enclaves. The enclave character of trade had an important part in the fact that the accumulation of commercial capital has not been transformed into industrial capital accumulation. Due to its being ethnically alien, it could not acquire political power, and thus, the economic functions of towns remained lastingly at the level of trade.

The level of trade in all Oriental formations is low anyway. The underdevelopment of the social division of labour, as well as the interior isolation of the mode of production entails necessarily *exterior isolation* and an underdeveloped international division of labour. Isolation was considered the main reason of Oriental stagnation by Marx, too. It was not by mere chance that among the South European countries it was Portugal, the country least exposed to Eastern effects and most exposed to Germanic influence, where the joining in the international division of labour started soonest and developed in the most intensive form.

One of the mainstays and foundations of the Oriental formations was the economic and social role of the state. The main sources of the central power were the concentration of landed property in the hands of the state, expropriation of the surplus product by the state, subordination of public works to the authority of the central government, and a class division depending on the central power. The greatest contradiction of Oriental class society founded on tribal common property was deduced by Marx from the role of the state as the representative of common landownership, and behaving as the proprietor of the means of production as well as a regular exploiter.

On the Iberian Peninsula the most important factors of centralizing state power were the late Roman and the Arab influences, the state of war lasting for centuries, the *reconquista*, the developing foreign trade and the colonizing expansion of Portugal which had started to develop earliest since overpopulation became manifest.

The powerful part of the state was justified, however, not solely by ownership relations and work organization, but conquest, too, had a significant part in it. Spain had pursued a policy of conquest and colonization for almost a thousand years, and had striven for world rule for centuries. Putting the energies of society at the service of exterior expansion consumed growth energies and hindered the emergence of the intensive factors of growth, and established a military and administrative bureaucracy which was parasitical from the point of view of growth.

Let us summarize the pre-capitalist development features of the Iberian countries that are common in their evolution. Of outstanding importance is the difference between the static character of the isolated Oriental models and the dynamism of the West European model receiving increasingly strong impulses from the national and international division of labour as well as from changes in technology. A difference still valid developed in the overall management and interest system of society. In the West European model the main driving force was profit, and its framework was private enterprise. In the Oriental model, i.e. in countries imbued with the elements of Oriental formations, the institutional framework of development was provided by the state, and thus the acceleration or structural transformation of development could receive stimulation only from above, i.e. from the central power. All changes in the economy started out of the power sphere, and it was neither the division of labour nor any market incentives that played determinant roles.

Owing to the indifference of the overwhelming majority of the population towards power and the underdeveloped states of economic incentives, power mechanisms or moral incentives (titles, ideological elements) had a much more important part in the operation of the Spanish model than in the long-term historical development of Western Europe. In a social environment where the source of authority and wealth is state power, energies of the dynamic elements of society are directed, quite naturally, at assuming first of all political power or at finding some means of cooperation with the state, and not at economic activities.

The social efficiency of the centralization of power is varying according to historical periods, and thus the evaluation of usefulness remains an insoluble problem. It is, however, illustrated by the main characteristics of development that the chief characteristic of Spanish development is by no means feudal backwardness but the incomplete nature of the evolution of feudalism. To cite Ortega[28]: "As regards Spain, the problem is not her being feudal, but her being not sufficiently feudal." To the school of historians reared on the experiences of the West European model all deviations from this model seem to be deformations. Yet, if we are objective, it becomes clear that Spain on the threshold of Modern Times witnessed not some kind of feudal distortion but a social development different from the West European model.

The criteria of the Spanish pattern of development explain the peculiar traits of *Latin American development*. The Chilean historian C. Veliz states as follows: "... the explanation of the historical development lies in the deep-rooted and characteristic centralist traditions of Latin America, this being also the pivotal question of understanding the future of our continent."[29] The dominant model in the colonized regions was quite obviously that of 16th century Spain, yet—compared with the metropolitan country—the ideal of the Spanish model played an even greater role in the life of the colonies. Thus, centralism taking shape in the colonies as the heritage of the ancient Mediterranean model, grew much stronger than in the metropolitan country. It is mentioned by Haring[30] that 16th century Spanish colonial policies among all the colonial ventures of Modern Times show the greatest resemblance with the colonial policy of ancient Rome based on military and institutional means. This similarity, however, is not a simple superficial phenomenon, but an expansionist activity following logically from a similar social formation.

[28] J. Ortega y Gasset: *op. cit.*
[29] C. Veliz: "Centralism, Nationalism and Integration." *Studies on the Developing Countries.* No. 25, 1969, p. 4. Published by the Afro-Asian Research Institute of the Hungarian Academy of Sciences.
[30] C.H. Haring: *op. cit.*

Despite of enormous geographical distances and transport difficulties, the power system developed in the colonies was characterized by a highly intensive centralism unparalleled at the beginning of Modern Times. The degree of *centralized public administration* is well illustrated by the fact that even the low-ranking officials of the colonial administration received their appointments directly from Madrid. The dependence of the administrative system on the capital used to be almost absolute in all personal matters, the decisions were made without exception in the metropolitan country, thus rendering local autonomy a negligible factor. When searching for the institutional sources of Spanish democracy, the famous liberal historian and philosopher, Salvador de Madariaga stated that the town councils *(cabildos)* represented democratic social organizations, nevertheless he fails to mention that members of these councils were appointed by the Crown or by the heads of the administration, whereas still later, offices became matters of business transactions. Thus, by the end of the colonial era, all offices had passed into the hands of the wealthy classes.[31] The *cabildos* can in no respect be regarded as representatives of city pressure groups, similar to those developed in the course of West European development as their tasks were limited to commonplace activities of city administration.

In addition to concentrating secular power in the colonies, the royal power obtained church control, the so-called patronate over the new territories, as stipulated by the Papal bull "Universalis ecclesiae regimini" issued in 1508.[32] The *ideological monopoly*, characteristic also of Spain, was further enhanced by the interpenetration of state and church. The possibility of religious nonconformism, the parallel development of different religions characteristic of West European and Central European development did not even arise as a problem. Right until recent times catholicism as a religion and a scale of values remained not only the dominant but the exclusive factor in Latin America.

The third element of the centralized Spanish model has been the *economy*. As a consequence of both the heritage of the Byzantine model and the practice of the *reconquista*, political power became the source of wealth. This phenomenon was even more pronounced in the age of colonization when the overwhelming part of the colonies' precious metals resources, metal mining, arable lands and even Indian population passed into the proprietorship of the Crown. The central power exercised foreign trade monopoly, controlled the development of colonial industries, domestic trade and finances, and transformed the West European interpretation of the social functions of property; at the same time, land property was divided among the conquistadors appointed as the commissioners of the central power. The degree of state interventionism exceeded by far the practice of European mercantilism (as a matter of fact, it did not originate in mercantilism). It is a well-known fact that West European economic development in the 16th to 18th centuries can be attributed but in a very small degree to interventionism practised by the central power. Economic development in the time of Spanish colonization of Latin America, however, had been entirely dependent on public interventionism.

In a peculiar way, the aim of economic interventionism was not the development of economic life, the securing of markets for the metropolitan bour-

---

[31] W.Z. Foster: *Outline Political History of the Americas.* International Publishers, New York, 1951.

[32] C. Veliz: *op. cit.*

geoisie, or the maximization of the growth of production but, in addition to providing for the safety of budgetary revenue, the performance of the defensive and security functions well known from the history of Oriental societies. In the name of the Spanish "global strategy" of that time, this policy set itself the goal of preventing any other power to gain foothold on this valuable continent, i.e. it strove to consolidate the colonial social system and to seclude itself entirely from the disintegrating effects of the external world.

Spanish colonization absorbed certain elements of the Pre-Columbian society, transformed them structurally, whereas elements which could not be integrated were physically liquidated or doomed to peripheric existence. As a result of these radical changes, there emerged a new, social organism.

According to Wittman, "...as the joint result of the development of the Pre-Columbian Indian societies and the Castilian colonization, there emerged a third, up to that time unknown economic and social structure".[33] Illustrating the unbelievably fast development of urbanization, he refers to the fact that between 1580 and 1610 the percentage of settlements with more than 500 taxpayers rose from 37 to 75 per cent.

The urban centres expanding rapidly in the course of colonization were superimposed on the former Indian centres and assumed a virtually administrative and commercial character. Owing to the competition of the goods from the metropolitan and the West European countries, the only function of industry was that of satisfying daily demand. Industrial production was based not on up-to-date European but on traditional Indian technologies. The rapid growth of urban population can be explained primarily by the fact that part of the colonial surplus value was used to finance the relatively high-level parasitic consumption of the colonialist population of the cities, while this high-level parasitic consumption resulted in a considerable demand for Indian labour force. Thus, there developed a concentration of population not stimulated by increased social division of labour or the advance of technology. The villages did not become the driving force of the structural development of the cities, and the relationship of town and village, industry and agriculture took a turn where neither of them could stimulate the other. Similarly to a number of Oriental formations, urbanization in Latin America had an unfavourable effect bringing about stagnation in both social and economic life.

The power colonizing this continent represented as an exception in Western Europe the factors of institutional conservatism, the historically most static and centralized late Roman-Byzantine model. This has left a lasting mark on the further development of Latin America. It follows from the application of this model under colonial circumstances that certain feudal elements could be transplanted only in a rudimentary form to Latin America. Feudal particularism became consolidated, oddly enough, exactly at the time when capitalism and liberalism started to unfold in Europe.

At the beginning of colonization, the distribution of landed property in Latin America took shape according to the Spanish model, not conforming to the characteristics of West European feudalism. The *encomiendas* are not related to territorial rights, and since they were not hereditary estates, their relationship with the Crown was not of a feudal nature.

The dominant role of the state in the distribution of land, however, has gradually weakened due to two exogenous factors. The economic and military

---

[33] T. Wittman: *op. cit.*

decline of Spain, a process which had been going on for two centuries, undermined the positions of the Crown in the colonies too, and gave an impetus to the emergence of alternative power factors opposed to the Crown. There ensued a struggle between the state and the landowner class for influence, authority, landed property and labour force. This process was increasingly furthered by the development of the world market, the sudden spurt of colonial trade, as well as the revival of West European interest in Latin American products and markets, going hand in hand with the penetration and the transformation of the whole system. Spanish power being of a rigid structure and protectionist nature, and the traditional Indian communities which were becoming increasingly its allies in the fight against the emerging class of big estate owners, could not play a part on the world market. This role had to be taken by the new-type big estates, the *haciendas*, developing beyond the narrow compass of the former *encomiendas*.

The *hacienda* is a peculiar product of the Latin American evolutionary process. Being a countervailing power developing parallelly with the weakening of central power, the *hacienda*, too, borrows from the synthesizing heritage of historic development. In its activities, similarly to the rival state power, economic, political, and ecclesiastic functions were merged in a complex way.[34] The internal work organization, the *peón* system, resembled in some of its aspects the conditions of slavery, and the autarchic character of the internal division of labour was linked with an export orientation of varying degree. The *hacienda* has been a form closed downwards and open upwards, inasmuch as it reserved the monopoly to market its economic products.[35] Following the War of Independence, the *hacienda* took possession of the political power, too, and its characteristics left their marks permanently on the further development of Latin America.

Let us summarize the specific features of the growth of the colonial institutional system: owing to the Neo-Byzantine colonial model and to its administrative restrictions, no comprehensive commodity market came into being. Likewise, there could develop no real labour market because of the use of forced labour, no real money and capital market because of colonial surplus skimming and no land market because of the regulation of landownership. Thus, no capitalist-type development could commence. In other words, in the Hispanian colonial model the Indian variant of the Asian mode of production has been alloyed with the Spanish model which itself had numerous elements of this very same Asian mode of production.

As a result of the administrative, ideological and economic centralism, in the development of the Latin American model during the era of colonialism, the parallel institutions of state power, political-ideological pluralism and an economy guided by market competition could not develop sufficiently. This complex centralism outlived the period of Spanish colonization and took root also in the capital of the new national states emerging after the War of Independence. The roughly hundred years of development from the period of the wars of independence up to the World Depression of 1929, the fact that the whole region came under the economic, ideologic and political influence of Western Europe, and finally, the homologizing effect of the era of liberalism— all this led to some kind of a break with the secular trends of development.

---

[34] P. Macera: "Feudalismo colonial espanol." *Acta Historica*. Szeged, 1971.
[35] Cited by Gy. Kerekes: *Kubától Chiléig* (From Cuba to Chile). Budapest, 1974. p. 68.

In a historical sense, the results of this developmental stage, however, were unable to strike deep roots and remained alien to the social, economic and ideological evolution of Latin America. The tendencies rooted in European historical development, i.e. industrialization, economic growth, institutions and political ideas, have not become integral parts of the Latin American scene. The socio-economic crisis beginning in the wake of the World Depression once again brought to the surface, sometimes in extreme forms, the specific features of historic development as well as the manifestations of economic, political and administrative centralism (protectionist *dirigismo*, military take-overs, ideologically oriented one-party systems). It is almost impossible to understand the characteristics of Latin American historic development and the present-day problems of the continent without seeing them in the context of history. Though all these phenomena are regarded as distortions and anomalies by English and American historians focussing on growth, their interpretation becomes rather simple if seen in the context of Latin America' s historical development.

After this short review of the institutional system of Spanish colonization, let us examine the problem as to what extent the appearance of Spanish colonialism in Latin America's social and natural environment influenced the long-term growth in the various regions of the continent.

The most important single motivating factor of Spanish colonization was the gold rush, as well as the lust for conquest and land acquisitions. From the point of view of the colonial power there were added several further elements to this, such as defence factors of global-strategic nature, military occupation of strategically important points of the region, and during the later stages of colonization the establishment of the productive as well as the export capacities of articles demanded by Europe (e.g. anil, cocoa, coffee, tobacco, sugar). As the resultant of individual and central objectives, the early colonial organization was superimposed upon the former Aztec and Inca power centres as it was here that the mines of precious metals were to be found, and labour force was also available for mining as well as for the colonists' necessary supply with food and retinue of servants required by the status and prestige of the conquerors. Thus, there emerged the Mexican and Peruvian viceregal centres built upon the former Indian empires.

The consumption and nourishment customs of the new colonialists necessitated, however, the acclimatization of new kinds of crops (wheat, rice, oat, sugar-cane) and animal breeds (cattle, sheep, pigs, goats, poultry). In Mexico and Peru which were situated partly in the tropical zone, this could not be solved and thus from the point of view of food supply of the viceregal centres the sparsely populated regions of the temperate zone became gradually more important (Chile and North Argentina).[36]

Several geographical units could be distinguished within the colonial economic system: first, viceregal centres; second, regions without mines, the economies of which were based virtually on agricultural self-sufficiency (present day Chile, North Argentina, Ecuador, Colombia, Venezuela, Central America, the Caribbean Islands); and third, the uninhabited, empty areas (the La Plata states of today) the occupation of which was effected solely for strategical reasons; their role in the early period of colonization was limited to the army and public administration. At the time of the development of the colonial

---

[36] O. Sunkel–P. Paz: *Desarrollo economico.* Ilpes, Santiago, 1969.

empire in Latin America, there evolved—in addition to the high degree of dependence on the metropolitan country—economic, political, cultural as well as military subordination and also a centre vs. periphery relationship between the Peruvian and Mexican viceregal centres and all the other areas of the region. In the viceregal centres a higher developmental level, an economic structure founded on mining, agriculture, rudimentary industries *(obrajes)* and trade was established. Here the percentage of the European population which had arrived not as settlers but as colonizers remained insignificant as compared to the population as a whole, while at the same time there ensued a strong ethnical mixing. The extremely unequal income structure (white European—Creole—Mestizo—Indian—Negro) is also somewhat related to this phenomenon.

The originally less populated regions which possessed no mines were initially characterized by a lower degree of development, an agrarian economy producing very little surplus being virtually self-sufficient, a much lower rate of ethnical mixing, more balanced income relations, and a stronger sense of local identity owing to the great distances from the viceregal centres. Due to the decay of the Peruvian mining boom becoming apparent from the end of the 17th century, and to the spread of mercantile policies, the formerly essentially self-sufficient tropical regions (mainly the Caribbean Islands and Venezuela) became gradually specialized in the production and export of tropical produces sought for on the international markets. Besides the traditional great latifundia producing mainly for self-supply and regional needs there emerged an export-oriented plantation system using slave labour. In the 18th century the formerly uninhabited areas of the moderate zone gradually became populated. Here the basic economy was animal husbandry, the surplus products of which were marketed from the middle of the 18th century in the viceregal centres.[37]

The development trends of the Spanish American colonial empire in the 18th century and the relationship of the region's individual areas were considerably affected by the partial exhaustion of the Peruvian precious metal deposits, the recession in mining, and the weakening of the external positions of the Spanish Empire. Due to the decreasing significance of the Peruvian mines making itself felt from the middle of the 17th century (a consequence partly of the exhaustion of the deposits, and partly of the primitive mining technology) a change took place in the power balance among the individual colonies, and the centre vs. periphery relationship within the colonial empire became less important. In 1739 the vast region comprising present-day Colombia, Venezuela, Panama was raised to the rank of the New Granada viceroyalty, while in 1776 the present Argentina, Paraguay and Uruguay were reorganized as the La Plata viceroyalty. Thus, there commenced the process of territorial disintegration and the taking shape of the later national states.

Against a steadily strengthening England demanding free trade and pursuing a policy of economic penetration in Latin America the declining Spanish Empire became less and less capable of defending its interests. It was especially in the former peripherical regions of the colonial empire, in the new viceroyalties that the policy of free trade and the smuggling of goods made ever greater breaches in the system of Spanish protectionism. Parallel with the decreasing economic significance of mining easily controlled by the central power, the positions not only of the groups interested in mining but also those of the

---

[37] V. Roel: *Historia de la economia colonial*. Lima, 1970.

bureaucracy in control of economic management were shaken. The positions of the great landowners, plantation owners, and ranchers interested in free trade gradually became stronger in relation to the typical institutional representatives of 16th and 17th century colonization (public administration, church, army).

In the last stage of the era of colonization, the conflicts between the new interest-groups forging ahead and the colonial administrative structure, and also those between the colonies and the metropolitan country have considerably sharpened. It was not by chance that the prime factors and bases of the Spanish colonies' independence struggle were New Granada and La Plata, the new viceregal centres, the former peripheric regions of the one-time viceregal centres which made fast progress in the last stage of colonization and had the closest contacts with the outside world. If, however, the War of Independence is examined from the point of view of the contemporary active social movements, the different tendencies striving for the overthrow of the colonial structures are easily discernible. As a result of European–American influence, the commercial bourgeoisie went to war not only to do away with the colonial past but also in order to liquidate the Pre-Columbian remnants maintained in or merged into the colonial system. At the same time, there also appeared those forces (first and foremost in Mexico) which wished to link the overthrow of the colonial system with the socio-political integration of the Indian population, moreover with the creation of an institutional framework required by the social and cultural heritage of the Indian population. These latter movements all foundered in the course of the War of Independence and their revival came about only in the historic period beginning with the unfolding of the Mexican revolution.

The dramatic history of colonization, the different character of the Hispanian colonial model as compared with North-American development, the extraordinarily strong foreign dependence usually makes us forget the fact that, as regards the development of production forces, the backwardness of Latin America had internationally by far been not so great as in our times. In the last years of the colonial era the production, foreign trade, urbanization and cultural level of the Latin American countries was considerably higher than for instance that of the United States.

According to W.Z. Foster, the exports of the 13 English colonies in 1783 did not exceed the five million dollar mark, whereas Latin America exported twenty-seven times as much. The countries south and west to the new-born American republic showed all the characteristics of mature economies: flourishing agricultural and industrial production, rich cities, highly developed arts and science, stately edifices as well as all the luxuries of the period. While Philadelphia could boast of only 84 coaches at the middle of the 18th century, Lima had 5,000, and Mexico City even more. Up to the 18th century there were no cobbled streets and pavements in English-speaking America, whereas in Latin America such roads had been in use for 200 years.[38]

---

[38] W.Z. Foster: *op. cit.*

1.3. THE PERIOD OF ECONOMIC LIBERALISM AND SEMI-COLONIAL "OUTWARD-LOOKING"

As a result of the successful wars of independence, the Latin American countries (with the exception of the Caribbean Islands) got rid of the monopolistic restrictions and exploitation imposed on them by the declining colonial power, and thus theoretically they were provided the opportunity for development in a new direction. The Creole landowner class, however, which came into power as a result of the wars of independence was interested at the very most in unfurling the flag of "free trade" but, at the same time, it was not willing to change the class structure of society and liquidate the pre-capitalistic production relations. Partly owing to their direct interests, and partly to their attempt to hand down the traditions of the struggle against the colonizing state, the new forces which had come into power sided with the idea of the liberal, weak state, and all attempts aiming at the establishment of a strong state were thwarted.

The victorious War of Independence resulted not only in terminating the subordination to the metropolitan country but also in the ending of dependence on the power sub-centres of colonialism, in the developing of new centres different from the traditional centre of growth, as well as in the more rapid spread of institutional forms and ideas characteristic of these new centres. In the wake of economic development there evolved specialization which in the long run influenced the socio-economic framework of growth. On the territory of present-day Mexico, Peru and Bolivia there developed mining, in the region of today's Argentina, Uruguay, Paraguay, Chile, Colombia and partly Venezuela animal husbandry and farming characteristic of the moderate zone, whereas in the Caribbean region tropical monocultures took root.

It follows from historical progress that adaptation to the system of political and economic liberalism, attempts to imitate the West European and North American development model proved to be most marked in Argentina, Chile and Uruguay. Here there took place considerable English economic penetration, Spanish and Creole elements were ousted from the foreign trade sector, and gradually there appeared more and more European immigrants.

In the former power sub-centres and/or in the countries the economies of which were built on tropical monocultures (in the latter ones the secession from the metropolitan country took place much later) the adaptation to the requirement system of laissez faire capitalism progressed rather within the limits of some kind of pseudo-morphologic evolutionary process, while the conservative forces put up stiff resistance. Following the attainment of independence, the conservative forces—comprising civil servants, the big landowners and the privileged traders' stratum—tried to preserve some elements of the colonial model by means of a mercantile-nationalistic policy. (It should be noted that on the strength of referring to the principle of national defense, industries supplying the armed forces developed within the framework of protectionism even in the period of liberalism.) However, common characteristic features of the development in the decades following the achievement of independence, were the compromise between realities and the traditional model, varying from country to country, and also the decisive and homogenizing role of exogenous factors of development (export orientation, capital import, immigration, adoption of foreign ideas and institutions). The termination of the administrative, trade and military role of the one-time viceregal centres brought about regional loss of positions, economic recession and a lasting political imbalance. In the countries with a tropical farming economy there survived the plantation system relying on the increased world market demand, but substantial changes were not effected here either.

In the second part of the 19th century, all Latin American countries were characterized already by the so-called "crecimiento hacia afuera" ("outward-looking" growth) type model, based on the joining the international division of labour. The magnitude of this process may be measured by the fact that, in the period between the middle of the last century and World War I, the value of Argentine exports increased more than 50-fold, and that of Brasil and Mexico 7-fold.[39] In spite of the homogeneous cause of this new stage of development the inequalities among the regions of the continent increased depending on the nature of specialization, the constancy of world market prosperity, the character of technologies adopted, the relations of ownership within the sector of external economy, and the strength of state power. Thus, there have developed the structural characteristics affecting even contemporary evolution in Latin America.

In the century following the attainment of independence, when growth was based on the policy of "outward-looking", it was Argentina, Uruguay and Chile, i.e. the least densely populated countries burdened with *the least colonial heritage*, which made the fastest development. The mass influx of European immigrants and capital, the adoption of West European institutions created favourable preconditions for joining the international division of labour. The nature of specialization connected with the favourable natural-geographical conditions also proved to be propitious for accelerated growth. The territorial demands of agriculture and animal husbandry led to the opening up of extensive regions, to the building up of an up-to-date infrastructure, whereas manpower shortage resulted in rapid technical development, a higher level of relative wage-standard as compared to other developing countries, and finally in the capitalization of economic life as a whole. Argentina and Uruguay progressed essentially patterned according to the British dominions (Australia, New Zealand and Canada), i.e. extensively from the point of view of making use of lands. Up to World War II, Argentina used to be called often "the fifth British dominion", and as regards *per capita* gross national product, accumulation and productivity, she has achieved a similar standard to that of the British dominions in the period between the two world wars.

The only deviation from the development of the white dominions was caused by the colonial heritage and by the big estate system which evolved in the era of colonialism. While settlers emigrating to the British dominions have been able to find physically and legally unowned lands for a historically long period, during the three hundred years of Spanish colonial rule, the propriety rights of the empty La Plata territories, though by and large left fallow from the point of view of economy, have been divided among the big landowners. Thus, for lack of unowned lands, the immigrants arriving in the La Plata region could not find any other alternative but to put up for sale their labour-power. Under these conditions agricultural development did not take the American but rather the Prussian course with a strong concentration of property rights and surplus value.

---

[39] W.P. Glade: *The Latin American Economies*. American Book, New York, 1969, p. 216.

The "input" necessary for the development of the agrarian sector (tools, machines, subsidiary materials, warehouses) is less import-intensive, and thus can be produced domestically, too. Therefore agricultural specialization also contributed to the diversification of the structure of production and services, to the early development of the manufacturing industries, to a relatively large-scale internal integration, as well as to a higher standard of social services and cultural development. The production foundations of the external economy (landed estates, livestock) virtually remained in national possession, whereas a considerable percentage of the manufacturing industry (big cold-storage plants, slaughter-houses), foreign trade and the connected services have come under the control of foreign interests. As a consequence of the great dependence on external market (especially on the English market) in the first thirty years of this century one third of the gross national product has been realized by export.

From among these type of countries it has been Chile which has shown different developmental characteristics from the last quarter of the 19th century. Mining industry which was very labour-intensive led to the concentration of the population around the deposits, to the emergence of an urban proletariat, and at the same time to an increased concentration of incomes due to the high mine rent. Though the copper boom replacing the salpetre boom—which had come to an end as a result of the appearance of synthetic salpetre— managed to uphold the dynamics of "outward-looking" growth in the twenties, the first disturbing factors made themselves felt already in the development of the national economy. Owing to the increased involvement of the North American big monopolies, the utilization of local materials, investment goods and manpower in the copper industry developed with capital intensive North American technologies remained negligible, and the greater part of the mine rent which was quite considerable because of the favourable extraction conditions, fell into the hands of foreigners. The key sector of the economy became "denationalized" and, as a consequence of the character of technology, it gradually became isolated, turned into an enclave and did not contribute to the growth of the domestic economy. And though prior to the World Depression Chile had been regarded as a country of higher income standard, there appeared an unfavourable trend in the development of the factors of economic growth.[40]

The progress of the *export-oriented tropical economies* taking shape in the Caribbean region present a markedly different picture. The production of sugar, coffee, cacao, tobacco and other tropical produce resulted in a considerable seasonal manpower demand. This demand had been met by the slave-trade in the 17th and 18th centuries, however, in the 19th century, in the wake of the liquidation of the slave import, there emerged, in addition to export-oriented plantations, the so-called minifundia providing for the livelihood of cheap manpower which supplied the big estates with seasonal labour force. At the middle of the century plantation farming based on slave labour switched over to making use of labour force of the agrarian proletariat concentrating in the neighbourhood of the big estates. The organizational framework of economy and society have been left untouched by the wars of independence.

Technologies of the export branches were modernized around the turn of the century, and at the same time there began the large-scale influx of North Amer-

---

[40] I.P. Pena: "Comparación de tres procesos globales". *El trimestre economico*, 1970 enero-marzo.

ican capital and technology resulting in an accelerated rate of the concentration of landed property, rural unemployment and pauperization. The streamlined export branches, however, exerted scarcely any influence on the development of other sectors of the national economy, their manpower and product demand remained negligible, the home market contracted relatively, and even the existent rudimentary handicraft and processing activities declined. The overwhelming part of the export branches was vertically seized by foreigners. The uncertainty of internal affairs, partly a result of increasing pauperization, social tension, as well as the ever growing political, military, etc. subordination to the regional great power, i.e. the United States, hampered the emergence of a strong state power, and the subsequent dictatorships did not prove strong enough to draw away the national surplus value realized in the export branches, from the foreign capital, or to effect a diversification of the production structure.

Finally, it is very illuminating to examine the "outward-looking" growth of the former *power sub-centres*. In South America the War of Independence had been also an armed conflict between the power centre and the periphery. The political chaos in the defeated viceregal centres entailed a considerable loss of territory. Upper-Peru (the present-day Bolivia) and Ecuador seceded from Peru, while her southern province was lost as a consequence of the war against Chile. Mexico was forced to surrender her sparsely populated northern territories (California, Texas, New Mexico) to the United States. The era of "outward-looking" was entered by both countries deprived of their growth energies and partly of their former territories. As regards the Peru of the 19th century, the only possibility of export extension has been the guano deposited all over the coastal islands. The relatively easy exploitation of the guano exerted virtually no effect on the other sectors of the economy, and provided only a source of revenue for the state budget. Given these circumstances the economic, technological and social development of the country has hardly been stimulated by the policy of "outward-looking". The guano boom lasting for about half a century made itself felt only in the slight development of the infrastructure and in the setting up of cotton and sugar-plantations in the coastal regions. Following the exhaustion of the guano deposits, there ensued a lasting socio-economic stagnation.

In the case of Mexico the loss of the role of power centre had by far not so unfavourable consequences. After nearly half a century of political anarchy and economic recession, a new export basis, that of mining and animal husbandry, has been established in the last thirty years of the 19th century. In the colonial period the modernization of mining was accomplished mainly by means of North American capital. Reliance on the near-by US market brought about a rapid and more diversified development, a considerable amount of infrastructure, and a modest industrialization, based on the processing of mining products. At the same time, however, it opened up foreign capital almost unlimited possibilities. The space requirements of animal husbandry resulted in the further increase of the already exorbitant property concentration, and also in the expropriation of Indian communal lands.

Due to the extreme policy of economic liberalism, by the turn of the century there came about a property concentration of unprecedented proportions in Mexico. Nearly all of the national wealth and the cultivated area were concentrated in the hands of three per cent of the population, and at the same time there appeared large-scale rural unemployment and pauperization. The Mexican revolution of 1911 was called forth by accumulated social tensions. As a consequence, the first land reform in Latin America was accomplished, and a constitution was passed which for the first time on the continent went beyond the limits of bourgeois revolution. The Indian village communities, the products of historical evolution, were restored, and a development model was approached which could be integrated in the long-range historical evolution. In the fifty years following the revolution, Mexico which had been the most developed country of Latin America in the early 19th century and one of the most backward ones at the beginning of the 20th, fought her way once again among the top-ranking states of the continent.

The above examples prove that the model based on "outward-looking", broadening markets, capital and technology imports, by being superimposed upon the colonial structure, brought about different results and development types, and increased the inequalities among the individual countries. A more dynamic capitalist development, though controlled from abroad, commenced in the formerly peripheric southern countries, which had no Pre-Columbian traditions and less colonial heritage. As regards *per capita* national product, these countries have fought their way into the international front-rank. The progress of the one-time viceregal centres and countries founded on the mining industry or tropical agriculture is much more unfavourable. Inheriting quite a few characteristics of the static, less dynamic Oriental societies, these regions progressing according to the Indio-Hispanic model were not prepared for the world economic relationships incidental to accelerated economic and technological development as well as increased social mobility, i.e. for "the economic environment of liberalism", therefore their development as a whole became much slower and rather distorted.

The development of all Latin American countries, and especially of those belonging to the latter category, was unfavourably affected by wrongly established institutional systems. The appearance of the ideology and practice of free competition resulted in the decline of the role of the state. This might have been the exaggerated reaction to the excessive historical secular-centralism, but might have also originated from the fact that, at the end of the 18th century, the concept of protectionist development flourishing for two hundred years began to decline in England, the new dominant power of the region and the growth centre of world economy. It was obviously largely under the influence of England that the 19th-century practical experiences of the European countries committing themselves, though belatedly, to the course of industrialization were only scarcely taken into consideration by the overwhelming majority of the Latin American countries in that century. In the non-Marxist literature on economic history the historical interpretation of the state's role in growth has been emphasized for the first time by Gerschenkron.[41] He pointed out that the growth of the relatively backward East European countries which belonged to the third wave of industrialization (taking place in the *laissez faire* period of capitalism) in the 19th century is distinguished from the West European pattern by vigorous state intervention. In Latin America, the imitation of the practical development policy of the most advanced country has become a serious obstacle to growth and the role of the state has weakened at the very time when the warding off the menace originating from approaching the world economy, the processes of complex modernization and moulding a nation would have demanded a stable and strong executive power.

---

[41] A. Gerschenkron: *Economic Backwardness in Historical Perspective.* Cambridge, 1962.

1.4. THE PERIOD OF "INWARD-LOOKING" GROWTH

The external preconditions of "inward-looking" growth were eliminated by the World Depression of 1929–1933, the drop in export prices (out of typical Latin American products the prices of wheat and rubber fell by 50 per cent, those of meat, leather and petroleum by 25 per cent), the narrowing down of markets and the drying up of the sources of capital import. The foreign trade consequences of the crisis can be seen from the UNO data in Table 1.

Table 1

Trade indices of some Latin American countries
(Changes on a percentage basis as compared to the annual averages of the period 1925–1929)

	Volume of exports	Terms of trade	Import capacity	Volume of imports
Argentina				
1930–1934	− 8	−20	−27	−32
1935–1939	−11	0	−11	−23
Brazil				
1930–1934	+10	−40	−35	−48
1935–1939	+52	−55	−32	−27
Chile				
1930–1934	−33	−38	−58	−60
1935–1939	− 2	−41	−42	−50
Mexico				
1930–1934	−25	−43	−55	−45
1935–1939	−11	−36	−39	−26
Latin America				
1930–1934	− 9	−24	−31	.
1935–1939	− 2	−11	−13	.

Source: ECLA, *Economic Survey of Latin America*, 1949.

The World Depression, however, proved to be not simply an economic crisis but assumed the proportions of a comprehensive socio-political crisis. The considerable decrease in the budgetary revenue based on foreign trade relations resulted in the weakening of the state power, while the collapse of economic balance in the upsetting of political balance.

The crisis affected all the Latin American countries, though to a different extent, and these reacted differently according to their historic development and to the extent of their affliction. Those ones which in the period of "outward-looking" growth have reached a higher degree of national economic *development* and diversification (Argentina, Brazil, Chile, Mexico, Uruguay and in a more modest way Colombia, i.e. out of the five countries of the former Spanish colonial empire four one-time peripherical regions) and have been in a more favourable position also from the point of view of socio-political development (national bourgeoisie, intermediate strata, more up-to-date political mechanisms) were compelled to switch over to a new source of growth, and have concentrated their resources in branches fit for import substitution.

As a result, instead of the international division of labour internal division of labour has become the main driving force of the growth of these countries.[42]

The economic, political, and social crisis brought about the modification of the political balance of power. In a number of countries the landowner-comprador bourgeois stratum, which as a consequence of the aftermath of the crisis established close ties with foreign powers and was highly involved in export transactions was obliged to hand over to, or share power with the representatives of the national bourgeoisie or intermediate strata interested in protectionist development. Due to the takeover which took place under the conditions of increased social tension, there emerged new political power centres much stronger than their predecessors and having a broader domestic foundation. These centres felt themselves strong enough to organize political power within the new framework of growth and to realize capitalist modernization. These socially heterogeneous political coalitions were linked by far-reaching compromises varying in contents from country to country in the different periods. The coalitions termed "populist" provided key roles and prosperity to the industrial bourgeoisie, positions in public administration to the intelligentsia, and a more favourable social policy to organized labour. The unity of the coalitions was often hallmarked by a dominant political personality (Vargas, Peron, Cardenas, etc.). Simultaneously with the switchover to import-substituting growth, a new social picture emerged which was marked by more or less interconnected elements of the strengthening social and cultural factors of "inward-looking". Liberalism, borrowed from West European development, considerably weakened in the course of the changeover from one growth period to the other, and there emerged ever more vigorously the outlines of the revival of traditional centralism.

The coming into prominence of the economic role of the state, protectionist customs and trade policy, foreign exchange restrictions, the extensive adoption of the elements of budgetary policy, and optimal allocations of means for import-substitution resulted in rapid initial successes. Due to these achievements, it became possible to survive the crisis, to reduce dependence on the world market, and finally, to surmount the emergency situation brought about by World War II, as well as to accelerate the growth process. Thus, the strategy of switching over to "inward-looking" in economic policy has been elevated to the rank of "revolution in economic policy", characterizing the Latin American countries as well as a broadly defined group of developing countries not only in the times of emergencies but all through the three decades following the big World Depression.

Beginning with the late fifties, however, here appeared ever more frequently the symptoms of the limitations of such a strategy in time and space, indicating the exhaustion of that growth period, and also the internal and external economic and political imbalance. All these factors resulted in the decline of the period of "inward-looking" in most of the Latin American countries by the middle of the sixties.

The reaction of the *less developed* Latin American countries to the consequences of the World Depression was rather diverse. Owing to the underdevelopment of the national bourgeoisie interested in the domestic market and/or protectionism, as well as to the survival of certain elements of the traditional model, there occurred no switchover from one power structure and growth

---

[42] CEPAL, *Estudio Economicó de America Latina*, 1959.

period to another. Because of the increasing enclave character of the export sector of the tropical and monocultural mining economies, the spill-over effects in the national economy were of lesser significance. While the incomes in the exporting branches were considerably reduced by the crisis, the manpower employed in this sector of the economy was forced to return to the traditional sector producing for subsistence. Owing to its reduced incomes, the export sector became insolvent, the effect of which is less important from the point of view of the less monetized domestic economy based essentially on subsistence production than from the aspect of further unfavourable modifications of ownership relations. The export sector which proved unable to pay off its earlier debts was bought up increasingly by foreign corporations and, at the same time, there did not occur any new capital imports. Parallel to the narrowing down of the export sector, reduced public revenue resulted in the decline of the standards of investments and social services. In the thirties and forties these countries were unable to change over to a new stage of development and to adapt themselves to the conditions of shrinking international division of labour.

The favourable tendencies on the world market following World War II had a favourable influence on the export sectors of the Latin American countries. Nevertheless, the scale of import-substituting industrialization launched on the basis of the negative experiences of the preceding decade, and the post-war boom were of a much more modest nature and hardly represented any kind of changes in the pattern. As a consequence of the *a priori* limits of their economies and/or the appearance of the first negative signs in the development of the more advanced Latin American countries pioneering in import substitution, it was only Venezuela relying on her oil resources, Bolivia in connection with the socio-political changes of the fifties, Cuba in the sixties, as well as Peru in the seventies which committed themselves to the course of growth based on import substitution. The economic progress of all the other countries fell into step with the new trend of an export orientation discernible from the midsixties which occasionally took the institutional form of increased subregional integration, as in the case of the countries of Central America.

As regards the evaluation of the "inward-looking" growth period of the Latin American economies, the time limits and structural constraints of import substitution are by now well-known facts. According to the experiences of development, the excessive prolongation of the phase necessary for overcoming emergency situations, and the transgressing of certain structural thresholds (e.g. heavy industry) brings about more and more negative features at a higher level of development. It is a well-known fact, too, that the proportions of the national economy, *ceteris paribus*, are also instrumental in shaping the efficiency of the growth stage. The reserves of relying on internal division of labour and import substitution, the dynamics and the degree of efficiency of growth are much greater in big countries than in small ones, as it is borne out by the cases of Brazil and Uruguay.

Historically established international specialization, the structure of external economy and strategy are by no means negligible factors either. In countries where export capacity was maintained, and sometimes even enhanced by external factors or development policy (e.g. Mexico, Venezuela), import substitution proved to be a lasting and vigorously dynamic period, whereas at the same time, countries with a decreasing export capacity very often had to face the problems arising from the fluctuations of prosperity and the slowing down of growth rates. Thus, considerable capital imports became necessary for counterbalancing the external disequilibrium.

Development problems, though not quantifiable ones arose also from the so-called pseudo-morphological development. These problems were caused by the radical break with the traditional development pattern, the simultaneous mass scale adoption of institutions formed in an entirely different socio-economic environment, by ideas, by technologies as well as by growth priorities, with one word by elements alien to the historical model. It is not by chance that in Latin America, taken as a whole, the acceleration of the rate of falling behind the world standards coincided with the changeovers from one stage of development to another entailing always a break with historical tradition. At the end of the colonial era the difference in the development level between the West European–Anglo-Saxon and Latin American development poles had been negligible. The widening development gap came about during the half century following the attainment of independence and the two decades after World War II, i.e. first by the adoption of West European liberalism, and later by the 20th-century adoption of 19th-century West European protectionism.

If we examine the development of the individual countries, a similar trend takes shape. The adoption of the liberal West European model proved most successful in the regions where the roots of the traditional Indio-Spanish model did not strike very deep (Argentina, Uruguay, Chile), while the development of countries most deeply rooted in this model (Peru, Bolivia, Ecuador, Mexico) lagged behind. On the other hand, the returning to the centralist model, to the requirement system necessitated by long-term historic development, has liberated considerable energies stimulating accelerated growth, but it has brought about an almost irretrievable time-lag and loss of advantages in the case of countries lacking such kind of centralist traditions (i.e. Argentina, Uruguay and Chile). The negative symptoms of growth being not in harmony with the historical characteristics of development were reflected both by post-war Chile with a centralized economy developing within the limits of liberal politics, and Peru with her deeply rooted centralization and a liberally patterned economic development lasting right till the mid-sixties.

The period of increased reliance on a new, more up-to-date international division of labour (bringing about not simply the export of raw materials but also that of industrial goods) will be analysed in a subsequent chapter. The development of this stage is connected with several internal and external factors. Out of the internal factors let us mention here the gradual exhaustion of the policy of import substitution, the emergence of an employment crisis, and consequently, the sharpening of political and social tensions. As regards the external aspects of the problem, the unfolding of the process was furthered by the change in world economic environment, increasing manpower shortage in the advanced capitalist countries, inflation, the gradual ousting of labour-intensive branches based on simpler technologies from the highly developed areas and countries, the rapid broadening of international division of labour and its enhanced role in the transfer of the achievements of scientific and technological progress as well as the differentiation in the international balance of economic power. This latter factor provides the weaker countries with more favourable possibilities of manoeuvring, which may also reduce the risk of the policy of "outward-looking" as well as its socio-economic costs.

Increased export orientation of the national economies, a phenomenon characteristic of most of the Latin American countries, depended not only on the degree of development, the scale and structure of economy or the socio-political system. The process is demonstrated by the successful economy of Mexico carrying through the bourgeois revolution in its own peculiar way; by Brazil which underwent a slight crisis of growth and is governed by a military dictatorship; by politically liberal Colombia showing only the first signs of "outward-looking" import substitution; and also by Chile of the first three years under a popular front government, trying to shape a socialist model open towards world economy. The most striking features of the unfolding new period of export-oriented growth are the variety of political systems and institutional frameworks, and the taking shape of multifarious approaches in the different countries. Thus, Latin America has become a veritable depository for analogies in economic policies and for the analysis of international economic policy.

1.5. THE SPECIAL FEATURES OF THE DEVELOPMENT OF BRAZIL

Brazil shares in the heritage and symptoms of socio-economic backwardness with all the other Ibero-American regions, nevertheless there appeared considerable differences as early as in the initial stages of her modern history. The differences originated partly from the special features of the historical development of the metropolitan countries. Though both Spain and Portugal underwent the Roman, Visigothic and Mohamedan conquests, the development of Portugal had started under conditions differing from the Hispanian model.

Stanislawski[43] pointed out that the northern part of present-day Portugal had been settled not by the highly Romanized Visigoths but by the more backward Swabians. Here the penetration of the Germanic mode of production was much more vigorous and the disintegration of the Roman model much faster than in other regions. Besides the former Roman big estates, reduced in numbers, many small holdings came into being. In the northern parts of the country the Moorish influence proved to be rather weak, and even in the central districts of Portugal it could not take roots. Contrary to Spain, the *reconquista* had been finished here as early as the middle of the 13th century. Thus, contrasted with the Spanish model, the disintegration of the Roman socio-economic infrastructure had taken place on a much greater scale, no large-scale transfer of elements originating from the Asian mode of production had been effected by Moorish influence, and due to the earlier ending of the *reconquista*, there had not evolved a historically prolonged emergency situation or a permanent state of war which could had served as a source of centralism.

The economic and political power of the Crown was consolidated by property rights of the deserted lands following the driving out of the Moors, and this power was put relatively early in the service of certain development objectives. The commercial treaty of 1294 concluded with England should be regarded in a certain respect as a marker, since it laid the foundations of English-Portuguese relations to become traditional later on.[44] The treaty also guaranteed Portugal's independence against Spain, and opened up new vistas for the broadening of relations with the countries of Northern Europe, and also for the

---

[43] A.K. Stanislawski: *The Individuality of Portugal.* Texas University, Austin, 1959.
[44] A.K. Manchester: *British Preeminence in Brazil.* The University of North Carolina Press, Chapel Hill, 1933.

adoption of institutional, cultural, etc. experiences. In subsequent years it resulted in transforming Portugal's position into a commercial vassal of England; this process was confirmed by the Treaty of Methuen in 1703, known as the first free trade agreement in the history of trade. The early and active joining in international trade had considerably influenced economic growth, urbanization, institutions and later the colonial policy of Portugal.

The development of the economy and foreign trade by the state served as a background to colonial expansion begun in the 15th century. Contrary to Spain, Portugal had established a prospering foreign trade by the beginning of the 16th century, and in addition to the state trading enterprise, a number of private merchants and private trading companies had also participated in foreign trade. While in Spain it was the military mentality and institutional system which grew strong, in Portugal the commercial attitude and institutional system struck deep roots. Consequently, the prospering commercial bourgeoisie became the counterbalance to feudal aristocracy, and business considerations played a major role in government decisions. Religious intolerance and striving for religious conformity was far less vehement than in Spain, and Portugal did not see any persecution of Jews which had been a factor of the decline of Spanish economic life. By the beginning of the colonial era, the outlines of an expansive model, open in relation to world economy and fit to direct centrally the different economic and social interest-groups, took shape in Portugal.

Owing to the dissimilar trends of historic development and also to some natural-geographic factors, the special features of Hispanian an Lusitanian *colonialism* have clearly stood out from the very beginning. The colonization of Brazil was begun in the northern, i.e. tropical coastal zones of the country, and thus the Portuguese colonies were much nearer to the routes of world trade than the centres of Spanish colonialism superimposed upon the former Indian highland empires. Settlement along the coastline has ensured from the very beginning increased commercial activity, closer relationship among the separated colonial trading posts and, last but not least, the possibility of getting cheap slave labour force from West Africa. On the other hand, as the task of incorporating the population and/or the Indian formations did not arise in Brazil which was sparsely populated by Indian tribes of a lower-grade civilization, the socio-cultural dualism—despite slave trade getting under way—was not as pronounced as in the Spanish regions. Though at first it seemed that by virtue of the treaty of Tordesillas Portugal has got the less valuable territories of Latin America poor in precious metals and population, the initial disadvantages have proved to be by far not unequivocally negative factors from the point of view of long-range development.

As a result of the factors mentioned above, the institutions of Portuguese colonization were different from those of Spain. Though the Eastern trade, being of a pivotal importance, was controlled by the Crown, the Brazilian colony regarded as worthless was virtually left in the hands of private enterprise. Colonization was effected by private persons, the so-called "donatarios" who, having acquired a concession, developed these regions at private expenses. As a result of the policy of "beneficial neglect", colonial organization was characterized by decentralization and local power centres (municipal councils, big estates).[45] Contrary to the structure of the Spanish colonial ruling class,

---

[45] C.R. Boxer: *Portuguese Society in the Tropics*. The University of Wisconsin Press, Madison, 1965.

it was not the representatives of state administration but those of the merchant-landowner strata which wielded a state power. It is indicative of the looser relationship with the central power of the metropolitan country that not a single town in Brazil was named Lisbon. As a consequence of the ideologically less rigid Portuguese model, the role of the Church was not so significant and it acquired less economic and political power in the course of colonization than in the regions of the Spanish colonial empire. According to Freyre's witty remark, the power aspirations of the Church have been restricted from the very outset, owing to the lasting conflict between the settlers and the Order of Jesuits as regards the problem of slave import. The presence of the Church was limited to the role of the plantation owner's domestic chaplain.[46]

At the beginning, sugar-plantations seemed to be the only practicable opportunity for the new colony. Because of the very small number of local population and/or manpower shortage, the growing of sugar-cane was accompanied by the introduction of slave labour. The less important source of getting hold of slaves, was the rounding up of the Indian population living in the interior of the country, or even in adjacent Spanish colonies, by slave-hunters' expeditions *(bandeirantes)*, the other, and more important one, was the slave import form Angola. Both forms always ended with buying in cash and thus, the monetization of the Brazilian economy was strikingly accelerated by the spread of slave-trade. Besides sugar-cane the basis of colonial economy was the export of Brazilian wood, hides, tobacco and cotton. In addition to the Portuguese merchants, English, Dutch and French traders have also participated in foreign trade from the very beginning, thus strengthening the openness of the colonial economy to world markets.[47]

Brazilian development, the pushing forward towards the western internal territories *(marcha para oeste)* has been determined for about one and a half centuries by the extraordinary export boom. Around the middle of the 17th century, however, concurrently with the rise of the British, French and Dutch colonial empires and the establishment of the practice of mercantilist protectionism, external markets shrinked rapidly, the traditional export sector declined and only at the end of the 18th century did it reach again its earlier level of 150 years ago.[48]

This process has resulted in the diversification in space of both the structure and mechanism of production. The centre of colonial economy has shifted towards Central and South Brazil around the gold and diamond mines opened up at the beginning of the 18th century (Minas Gerais, Matto Grosso, Goiás), and the gold rush has brought about accelerated immigration, the founding of administrative centres, as well as increased internal division of labour. The recession in mining at the end of the 18th century left behind a more diversified production mechanism and social structure. In the northern coastal regions there survived the institutionally rigid remnants of the one-time dynamic export economy based on plantations, whereas in the southern regions of the country there developed the model of a socially more dynamic economy rather similar to the European pattern, based on agriculture originating from the mining boom and to a lesser degree on industry.

As opposed to the disintegrating Spanish colonial empire, the colonial empire

---

[46] G. Freyre: *The Masters and Slaves*. A.A. Knopf, New York, 1946.
[47] A.K. Manchester: *op. cit.*
[48] C. Furtado: *Economic Development of Latin America*. Cambridge University Press, 1970.

of Portugal was characterized at the beginning of the 19th century by a more decentralized control, a more diversified structure, a higher degree of internal integration, stability and less obstacles to capitalist development. At the same time, however, cultural development was much more backward.[49]

The termination of Spanish colonial rule resulted in territorial disintegration. A new power centre was established in Brazil in 1807 by the Portuguese royal family escaping from the armies of Napoleon; later, in 1822 Brazil seceded from the metropolitan country and an independent empire was proclaimed. Thus, independent statehood was attained peacefully, without any changes in the power, social and economic structures, and also without disintegration of the country. The destruction of the means of production during the Spanish American War of Independence turned out to be indirectly even beneficial to the development of Brazil, as combined with a political stability, it gave a great impetus to the exporting industries. The main commodity of the new export boom was coffee which could be produced on small plantations, too. This sector of the economy which developed outside the bounds of the latifundia, further extended the internal frontiers of the economy towards the almost unexploited southern territories. The rubber boom in the Amazonian regions had a similar, though less vigorous impact.

The stability of political power resulted in more favourable negotiating positions in foreign trade, too. Contrary to the almost boundless liberalism of the former Spanish colonies, there survived the ends and means of interventionist policy established in the course of 18th-century Portuguese reformism. Thus, despite British pressure, a moderate protectionist customs policy (20 to 60 per cent) was introduced in 1841, serving not only fiscal objectives but also the aims of industrial development. The state took a part in the development of education, training specialists for the technical branches and public administration, in furthering the development of industrial branches of strategical significance (e.g. metallurgy), agricultural research institutes and the infrastructure as a whole, as well as in stimulating immigration. The imperial decree of 1848 abolishing slavery also served the acceleration of capitalistic development.

The manpower demand of the new sector representing the foundation of the boom, the simple tools used by this sector, the fact that the necessary equipment could be produced locally—all these factors in themselves were favourable to the acceleration of the internal division of labour. Due to the stability of political power, and the growing mercantile mentality, the purchase, the foreign trade and financial relations of the coffee sector were centralized almost entirely in the hands of the national capital. Thus, contrary to the export oriented development of Spanish America, controlled by foreign monopolies, a much higher percentage of surplus value realized in the external economic sector remained in the country and helped accelerate the process of domestic accumulation of capital.

Population growth speeded up by immigration (three million European immigrants between 1875 and 1914), the higher degree of the monetization of economic life, a broader domestic market, the greater weight of labour-intensive branches, the policy of the state and the stability of its institutions, the lack of Pre-Columbian vestiges, and also a less burdensome colonial heritage—all these elements contributed to the fact that the Brazilian economy reacted

---

[49] P. Calmon: *Historia social do Brazil*. Cen, São Paulo 1941.

more favourably to impulses from the dynamic export sector and forged ahead faster on the road of structural progress. More than hundred textile mills were in operation by 1889, and in 1913 the first Brazilian big slaughter-house for domestic supply was opened. There also emerged the Brazilian big capitalist groups the most outstanding representatives of which were Baron Maua, Matarazzo, Jafet and Rodovalho. True, the process of internal economic integration was interrupted towards the end of the colonial era, but alongside the northern part of the country representing one of the rock-bottoms of Latin American backwardness there gradually emerged a new, developed region in the south. As regards *per capita* indices, Brazil as a whole still considerably lags behind Argentina, Uruguay or Chile, her dynamic southern regions, however, had become outstanding in the economic and technological development of Latin America, even prior to the World Depression.

The well-known consequences of the Depression have affected Brazil even more seriously than the other countries of the continent, since the income-flexibility of demand in coffee and other tropical produce is higher than those of food-products of the moderate zone. Even a slight decrease in demand may trigger a disastrosus slump in prices because production capacities are not convertible like those of the produce of the moderate zone.

In the first Depression years, the state, being under the influence of the coffee producing sector, strove to maintain the income and employment level of the coffee sector representing the basis of national economy, by financing supplies. In addition to recently introduced protectionist measures, this initiated a spontaneous process of industrialization which, compared to the neighbouring countries, received only belatedly top-priority in public development policy. The rate of Brazilian industrialization, as compared to the other Latin American countries, was accelerated by considerable public subsidizing, deficit financing, immense raw material resources, broad markets and, last but not least, by the abundant supply of cheap manpower released from the labour-intensive but stagnant coffee sector. However, by the beginning of the sixties, all reasonable possibilities of import substituting have been exhausted. More than 90 per cent of Brazil's consumption of industrial goods has been supplied by the domestic industry, thus all further import substituting industrialization has become even more difficult and expensive. The process of import-substituting industrialization bogged down on account of increasing internal and external imbalance, and finally the increased social tensions resulted in a political crisis and a military take-over in 1964.

Though the balance of import-substituting growth seems to be more favourable in Brazil than in the other Latin American countries, the recovering from the consequences of the growth crisis was rendered possible only by relying on administrative constraining measures, by enhancing the extremities of income division and by opening the doors to foreign capital. The buoyant expansion of the late sixties, however, commenced under the conditions of a new growth period. The Brazilian economy returned to the four-century-old policy of export-oriented development, and realized this concurrently with an unprecedented degree of centralization.

The comparative analysis of the main stages of Latin American development and the specific evolutionary traits of the most important country types cast light upon the manifold character of the trends and causal interrelation of historic development, as well as on the still perceptible influence of historic socio-economic heritage on growth. Without the knowledge of these peculiar traits it would be hard to understand the historic failures of the adoption of experiences gained from the political institutional system, scale of values, etc. of the so-called West European-Anglo Saxon type of development. The present trends of progress, the exploring of possibilities—based on national development—for the acceleration of the economic and social progress of the individual countries and the efforts for the termination of economic dependence cannot be understood without taking into account history.

CHAPTER 2
THE MAIN TRENDS OF DEVELOPMENT OF THE LATIN AMERICAN ECONOMY

2.1. A COMPARATIVE ANALYSIS OF THE GROWTH PERFORMANCE OF THE LATIN AMERICAN COUNTRIES

After having reviewed the main development trends in the Latin American countries, let us examine how this evolutionary heritage is reflected in the development of the continent's economic processes. The outstanding characteristics of the growth dynamics, structural development and economic balance of the continent will be surveyed in this chapter.

Let us begin the presentation of the main development trends with an analysis of the rate of development. (See Table 2.)

A number of interesting conclusions may be drawn from the above data. *The acceleration of the rate of growth* in recent years is a noteworthy and unambiguous phenomenon. While the average rate of regional growth has remained

Table 2

The rate of growth of the national product in Latin America
(average annual growth, per cent)

	1950/60	1960/70	1960/65	1965/70	1970	1971	1972	1973
Argentina	3.1	3.7	3.5	4.0	4.1	3.8	3.8	4.8
Bolivia	0.4	5.6	4.9	6.0	5.7	3.9	.	3.5
Brazil	6.8	5.8	4.5	7.5	9.5	11.3	11.5	11.0
Chile	4.0	4.3	5.0	3.7	3.2	8.6	5.5	2.0
Costa Rica	7.1	6.8	6.3	6.9	5.4	4.9	5.0	.
Dominica	5.7	3.7	1.0	6.6	6.5	7.4	12.5	.
Ecuador	5.0	4.7	4.4	6.2	6.0	6.5	7.1	9.9
El Salvador	4.7	5.8	6.9	4.7	4.6	5.1	4.5	5.4
Guatemala	3.8	5.2	5.3	5.0	5.7	5.6	6.4	.
Honduras	3.5	5.2	5.3	5.1	3.1	4.3	4.3	5.0
Haiti	1.9	0.5	1.1	0.1	4.7	5.7	.	.
Colombia	4.6	5.2	4.6	5.8	6.7	5.5	7.1	8.0
Mexico	5.8	6.8	6.9	7.2	7.5	3.7	7.5	7.6
Nicaragua	5.2	6.7	8.1	4.8	5.0	5.7	5.1	.
Panama	4.9	8.0	8.2	7.8	7.0	8.6	.	.
Paraguay	2.7	4.7	4.8	4.1	6.1	4.5	.	.
Peru	5.3	4.9	6.6	3.3	7.5	5.2	5.9	6.0
Uruguay	2.1	1.3	0.9	1.5	5.0	—0.5	—1.0	.
Venezuela	7.7	5.8	5.0	6.4	4.4	4.5	4.5	7.0
Latin America	5.1	5.5	5.1	5.7	6.8	6.1	7.0	7.8

Source: ECLA, *Economic Survey of Latin America*, 1970, 1974.

unchanged in the fifteen years following 1950 (within this, however, it has markedly decreased at the beginning of the sixties), a significant quickening in the rate of increase is under way since 1967. In the second part of the past decade the rate of growth has increased in almost all the countries of the continent, with only a few exceptions.

Uneven regional development is indicated by the growth rates of the individual countries. The considerable acceleration of Brazilian growth plays undoubtedly an important role in the upswing of regional growth averages. Brazil not included, the growth rate of the region is lower by almost one per cent. In the period under review an increasing growth rate is shown by Brazil, Colombia, Mexico, the Central American states, Ecuador and Bolivia, whereas Venezuela and Chile have experienced a somewhat slowed down growth rate in the examined last ten years. The changes in the rates of increase are quite conspicuous and indicate the fact that the individual national economies are highly responsive to the movement of the external or internal factors of development.

Besides the changes of growth rates, the problem of the average economic rate, characteristic of the long-term development of the individual countries, is also a noteworthy factor. The members of the group with a long-term fast development were the two big countries, i.e. Brazil and Mexico, oil producing Venezuela, Panama the economy of which is based on the service industries, and also Costa Rica and Colombia, almost the most stable countries in the region with the most democratic political institutional systems. Moderately dynamic countries are the Central American republics, as well as Peru and Ecuador. Among the countries with slower growth rates Argentina and Chile are the most important ones. The economic life of Haiti and Uruguay, two little stagnant countries of the region (quite an exception internationally, too) has essentially been paralyzed between 1950 and 1970, mainly for political reasons and/or because of considerations of economic policy.

It is worth-while to scrutinize these growth rates from the point of view of *demographic* data as well. Owing to the general development and especially to progress of public health conditions, the already high rate of increase in population has intensified in the past three decades. Whereas the average regional rate of the increase in population had been 2.3 per cent in the forties and fifties, it increased to 2.9 per cent by the beginning of the seventies, and within that it amounted to 4 per cent in Costa Rica, and approximately 3.5 per cent in the other Central American republics, i.e. in countries with an overwhelmingly Indian population. Only Argentina and Uruguay have a lower rate of population increase (1.2 and 1.5 per cent).[1] Such a high rate of increase in population is not to be found in any other major geographical region of the world economy, and as regards those ones with a similar level of economic development, they hardly reach one third of the Latin American growth rate. Under the present conditions, the stagnation of the economy means a decrease in the level of *per capita* incomes, and an annual growth of about 3 to 4 per cent makes it possible only to maintain the standard at the very most. It is not by chance that the sharpening of socio-political tension is characteristic first and foremost of countries with permanently slower rates of development.

The comparison of the growth rates of population increase and economic growth also indicates, though in a simplified form, the extent of the increase of productivity on a national economic scale, and the level of modernization and structural development. In case of the stagnation of the *per capita* national product or a 1 or 2 per cent annual growth, an accelerated technological progress, the liquidation of backward structures, and the historic heritage of unemployment could hardly be excepted. Under such circumstances, up-to-date

---

[1] *Monthly Bulletin of Statistics*, 1974/6.

development is but of an enclave character; it occurs to the detriment of other sectors, and uneven development, in case of quasi-stagnation, brings about very serious economic, social and political tensions.

High population increase itself renders the efficiency of accelerated growth rather doubtful. Taking into consideration the *proportions of growth*, too, the number of negative factors of growth increase. Accelerated growth did not cover all the sectors of the national economy; it was based on the industry and/or the service sector established around the industrial centres. Between 1960 and 1970 the average increase in Latin America has been 2.6 per cent in agricultural production, 7.3 per cent in the manufacturing industries, 7.1 per cent in the services.[2] The differences in the dynamics of the individual sectors, viz. the lagging behind of agricultural production, has become a source of serious difficulties. It should be indicated only, that in the majority of the region's countries the slow progress of agricultural production falling behind the increase of population, did not allow for the improvement of the rather unfavourable nutrition conditions and a better utilization of the human factor of growth. *Per capita* calory consumption has increased barely by 5 per cent, *per capita* protein consumption by 8 per cent,[3] i.e. it has remained stagnant, and has even deteriorated in certain countries. In the more backward countries and/or in case of a lower income level, the income flexibility of food demand is extraordinarily high (50 to 90 per cent as against the 25 to 35 per cent of the developed countries). Thus, if the progress of agriculture does not keep abreast with the increase in national income, internal or external imbalance cannot be averted.

A further problem arises from the fact that a slowly developing agriculture is unable to ensure the appropriate employment of the rapidly increasing population. The process of the influx of redundant manpower to the cities is speeded up, and only in a handful of countries does industrialization make such progress as to be in the position to absorb efficiently the manpower supply originating from accelarated population flow to the cities. Therefore, the service sector is instrumental in the relative absorption of manpower supply, especially in the less dynamic countries. The service sector, however, is able to increase rapidly only the number of fictitious and socially superfluous working places, whereas its excessive burdens, owing to employment tasks, results in the stagnation and deterioration of the sector.

This process is clearly illustrated by the development trends of the *branch productivity* in the individual countries. On an average (taking Latin America as a whole) the service sector is a relatively modern branch; in 1970 its productivity surpassed that of the average of the national economy by 57 per cent, i.e. to the same extent as the industry. Within the regional average, however, the service sector turned out to be increasingly a sector lagging behind in the case of countries with a low rate of development: in Argentina it fell by 30 per cent, in Chile by 16 per cent, in Uruguay by 14 per cent as compared to the average national economic productivity rates of 1970. This was mainly due to the low and practically negative productivity of artificially created, socially superfluous or unjustified work places.

The widening of the gap between the development levels of industry and agriculture naturally hampers the process of interior socio-economic integration

---

[2] *IADB, Annual Report*, 1972.
[3] *U.N. Statistical Yearbook*, 1972.

as well. The diverse dynamics of sectoral development has altered the structural proportions of the individual national economies. As a consequence of economic growth, the share of agriculture in the gross national product decreased between 1960/61 and 1970/71, taking the regional average, from 19 per cent to 15 per cent, that of the extracting industries from 4 to 3 per cent; the share of the service sector remained unchanged at 44 per cent, whereas that of the manufacturing industries increased from 23 to 26 per cent. In some of the more industrialized countries, such as Brazil, Argentina, Chile, Mexico or Venezuela, this share is even higher.

The *structural indices* are much more favourable than in the average of the developing countries: the average ratio of industry in Latin America, taken as a whole, is essentially equal to indices of the European Mediterranean countries and of Norway, Finland and Ireland. Thus, examples taken from the Latin American countries do not corroborate the interrelations between structural development and *per capita* national income, since the ratio of industry in the Latin American countries (as mentioned above) has been achieved with half or one-third of the *per capita* national incomes of the small European countries which have been lagging behind in industrialization. The out of the ordinary indices of structural development are to be explained mainly by the development strategy of the Latin American countries as well as by factors related to their growth characteristics. Latin American economic development between 1930 and 1965 has taken place mainly within the bounds of protectionist policy, and this has been instrumental in the shaping of dissimilar trends of structural development and the growth process as a whole.

The dissimilar trends, the low rate of modernization and rise of efficiency explain to a certain extent the fact why accelerated growth has not been accompanied by a more rapid increase in the accumulation capacities of the individual national economies. The acceleration of growth following 1965, originated in the changeover from one stage of growth strategy to another, and also in making use of the advantages of joining the international division of labour. At the same time, no signs of a major increase in investments can be detected (see Table 3).

The above data call attention to phenomena not so astonishing, knowing the trends of economic growth. The development of the investment ratio does not back up the views of the structuralist school, as the correlation between the degree of economic development and the ratio of investments cannot be pointed out unambiguously. It is rather the development strategies and the economic size of the individual countries which provide some kind of orientation. In the formerly developed regions which have been carrying out a policy of protectionism for a long time, investment activity is less intensive than the average, the ratio of investment is on the decline, while the specific capital intensity of growth shows a tendency of increase. In the case of Argentina, the maintenance of a modest rate of growth seemed to be attainable only by increased investments. At the same time, a more favourable ratio of investments and their degree of efficiency become evident in the major countries which have joined more intensively in the international division of labour.

The smaller countries of Latin America present an interesting picture coinciding with the trends of the small European countries. In all the smaller countries, with the exception of Haiti and Dominica, the trend of increasing investment ratios concurrently with a frequent increase in capital intensity may be regarded as a general phenomenon. This indicates a relationship on the

Table 3

The development of the ratio of investments in the national product in some Latin American countries
(per cent of the national product)

	1955	1960	1970	1973
Argentina	18	23	21	21
Chile	17	17	17	14
Uruguay	20	18	16	.
Brazil	17	18	20	23
Mexico	19	20	21	21
Venezuela	29	18	20	28
Colmobia	26	28	18	23
Peru	24	22	18	13
Costa Rica	19	19	20	23
Panama	15	16	24	29
Honduras	17	14	21	16
Paraguay	12	17	24	.
Latin America, total	19	19	20	22

Source: U. N. *Statistical Yearbook*, 1958, 1972.

international scale, according to which a faster rate of growth in small countries can be achieved only by means of considerable investments and high specific capital intensity.

Naturally, the interrelationships between investments and the rate of growth should be treated with due circumspection, as a direct stimulating effect on growth can be expected only from investments influencing the means of production. Owing to statistical restrictions, no breakdown according to investments in Department I and Department II can be demonstrated, however, we may infer from partial data available as regards the breakdown of investments in the building trade, machinery and equipment. In the average of Latin America, 56 per cent of all investments in 1969 were allotted to machinery and 44 per cent to building investments. As to Brazil, the 81 per cent share of machinery investments, unparallelled in international economic life, throw some light on the relatively extremely low capital intensity of the acceleration of economic speed. As regards, however, Argentina and Chile, indices conforming to the regional average indicate that these countries are not facing problems rooted in investments directly expanding or not expanding production, but efficiency problems and disorders of economic circulation.

The examination of the *sources of investments* is of considerable interest, too. Taking the average of the past decade, 93 per cent of all investments in Latin America came from internal sources, i.e. measured by international standards, the share of external sources was modest. No major shifts could be detected in the ratio of internal and external sources. Noteworthy changes took place, however, in the shares of investments by the public and private sectors, i.e. in the development of the socio-economic role of the state. Data of Table 4 demonstrate the breakdown of fixed capital investments by the public and private sectors, as well as the ratio of budget expenditures as compared to the national product.

The above data are a striking proof for the fact that a more up-to-date approach concerning the role of the state, and the *lengthening of the operational*

Table 4

Some indices relating to the economic role of the state

	The share of the state in fixed capital investments (per cent)		Budget expenditures in the percentage of the national product	
	1959/60	1968/69	1960	1970
Argentina	24	37	20	27
Brazil	40	39	25	27
Chile	46	59	31	36
Costa Rica	19	18	17	19
Colombia	15	29	10	15
Mexico	43	41	14	17
Peru	13	35	19	20
Uruguay	16	17	25	28
Venezuela	46	37	23	22

Source: ECLA, *Economic Survey of Latin America*, 1970.

radius of the state are general phenomena; an ever growing percentage of the national product is distributed through the state. Latin America is ranking high internationally as concerns the budgetary distribution of the national income. The increased part of the budget in itself does not, however, provide for more favourable conditions as regards the growth process, since the budget (especially in the one-time developed, southern countries) frequently takes an active part in financing socially useless activities and parasitic consumption. The role of the public hand in financing investments is a far more important means of influencing the structure, efficiency and patterns of the growth process. In the period under review, the part taken by the state in financing investments has also increased, nevertheless this process has not been quite unambiguous as it has reflected evolutionary and political factors as well.

On the evidence of the vigorousness of public activity in this field, two groups of countries may be distinguished: those which interpret this role in a liberal, and others which interpret it in a "dirigist" way. The emergence of these groups was by no means due to identical factors. It could be stated in an oversimplified way that in the bigger, more developed countries, adopting a protectionist strategy, the role of the state is more dominant than in the smaller, less developed ones which are open in relation to the world economy. The partial exceptions are, however, not few in number. The historically more developed countries which are non-dynamic as regards the world economy and set out from the functions of the so-called welfare state, such as Argentina, Uruguay and Chile, have enhanced the re-distributional role of the budget going beyond the Latin American average. However, as regards investments reflecting the direct economic participation of the state, the public sector of Uruguay, a dwarf economy related to the world economy, scarcely participates at all in such activities. Colombia, Costa Rica, and partly Uruguay and Peru are minor states even within Latin America. In the cases of Venezuela and Peru the consequences of political events and changes are also clearly reflected in the development of the role of the public sector. The co-existence of economic centralism and political liberalism is a frequent phenomenon; it is highly questionable, however, whether this proves successful.

The *directly state-owned sector* which has rapidly expanded since the thirties, is instrumental in increasing the economic influence of the state. Unfortunately, no reliable statistical data are available necessary for the quantification of direct public participation in the economic life of these countries. On the strength of partial surveys, however, it may be stated that the public sector represents not only the most dynamic but also the most concentrated economic factor as well. According to a survey of the early sixties, out of the biggest 183 enterprises of Argentina, Brazil, Chile, Colombia, Mexico and Venezuela, 60 were wholly and completely owned by the state, in 11 the majority of the shares were held by the state, whereas 71 were controlled by domestic private capital and 41 by foreign capital.[4] Out of the aggregate capital of these 183 corporations two-thirds were controlled by public enterprises.

It should be noted that the public sector is essentially confined to big corporations, while thousands of medium-sized and small enterprises remain privately owned. Out of the 183 big corporations mentioned, 19 were banks, 13 were active in electric energy production, 7 in railway transport, 6 in the iron and steel industry, 4 in the petrochemical industry, 4 in communications, whereas the rest were active in mining, various branches of transport, and the processing industry. Thus, the public sector struck roots in the infrastructure demanding huge capital investments and a well-founded organizational background, and also in the key industries. Over the past decade the public sector has continued its expansion partly to the detriment of foreign capital, and partly to that of domestic private capital. According to a survey in 1969, out of the 10 biggest corporations of Argentina 5 were publicly owned, while out of the 10 biggest of Brazil 7.[5]

The increased economic role of the state was connected in a peculiar way with the internal imbalances and inflation. At the beginning of the modernization and the industrialization process, the financing of public investments could have relied on the revenue, i.e. on tax income. However, in underdeveloped countries tax revenues are extremely low, and the major part of taxes is used for covering expenditures of a non-investment nature. In addition, income structure was highly susceptible to any changes in the world economy, as 30 to 50 per cent of such income came from import or export customs duties. Any set-back or stagnation of foreign trade turnover had a highly detrimental effect on state revenue as well. Under the given historic conditions, the Latin American countries steering the course of accelerated import-substituting industrialization adopted the Keynes remedy and resorted to the policy of budgetary credits. Industrializing Latin American economies have been for decades characterized by a policy of financing by budgetary deficit and co-existence with inflation. The initial mobilizing effect of deficit financing is beyond argument, however, the macro-economic efficiency of import-substituting capacities newly founded and developed in an atmosphere alien to competition turned out to be low, and up-to-date technologies were adopted mainly on a sub-optimal factory size. Thus, the cost level of the domestically manufactured new products frequently rose to many times of that of the former import prices and went hand in hand with the intensification of the inflationary process. Table 5 shows the annual growth rates of the consumer's price index broken down according to individual countries.

---

[4] F. Brandenburg: *The Development of Latin American Private Enterprises*. Washington, 1964.
[5] "Public Enterprises" ECLA, *Economic Bulletin*, New York 1971, Vol. XVI/1.

Table 5

Development of the consumer's prices
(percentage of average annual increase)

Country	1961/65	1966/70	1971	1972	1973
Large-scale inflation					
Uruguay	30	60	24	77	97
Brazil	62	28	20	17	16
Chile	27	26	20	78	several hundred per cents
Argentina	23	19	35	59	58
Moderate inflation					
Colombia	12	10	9	14	21
Peru	9	10	7	7	7
Relative stability					
Mexico	2	4	3	8	15
Costa Rica	2	3	3	5	11
Panama	1	2	2	6	7
Venezuela	0	2	3	3	6

Source: *Monthly Bulletin of Statistics*, 1974/76.

The rates of inflation clearly demonstrate the correlations of internal imbalances and import-substituting industrialization and/or deficit financing. As regards the rate of inflation, here, too, the former more developed countries rank among the first. It is interesting to note that the smaller countries which are forced to adopt an export oriented economic policy, as well as Mexico and Venezuela have developed under the conditions of a relatively more balanced internal economy. In Brazil and Colombia, which have switched over to export orientation about the middle of the last decade, the speeded up joining in the international division of labour has not proved to be a factor of inflation—owing to an improvement in macro-economic net efficiency—although inflation characterizing the internal economies of their main partners, i.e. the developed capitalist countries, has expanded ever more to the sector of external economy since 1967/69. Related to the rate of world economic inflation, the increase in the rates of inflation of the export-oriented Latin American countries in the early seventies may be considered moderate. The wide gap which formerly existed between the rates of inflation of the developed capitalist countries and the Latin American republics has been fading away in the capitalist world economy. The high rates of the one-time developed South American countries do not reflect economic factors any more, but primarily political ones.

The disproportions between sectoral and regional development, the negative factors due to the evolution of the human elements of development, the insufficiency of social development, growing unemployment, the problems manifesting themselves in the quality of social services, education, public health, all reflect the symptoms of *forced, unbalanced economic growth*. In the countries which have effected a switchover from one stage of economic growth to another one, structural disproportions and the extent of internal and externa imbalances have obviously decreased, and the competitiveness of the individual economies has increased. The growth process itself, however, has continued to be constrained and unbalanced, and has forged ahead by leaving behind huge enclaves of backwardness. The sharpening of political tension has also indicated a social criticism of growth.

The narrow range of social insurance, too, reflects the slow rate of social development. Two-thirds of wage-earners in Argentina and Costa Rica, two-fifths in Panama, one-third in Chile benefit from social insurance, whereas the respective statistical indices of the other countries fluctuate between 5 and 25 per cent.

Although the expansion of public education has accelerated over the last quarter of a century, illiteracy is still considerable: taking a continental average, it amounted to 26 per cent, at the end of the sixties, and within this it reached 9 to 10 per cent in Argentina, Uruguay and Chile, 15 per cent in Costa Rica, 50 per cent in the Central American republics, and more than 80 per cent in Haiti. From the point of view of future development it is a highly critical fact that at the end of the sixties a mere 60 per cent of the population obliged to go to school received regular education, and even in the more developed countries the situation is not much better in this respect. The unfavourable quantitative indices of education are coupled with increasingly serious qualitative problems. Political aspects were involved in the development of this sector in almost all the countries (securing financial aid for universities and high-schools, local points of view of prestige, postponement of employment problems, etc.). The co-ordination of the development of education with the requirements of growth, substantial, *in merito* changes in education, the differentiated development of regions, towns and villages according to needs can be observed only in exceptional cases.

Even these brief outlines in the indices of social evolution indicate the fact that improvement in social conditions did not keep abreast of accelerated economic development, and economic disequilibrium went hand in hand with increased social imbalance. By the middle of the sixties it has become more and more obvious that, from the point of view of economic growth, the deficiencies and unevenness of social evolution are posing the most serious troubles and difficulties.

2.2. PROBLEMS OF INCOME DISTRIBUTION

One of the most debated fields of the literature on economic growth and also of practical development policy are the alleged or actual contradictions concerning the growth rates as well as the proportions of income distribution. It is a well-known fact that disproportionate income distribution is considered a growth stimulating factor by the traditional schools of bourgeois economics, arguing that the saving capacity and propensity of social strata with higher incomes is greater than those with low incomes. At the same time, the progressive and even the reformist economists take a stand against the expediency of income concentration and—on the strength of both political and economic reasons—demand an equalization of the income distribution structure.

It cannot be doubted that the structure of the growth process is to a great extent directly influenced by income distribution relations and these relations indirectly affect the economic processes by influencing the socio-political atmosphere of economic development. Owing to their synthetic nature, they are suitable for the indirect demonstration of the production relations of a given economy. Thus, no outline of the development trends of the Latin American countries could be complete without a survey of the income distribution relations. Unfortunately, though for reasons easily accounted for, the majority of the Latin American countries have not paid much attention to the checking on income distribution from a statistical point of view, and this makes it rather difficult to gain a clear picture of these relations. We are going to try to examine the characteristic features and causes of income distribution in the Latin American countries, their effect on the process of economic progress, using the scarce statistical material available. Table 6 shows the structure of income distribution in the individual Latin American countries.

Table 6

Percentage share of the individual income categories in total incomes in 1965

Country	Lower 20 per cent	Medium 60 per cent	Top 15 per cent	Top 5 per cent
Argentina	5	41	24	31
Brazil	4	35	22	39
Chile	5	34	25	33
Costa Rica	6	34	25	35
Colombia	6	37	27	30
Mexico	4	38	29	29
Panama	5	39	22	35
Venezuela	3	39	31	27
El Salvador	6	34	28	33
Average of Latin America	3	37	27	32
USA	5	50	25	20
Great Britain	5	50	25	19
Norway	5	55	25	15

Source: ECLA, Economic Survey of Latin America.

The data of Table 6 reflect in a condensed form the peculiar features of income distribution and the extent of income concentration in Latin America. Even in the percentage share of the lowest income categories there are no manifest substantial differences between the developed and the Latin American countries. However, the share in total income of 60 per cent of the population amounts to 50–55 per cent in the highly developed countries, whereas only to 37 per cent in Latin America. Even more conspicuous is the trend in the top 5 per cent bracket: a 30 per cent share in total income in Latin America as against 15 to 20 per cent characteristic for the developed countries. Thus, marked differences manifest themselves in the structure of income distribution even on the basis of UN data presumably more favourable than the actual situation.

The main ratios of the income structure also help us draw adequate conclusions as to the extent of social inequalities. 80 per cent of the population in Latin America do not reach the average *per capita* income calculated, whereas the highest 5 per cent can afford a living standard about seven times that of the average. The deviation from the average is much less in the developed capitalist countries where the bulk of the population has incomes approaching the average standard, while the incomes in the highest bracket exceed only three or four

Table 7

Relation of the individual categories to the national mean values
(national average = 100)

Country	Lower 20 per cent	Medium 60 per cent	Top 15 per cent	Top 5 per cent
Argentina	26	68	153	620
Brazil	17	58	145	790
Costa Rica	30	57	167	700
Colombia	30	62	175	610
Mexico	18	63	197	590
Panama	24	64	146	690
Venezuela	15	65	210	520
Chile	25	57	290	660
El Salvador	27	56	188	650
Average of Latin America	23	62	175	650
USA	23	85	170	400
Great Britain	25	86	167	380
Norway	22	92	166	300

Source: ECLA, *Economic Bulletin of Latin America*, 1969.

times the average. Thus, in Latin America the statistical mean value is a result of extreme values which hardly reflect the real economic situation. The deviation from the average is shown by data in Table 7.

The extent of extreme income distribution and *social polarization* is clearly demonstrated by the following data of 1965. The deviation between the average *per capita* incomes of the lowest 20 per cent and the highest 5 per cent is 25-fold (in the case of developed countries this amounts to a 13 to 18-fold deviation), in certain Latin American countries, however, values are even more extreme than the above. Thus, in the case of Brazil the gap is 46-fold, in Mexico 33-fold, in Colombia 20-fold, in Venezuela 37-fold, and in Argentina 24-fold.

The political and economic effects of the inequalities of income distribution are even more manifest if the extent of polarization is analysed not for the lowest and highest income brackets but for the deviation between the medium 60 per cent representing the bulk of the population and the top 5 per cent. Contrary to the developed capitalist countries, in the developing states the 20 per cent of the population in the lowest income category generally live outside the limits of monetary economy, pursue a subsistence economy, and is mostly isolated socially, politically and often even geographically as well. Thus, the evaluation of their income standard cannot be based firmly on statistics.

To point out the extent of polarization in the individual countries as regards the population living under organized socio-economic conditions, it is much more convenient to compare the income ratios of the bulk of the population partly or as a whole within the limits of modern economy and the 5 per cent in the highest income bracket. In developed capitalist countries the gap between these two categories amounts to a 3 to 5-fold deviation, whereas in the case of Latin American average it amounts to a more than 10-fold deviation. Within the Latin American average there are some countries with more favourable averages, viz. Argentina with a 6-fold, Venezuela with an 8-fold and Mexico with a 9-fold deviation; at the same time Brazil has a 14-fold, the small Central American republics and Chile a 11–12-fold deviation.

40 per cent of manpower is employed by backward, primitive agriculture essentially of a self-supporting nature, whereas a mere one-eighth of manpower works in up-to-date industrial plants, mines or plantations. According to calculations by UN experts, about half of the national product is produced by the manpower employed in the modern, capital-intensive sector, while the 40 per cent of labour force employed in the primitive sector produces scarcely one-tenth of the GNP. Thus, in Peru and Chile one per cent of the manpower employed in mining and related activities produces the bulk of exports and 15 to 20 per cent of the national income. The reasons for differences in incomes are by and large indicated by the immense differences in the level of sectoral productivity. Relationship between structural and income inequalities is of a rather direct nature. In countries where a considerable part of the national income is produced by a sector of high productivity, by some kind of extractive activity employing but a fraction of total manpower, the degree of polarization and income concentration exceeds the average. The data in Table 8 indicate the differences in productivity levels of the various sectors.

Table 8

The level of productivity in the different sectors
(average of non-agricultural activities = 100)

	Agriculture	Mining	Big industry	Small-scale industry	Manufacturing industry total
Latin America	35	329	211	21	128
Brazil	49	80	223	22	140
Argentina	111	268	239	28	158
Chile	41	244	216	30	133
Colombia	60	224	270	28	119
Peru	25	206	163	18	81
Mexico	19	298	175	17	123
Central America	29	140	173	18	84
Venezuela	27	154	244	24	94

Source: ECLA, Economic Survey of Latin America, 1970.

Due to historic, geographical and political reasons, the effects of sectoral inequalities are enhanced by differences in the level of development. In Brazil for instance, *per capita* income in the state of Sao Paulo is three times higher than in the north-eastern parts of the country. Similar ratios are to be found in Mexico as regards the Federal District and the south-western part of the country, and in Venezuela regarding the district around Lake Maracaibo and the coastal Cordilleras as compared to all other regions of the country.

The organization of the labour force is reflected to a certain extent by the development of income structure. In countries where immigration from Europe has taken on a mass character, the immigrant workers have brought with them their traditional experiences of the workers' movement and organization, and could make use of them in order improve their income situation, especially in growth phases characterized by considerable manpower demand.

The dissimilar income structures of the more developed Latin American countries are enhanced by factors like the liquidation of the primitive sector of economy, the industrialization of agriculture, and a more levelled progress

Concentration and polarization ratios which may be regarded as simplified indices of socio-economic tension make it possible to draw some interesting conclusions. The most extreme income distribution relations are to be found in Brazil adopting a policy of forced growth, as well as in the smaller Central American countries and Chile. At the same time, in Argentina, Mexico and Venezuela, i.e. in the bigger countries with more developed institutional structures, both the concentration of incomes and the extent of polarization is more favourable than the Latin American average.

Unfortunately, only the above data characterizing but the present state of affairs are available for conveying the differences between the individual countries, since dynamic time sequences as to the progress of the process of concentration and polarization are not ready to hand. Partial data, however, do not render it possible to reach unambiguous conclusions. In Mexico, for instance, the share of the top 5 per cent in total income fell from 40 to 29 per cent in the period 1950/64, and at the same time the share of the 40 per cent of the population in the lowest income bracket also decreased from 14 to 11 per cent. While the extent of concentration has decreased and that of polarization has remained stable, the income position of strata in the medium bracket, especially that of the skilled strata has improved. Quite naturally, the favourable trend of decreasing deviations from the income average does not obliterate the negative fact that the results of the dynamic Mexican economy, outstanding even internationally, have not got through to the broad masses with the lowest living standard.[6]

While socio-economic development in Mexico has improved the income situation of the medium strata, the development in Brazil between 1960 and 1970 has resulted in an increased concentration of incomes.[7] Due to the political character of the military system and to the inherent nature of forced economic growth, the achievements of economic growth were monopolized first and foremost by the 5 per cent of the highest income bracket, the position of the medium strata remained essentially unchanged, and that of the broad masses deteriorated. Between 1960 and 1970 the share of the highest income 5 per cent increased from 27 to 37 per cent, that of the subsequent 15 per cent from 27 to 28 per cent, whereas the percentual share of the 50 per cent in the lowest income bracket dropped from 18 to 14 per cent. Thus, the Brazilian model demonstrates not only the progress of income concentration and polarization but also its intensity unparalleled on an international scale.

Let us now examine the factors underlying the considerable differences in the structure of income distribution in the individual countries. There are a number of reasons for the diverse extremities of income ratios in the various countries, owing to uneven development, backwardness, the development strategies of the individual countries, and to the economic potentials which all have a direct or indirect effect on income relations. Partial statistics available prove the fact that here property, too, is much more concentrated than in the advanced countries. The concentration of landed propriety is a well-known fact. Lack of capital, however, has generally resulted also in industry and commerce in the emerging of a small number of enterprises and owners. The inequalities within structural, sectoral and regional development, too, have a decisive part in that process. Taken Latin America as a whole, about

---

[6] CEPAL, *Estudio sobre la distribución del ingreso.* 1967.
[7] E.F. Cardoso: "Income Distribution in Brazil." *Financial Times*, Sept. 13, 1973.

of the individual sectors and geographical regions which is also due to intervention on the part of the economic policy of the state.

The process of income concentration in Latin America was strengthened by the capital-intensive, labour-saving and mainly skilled labour demanding nature of import-substituting industrialization. Owing to the lack of experts, a fairly general phenomenon, and/or the greater mobility of specialists, their income level has developed not according to the actual domestic possibilities but according to international patterns, and this contributed to the further polarization of income distribution. As a result of extreme income distribution, the purchasing power of the majority of the population is almost negligible and is absorbed 50 to 60 per cent in the buying of food. Even in countries with larger populations the internal markets are rather narrow. Taking the given income distribution, the up-to-date branches of industry demanding mass buyers' markets and mass production cannot be developed economically. Mass markets set off from the assumption of a great number of individual consumers. From the point of view of mass production, a small number of consumers even with high incomes cannot be substituted for a great number of consumers, since according to experiences, the demand of the first category is much more diversified and is aimed first and foremost at top-quality goods which often have to be imported. Even the aggregate demand of all high-income strata of Latin America would not provide a mass market similar in size to England. The income structure is instrumental in the high ratio of consumption of high-quality or luxury articles, and this triggers off either a buoyant demand for imported goods or makes it possible to operate local productive units turning out but small series at a very high cost.

True, a number of countries demonstrate the fact that, within limits, more differentiated income relations exert a beneficent influence on the development of saving capacity. However, owing to the extent of inequalities in income distribution in Latin America, this stimulating effect is hardly to be felt. Incidentally, the consumption structure of the higher income strata with a possibility for saving does not conform to the consumption structure of the overwhelming majority of the country's population but to that of the countries with a high standard of consumption. The national cost and price levels of goods consumed in the developed countries, however, are much higher in the Latin American states. Thus, assuming an identical level of income and consumption, the percentage to be saved of incomes of high-income Latin American strata is lower, and the rate of savings is less than the possible ratio.

As a consequence of inequal income relations, the life style of the strata belonging to the highest income bracket undergoes a transformation. These strata get isolated from the broad masses, and inequalities become the sources of class distinctions and antagonisms. They impede internal integration and the strengthening of national unity. Directly and often indirectly, too, they tend to contribute to the preservation of the various elements of dependence appearing in external economic relations. Though the inequality in incomes partly arises from the structural distortions of the economy itself, it also affects production structures and impedes the building of an up-to-date economy which requires an expanding domestic market as well as a more balanced income distribution.

2.3. THE DEVELOPMENT OF AGRICULTURE

Within the general development trends of the Latin American region it is worth-while to devote special attention to the situation of agriculture. The special significance of agriculture is explained partly by its role played in general in growth, employment and exports, and partly by the fact that its nature is much more contradictory than that of other sectors. Due to out-of-date agrarian relations and to the slow increase in production, agriculture is the chief internal obstacle to present-day development in Latin America. The acceleration of agricultural development is by far not merely an economic problem but also the main objective of all ideas and movements aiming at social transformation and stepping up the rate of progress. In his recently published book, György Kerekes points out that the agrarian question is primarily a political and social problem.[8] The work quoted gives a highly competent survey of the socio-political aspects of the agrarian problem in Latin America. Our aim is to scrutinize the problems raised by economic growth.

In the course of the past twenty-five years the most important structural change in the economy of Latin America *has been brought about by the fading of the crucial role of agricultural growth.* Around the middle of this century it has been still agriculture which has supplied more than a quarter of Latin American gross national product; this fell to one-fifth by 1960, and to a mere 17 per cent by 1970. In the more developed major countries of the region, such as Venezuela, Chile, Argentina and Mexico, the share of agriculture is much less than the Latin American average, whereas in the smaller Central American republics it amounts to about 25 to 45 per cent. The decline of the macro-economic significance of agriculture shows an intensifying trend all over the world, being the result of structural development. In the case of Latin America however, the problem has not been simply that of keeping up with international trends, but, to a considerable extent, that of the agricultural sector which is incapable of progress and has a low degree of flexibility.

Agriculture is still the main source of employment. About two-fifths of the wage-earning population of Latin America, i.e. 35 million people in 1970, were employed in agriculture. The absolute number employed in agriculture has not decreased at all, and even the relative decline in the ratio of employment structure has been rather moderate. With the exception of the three formerly developed southern countries and Venezuela, agriculture is the means of subsistence of 40 to 60 per cent of all employed, and this ratio has dropped by only a few per cent over the past decade. It is quite obvious that under actual conditions, productivity in agriculture has hardly increased at all, whereas the development gap between agriculture and the up-to-date sectors of economy has further widened.

The lagging behind of agriculture is demonstrated most tangibly by the slow rate of progress which is falling behind the annual increase in population. Only Mexico and Venezuela show a dynamic growth of agriculture, while the rate of development in the one-time highly developed southern states with agrarian exports does not reach even the average level of the Latin American continents. (See Table 9.)

---

[8] Gy. Kerekes: *Kubától Chiléig* (From Cuba to Chile). Budapest, 1974. p. 74.

Development Trends

Table 9

The development of total agricultural production
(1961/1965 = 100)

	1965	1970	1971	1972	1973
Argentina	95	112	107	108	106
Brazil	115	128	136	143	149
Chile	102	118	114	108	106
Colombia	103	131	140	151	.
Mexico	113	122	129	128	133
Peru	104	120	122	121	125
Uruguay	102	101	93	86	89
Venezuela	112	144	147	146	158
Latin America	107	125	127	129	131
World total	105	120	124	124	129

Source: FAO, Production Yearbook, 1972.

The rate of development has been stepped up in the past decade, nevertheless it has not proved sufficient to curb the increasing problems caused by growth.

As a consequence of the uneven rate of development of the individual countries, agricultural production has been concentrated to an ever increasing extent in Brazil and Mexico. In the early seventies, half of the Latin American agricultural production has been supplied by these two countries, whereas the share of Argentina amounted to only 14 per cent, and that of Colombia to 10 per cent.

Examining the sources of this moderate rate of growth, the backwardness of the agrarian sector becomes even more conspicuous. According to C. Furtado,[9] more than two-thirds of the moderate increment of agricultural production between 1948 and 1965 derived from an enlargement of the area under cultivation, and merely one-third could be explained by increased productivity. Thus, *up to the present date, the growth of agricultural production in Latin America has been due to essentially extensive methods*, and at the same time, has left behind enclaves of anachronism. Agricultural yields which are extremely low as compared to international standards, have increased—with some exceptions—rather slowly. (See Table 10.)

Table 10

Yield of the most important agricultural produce
(q/hectare)

	Wheat		Corn		Rice	
	1958/62	1972	1958/62	1972	1958/62	1972
Argentina	13	16	18	31	31	35
Brazil	6	18	13	14	16	15
Chile	13	17	20	34	29	33
Colombia	9	13	11	12	20	36
Mexico	16	27	9	12	18	27
Peru	10	10	13	18	39	42
Uruguay	9	10	8	8	34	41
Venezuela	5	4	11	18	11	15
USA	16	22	39	61	39	53

Source: FAO, Production Yearbook, 1972.

[9] C. Furtado: *op. cit.*

Both the level and the dynamics of agricultural productivity is extremely varied. Chile achieved the highest level, whereas the fastest rate of development was experienced by Colombia. The results of the Green Revolution are represented by Mexican average wheat yields.

The development of yields reflects the effects of a number of factors. In the case of Mexico which in old times had the most backward agriculture and unfavourable natural endowments, the effects of the land reform and an economic policy aiming at the development of agriculture are clearly discernible. In Colombia a policy based on diversification, in Venezuela the efforts of a government policy aimed at the technical development of agriculture are the most outstanding features of agrarian policy. Yields in general show a close connection with the efficiency of efforts aimed at the modernization of agriculture, and also with the increased use of fertilizers and stepped up mechanization. In countries stepping up the use of chemical fertilizers at a high rate (e.g. Venezuela, Mexico, Argentina) the increase in yields is much faster. Since cheap labour force available impedes mechanization, the degree of mechanization of Latin American agriculture is rather low as compared to international standards. Around the middle of the past decade the number of tractors per thousand hectares of arable land has been 1 in Brazil, 3 in Argentina, 6 in Chile, whereas 9 in Australia having an extensive agriculture, 46 in France, and 52 in Great Britain. Thus, the lagging behind, as regards the level of technology, has been much greater than it could have been assumed on the basis of the developmental and structural indicators of the individual Latin American countries.

Within agriculture the situation of animal husbandry is especially unfavourable, though the endowments of the continent as regards animal husbandry

Table 11

Development of livestock
(million)

	Cattle			Pigs		
	1961/65	1969/71	1972	1961/65	1969/71	1972
Argentina	43	50	54	3.5	4.3	4.4
Brazil	79	98	99	53.1	66.4	67
Chile	3	3.1	3.1	1.0	1.2	1.2
Colombia	16	21	22.4	1.6	1.5	1.5
Mexico	21	25	26	9.2	11.7	12.3
Peru	3.4	4.1	4.3	1.8	1.9	2
Uruguay	8.6	8.2	9.3	0.7	0.5	0.5
Venezuela	6.7	8.5	8.5	1.7	1.6	1.7

Source: *FAO, Production Yearbook*, 1972.

are excellent. In most of the countries the number of livestock does not keep abreast with the increase in population, and in certain countries it is even stagnant (see data in Table 11).

Again the rate of growth was satisfactory only in the two big countries, i.e. Brazil and Mexico. The slow development of animal husbandry was primarily due to mistakes made in economic policy, excessive income regulation and neglected development. The ousting of animal husbandry from the more developed regions and related cost increases were also responsible for this trend. As a consequence of practically stagnant livestock, low yields and decreasing rentability, beef production out of cattle breeding being of outstanding importance in the region has increased from 5.4 million tons in 1961/65 to but 6.4 millions tons in 1972. This had a considerable part in both internal and external imbalances.

In addition to difficulties in the process of growth, the slow progress of agriculture has given rise to negative developments primarily in the external economy. Agriculture used to be traditionally a foreign exchange earning sector in Latin America, and even at the end of the sixties it supplied 54 per cent of the total foreign exchange revenue from exports and one-seventh of agrarian world exports. The volume of commodity stocks to be exported was affected both directly and indirectly by slow development. On the one hand, the supply of export goods could be increased only to a very modest extent, and on the other hand, domestic demand which had grown as a result of stepped up development, had to be satisfied, not at least for political reasons, out of export commodity stocks. In 1970/71 the share of exports in regional meat production decreased from 18 to 16 per cent, whereas that of wheat production from 23 to 15 per cent as compared to ten years earlier. As a consequence of insufficient commodity stocks, the role of imports has increased in satisfying domestic demand. The value of agrarian imports had been only 970 million dollars in 1960, but reached 1.9 billion dollars in 1970 and 2.3 billion dollars in 1972, and has grown at a much faster rate than that of agrarian exports.

Thus, neglect of agricultural development has unfavourably affected external economic relations in addition to bringing about socio-political tension and internal economic disequilibrium.

2.4. THE CHARACTERISTICS OF INDUSTRIALIZATION

The industrial development in Latin America has undoubtedly a longer history than in other developing countries, in fact a longer one than in some of the less developed European countries as well. Already in the last century, the joining in the international division of labour has had a stimulating effect on branches of industry processing certain export commodities, and also on branches turning out building materials and simple consumer's articles for domestic consumption. Prior to the World Depression, the share of industry in the national economies of Argentina, Mexico, Chile and Brazil was already significant but its development was closely linked to the expansion of the external sector of the economy. Liberal economic policy, a relative abundance of foreign currency, dependence on foreign financiers, the high wage level of skilled labour (even compared to international standards) generally did not render it possible to cross the structural thresholds and to step up the development of branches turning out investment goods.

The outbreak of the World Depression, and later on the changeover to the so-called "inward-looking" growth, initially reduced considerably the dependence of industrial development on the external sector; industrialization was based almost entirely on making the most of the opportunities provided by import substitution. Beginning with the thirties, industry became the basic source of growth, and by the sixties it was already the dominant sector of economy in the majority of the Latin American countries. The outstanding characteristics of industrialization in Latin America in the past decade have been closely linked to the establishment of the dominant position of industry in the national economy on the one hand, and to the joining in the international industrial division of labour on the other.

The establishment of the dominant position of industry in the national economy is rooted partly in the development of industry exceeding the rate of development of the national product, and partly in the acceleration of the process of industrialization itself. In the period 1938/50, the average annual

Table 12

The share of the manufacturing industries in the gross national product (per cent)

	1929	1947	1960	1965	1970	1973
Argentina	23	31	30	33	36	37
Brazil	12	17	19	19	20	22
Chile	8	17	23	26	28	29
Colombia	6	12	19	19	19	21
Mexico	14	20	21	21	23	25

Source: C. Furtado: *Economic Development in Latin America*. Cambridge University Press, 1970; QER 1974.

rate of growth of industrial production amounted to 5.2 per cent, between 1950 and 1960 to 7 per cent, between 1960 and 1965, as a result of the exhaustion of the process of import substitution, to 5.5 per cent, in the period 1965/70 to 7.3 per cent, and between 1970 and 1973 to 9 per cent. The rate of industrialization in Latin America was faster than the world average, and therefore, the share of the continent in total world industrial output increased.

The results of accelerated progress in this field are to be seen in the trends of *structural development*, too (see Table 12).

The exhaustion of growth based on import-substituting industrialization in the five countries supplying 85 per cent of the production of Latin American manufacturing industries is reflected also in the practically stationary state of their macro-economic structures in the period 1960/65. The reasonable possibilities of import substitution have been exhausted and the rate of import substitution has decreased. Structural development got new impulses only from the joining in the international division of labour and from the launching the export of industrial articles. Structural development, however, was remarkable not only within the national economy as a whole but also within the industrial sector. The phase when light industry dominated in import-substituting industrialization had ended already in the fifties, and in the sixties efforts for development concentrated increasingly on heavy industry. Following the development of metallurgy, the chemical and engineering industries, the share of the branches of heavy industry in 1970 was 54 per cent in Brazil, 56 per cent in Argentina, 48 per cent in Chile, 55 per cent in Mexico (calculated on the basis of value added of the manufacturing industries).[10] The characteristics of the industrial structure are much more favourable in Latin America than in the other developing regions or in the smaller European capitalist countries, and are approaching those of the developed major countries.

[10] U.N. *Patterns of Industrial Development in Latin America*, 1966.

Development Trends

The adoption of the structures of the developed countries at the present stage of economic development seems to demand disproportionately great sacrifices and to result in losses. Thus, *oversized structures have become prematurely old* in countries with more modest resources. In the sixties the industrial dynamics of Argentina, an industrial country from the point of view of structural development, and that of the small Chile already fell behind regional averages, thus illustrating the impeding effect of too early and over-developed structures.

In the heydays of import-substituting industrialization the rate of development of the branches of heavy industry was more than twice that of the light industries. In the early seventies the gap in the development rates of the two sectors narrowed considerably. The present difference of level is for the most part due to the slower progress of the food industries based on a stagnant agriculture, whereas the development of the textile and the clothing industries is hardly lagging behind the average.

Recent events in structural development and the acceleration of industrialization in the past years reflect first and foremost *the reassessment of the development strategies of the Latin American countries and/or the launching of industrial exports*. Exports never had a significant part in the history of Latin American industries, and with the exception of Mexico, industrialization was mainly of an import-substituting nature at all times. According to calculations of the ECLA (Economic Commission for Latin America) in the case of the five leading industrial countries mentioned, 36 per cent of the increase in industrial output in the period 1929/60 was due to import substitution. As regards Argentina, Chile and Colombia, more than 50 per cent of the growth of industrial production[11] was directly stimulated by import substitution and only a minor percentage by the broadening of the domestic markets.

The radical turning-point which had not been anticipated neither by the competent economists nor by the planners came about in the late sixties. In 1960 only 5 per cent of the industrial output was exported, and the value of industrial exports amounted to a mere 780 million dollars. By 1972 exports of industrial articles has risen to 10 per cent of the production, amounting to the value of 4 billion dollars. In the case of countries developing under the conditions of a constant protectionism, such results are regarded as extremely favourable, especially if it is taken into consideration that one-sixth of the 24 billion dollar increment of the industrial output of the period 1960/72 (calculated at 1970 prices) was realized by exports. The gradual switchover of an industry traditionally based on an internal division of labour to the international division of labour has undoubtedly opened up a new stage of growth. It would not be entirely without interest to examine the internal and external factors shaping the change in the patterns of industrialization.

The socio-economic emergency situation which came about in the wake of the exhaustion of the "inward-looking" model had a major share in the switchover from one phase to another. Import-substituting industrialization had been of a capital intensive and manpower saving character, and thus could not do away with traditional unemployment; it could provide job opportunities for only a fraction of the new labour supply. (It absorbed but one-sixth of the new labour force in the period 1948–65.) Great differences between the living standards of people employed in the modern sector of the economy and the increasing number of workers excluded from there have

---

[11] U.N. *The Process of Industrial Development in Latin America, 1966.*

resulted in major tensions of social and domestic policy. The unemployment crisis necessitated development variants and strategical shifts in priorities which could ensure the maximal utilization of unemployed manpower. Turning out labour-intensive goods for export enabled the employment of redundant manpower and, at the same time, opened up possibilities for easing the foreign exchange disequlibrium as well as for accelerating the process of structural development.

In certain countries of Latin America the emergency situation by itself would not have ensured the success of changeover, if new international trends, favourable from the point of view of export-oriented industrialization, had not come into being. In connection with this, we have to mention, in the political sphere, the modification of the global strategy of the United States, the reduction of direct American presence, more differentiated economic relations between the US and Latin America, the increased interest of Western Europe and Japan in the continent, as well as the strengthening of the international positions of the socialist countries.

The economic reasons are of a similarly varied character. Under the influence of the increasing cost inflation in the developed capitalist countries, the deterioration of the competitiveness of labour-intensive branches of industry not in the vanguard of technological progress, and their gradual retreat from higher income-level areas have become increasingly conspicuous in the past decade. The demand in products turned out by labour-intensive, technologically simple industries lagging more and more behind in the developed capitalist countries is met increasingly by imports from countries with low wage levels. As a result of these shifts, there has developed the trend of locating labour-intensive industries in countries with low wage standards. In the early sixties this process has covered only the countries of Southeastern Asia, in the second part of the decade, however, it has extended also to the Latin American countries. Joining in the industrial division of labour is primarily based on comparative wage benefits, recently, however, the trend has been strengthened by the industrial overcrowding of the developed countries and also by some considerations regarding the problems of environmental protection.

The import of industrial articles from the Latin American and other countries with low wage levels serves the business interests and the political aims of the more developed countries and primarily those of the multinational corporations. Import also eases tension on the labour market; it puts a brake on wage-demands of the workers in the developed countries; it weakens the bargaining positions of the trade unions, and slows down the rate of cost inflation.[12] It may also ease environmental problems, and in the long run may contribute to the balancing of the economic disequilibrium and to the stabilization of the political scene in the developed countries with market economies.

The fact that in the Latin American countries the changeover from one phase of growth to another coincided with the emergence of new international trends, quite obviously set in action socio-economic trends which were not confined to the industrial sector, and had a part in the weakening of the so-called populist coalitions including the national bourgeoisie, the intelligentsia and organized labour. At the same time, the comprador features of the bourgeoisie have become enhanced. Though anti-American sentiments have not been

---

[12] B. Kádár: "Új írányzatok a fejlődő országok iparosodásában." (New Trends in the Industrialization of the Developing Countries). *Világgazdaság*, April 6, 1973. Supplement.

mitigated by the new global strategy of the United States, the attitudes of the reformist parts of the bourgeoisie are characterized by increased prudence, as compared to the past decade.

The establishment of an export-oriented industrialization stimulated industrial differentiation within the region. The intra-regional industrial positions of countries being able to make the most of the advantages gained by joining the international division of labour and the new global strategy of the United States, have strengthened considerably. The share of Brazil in the industrial production of the Latin American region has risen from 24 per cent in 1960 to 28 per cent in 1972, that of Mexico from 20 per cent to 24 per cent, whereas that of Argentina has dropped from 28 to 23 per cent. The development of foreign currency earning functions (especially in the bigger countries) has ensured industry a major and more active role in external economy. Recently, the interests of the industry started to play an increasing role in the motivation of the regional and international actions of the individual Latin American countries, and in the sphere of external economy and diplomacy, too, the shift in the industrial balance of power has made itself more and more manifest.

2.5. THE SERVICE SECTOR

We have various reasons for examining separately the position of the service sector in Latin America. It is a well-known fact that the most dynamically developing branch of the capitalist world in the past twenty-five years, as well as the pioneer of intensifying the international division of labour has been the service sector. The share of the services in the national product of the developed capitalist countries has reached 45 to 55 per cent, and 10 to 15 per cent in the international flow of payments. The importance of the service sector is especially significant in the economic and external economic development of the recently industrialized and smaller advanced capitalist countries. Services represent to an increasing extent the most important foreign currency earning branch of the national economy in almost all the small and industrially "latecomer" capitalist countries.

The progress of the service sector in Latin America, however, differs from the international trends. People, who were not needed in agriculture and were not employed by industry either, have increasingly sought for job opportunities in the service sector. Thus, this sector plays the important role of an employment "safety valve", whereas in the range of development policy it is regarded as a rather neglected field. In the period 1960/71 the service sector developed at a slower rate than the average of the national economy, this being an unparalleled phenomenon in international economic life. The relatively slower growth of the sector of services has brought about serious problems in the domestic economy, too, and has become an important factor of the increase of tensions on the labour market, of the decrease in the productivity of the services, and of the declining living standards of urban population in the lowest income bracket.

The fact that insufficient attention had been paid to the service sector resulted in grave problems in both the external and the domestic economy. Thus, many opportunities on the world market could not be used. Unfortunately, statistical data published by the Latin American countries pay little attention to the service sector and, therefore, quantitative trends of development can be pointed out only concerning international tourism. In the period 1950/70 is was international tourism that proved to be perhaps the most dynamic branch of international economic relations. The number of tourists rose from 25 million to 158 million, and profits gained from tourism increased from 2.1 billion dollars to 17.4 billion dollars. In a number of belatedly modernizing European countries, such as Spain, Greece, Portugal, or Ireland tourism had a decisive part in economic expansion.

Owing to both its climate and abundant manpower supply, Latin America has excellent possibilities for the development of tourism, however, because of neglecting development strategy, this sector is almost entirely controlled by private capital. The quality of services and the price-level is scarcely competitive, and organization is inefficient. In 1971 the share of Latin America in the total earnings of international tourism amounted to a mere 12 per cent. However, regional gross data are extremely misleading, since two-thirds of the tourist traffic of the region is directed to a single country, Mexico. In 1971 the share of Mexico in the continent's total earnings from tourism (2.4 billion dollars) amounted to 1.6 billion dollars, that of Jamaica to 109 million, of Argentina to 93 million and of Panama to 80 million dollars. Within the current headings of the balance of payments, revenue from tourism reached 48 per cent in Mexico, 18 per cent in Panama and Uruguay, while in all the other countries it had no significant share in foreign exchange incomes. Contrary to other small countries, the overwhelming majority of the Latin American states did not rely on the development of tourism which could have made it possible to improve foreign exchange positions and would have served as an effective means of making use of mainly domestic material and human resources, and also of easing problems arising from differences in regional development.

Besides tourism, exports of services are of major economic significance in the case of two countries: Panama and Uruguay. Whereas the sector of services and the national economy of Panama—based on canal revenues, transit trade and shipping—develops dynamically, the role of Uruguay, the one-time "Switzerland" of Latin America as a trader and banker, has gradually withered away. In connection with the phenomena mentioned, the service industries do not fulfil the functions of the foreign exchange earning branches of the economy, and the external balance of services in the majority of the Latin American countries shows a deficit and has become the source of tension in foreign exchange policy.

2.6. THE DEVELOPMENT OF ECONOMIC POWER RELATIONS IN LATIN AMERICA

Let us compare the performance of economic growth in the Latin American economies with world development trends. The rate of economic progress has undeniably been stepped up in the countries of the region as compared to the period prior to World War II. Acceleration of growth, however, has been a world economic phenomenon being a consequence of the unfolding of the scientific and technological revolution, the intensification of the international division of labour, and the increasing rate of socio-political progress. In the period 1950/70 the world average of the growth rate of the national product was higher than that of Latin America. The growth rates of the Soviet Union, but also those of Japan and the OPEC countries, even of some countries of Southeast Asia, as well as of Europe exceeded those of Latin America.

Development Trends

Although the rate of industrial development was somewhat higher than the world average, all the regions mentioned above overtook Latin America. The decreasing role of the region in world trade is particularly considerable. World trade has developed twice as rapidly as that of the Latin American region, and with the declining share of the continent in the world economy as a whole, its significance as an international importer of capital has decreased, too. Table 13 shows the trend of changes in the world economic significance of Latin America.

Table 13

The weight of Latin America in world economy
percentage of world data)

	1950	1965	1970	1973
National product	5	4	4	5
Industrial production	4	4	4	5
Exports	11	6	5	5
US capital investments	38	19	16	14
Gold and foreign exchange reserves	5	5	6	7

Source: ECLA Economic Survey 1971; Monthly Bulletin of Statistics; Survey of Current Business (various issues).

The development of percentage shares clearly indicates the decline of the world economic importance of Latin America in the period of "inward-looking" growth. Only the establishment of heavy industries within the framework of import-substituting industrialization has kept abreast with the rate of international growth. In all other fields there was a marked weakening of Latin American positions. After 1965, due to the growth-accelerating effect of the increased participation in the international division of labour, the lagging behind world economic growth was slowed down to a certain extent, but the weakening of Latin American positions has not been stopped. In the period under review, the process of the weakening of world economic positions was a characteristic feature of the countries of the Third World, nevertheless, taken as a whole, the position of the Latin American states has deteriorated to a greater extent than the Third World average.

Despite the regression mentioned above, Latin America has a rather important part in the world exports of petroleum, cotton, copper, lead, tin, zinc, coffe, fish-meal, sugar, and in a number of other products, too, and its positions as a supplying region are rather strong. Its importance as an international importer of chemicals and engineering goods is also significant.

In the mid-sixties the share of Latin America in the aggregate national product of the developing countries amounted to 38 per cent, in industrial production to 48 per cent, in exports to 30 per cent, in investments to 35 per cent, in capital import to 41 per cent, and within the latter, in capital import channelled into industry to 61 per cent. This clearly shows that, despite the loss of importance of Latin America in the world economy, this continent is the most advanced region of the developing world, owing to its significant economic potential. Prior to the outbreak of World War II, certain Latin American countries, such as Argentina, Uruguay and to a lesser extent Chile, figured among the leading countries as regards the level of *per capita* national product;

presently they have dropped back into the group of medium developed countries.

In the period following World War II, Latin America was unable to keep abreast with the accelerated growth rate of the world economy and—owing to the policy of "inward-looking" growth—could not exploit its benefits. Thus, the peripheral character of the region was not mitigated but strengthened by recent development trends.

In the course of the past quarter century the economic balance of power has shifted in the region, and the influence of certain countries within the region has undergone considerable changes. Prior to World War II, Argentina had been the most developed country of the region, with the most significant economic potential. Owing to an exaggerated and lasting policy of import substitution, however, the rate of growth declined markedly, and thus the influence of Argentina within the region rapidly decreased. Similar was the lot of Uruguay, also regarded as a "developed" country before World War II, and that of Chile, too. Parallelly to the regional eclipse of the one-time relatively more developed "southern white states", the process of the emergence of Brazil and Mexico as regional growth centres took increasingly distinctive outlines. Let us consider now the development of the main indices indicating the trends of intra-regional economic balance of power. (See Table 14.)

In addition to the shifts in the hierarchies of economic power, the development of the balance of power within the region casts a light on the considerable

Table 14

The economic positions of some Latin American countries within the region
(in percentage of the region)

	Argentina	Uruguay	Chile	Brazil	Mexico	Venezuela	Colombia	Peru
National product								
1950	25	3	6	26	20	6	6	4
1960	20	3	6	24	20	7	5	4
1974	15	1	5	32	23	8	5	4
Industrial production								
1950	32	4	7	21	18	.	.	.
1960	28	3	6	24	20	6	4	3
1974	22	1	5	32	24	6	4	3
Exports								
1950	22	3	5	16	6	14	4	2
1965	12	2	5	13	9	22	4	5
1974	8	1	5	16	7	31	3	3
Foreign exchange reserves								
1965	7	5	4	14	15	24	3	5
1974	7	1	.	29	8	36	2	2
US capital investments								
1950	8	1	11	13	9	21	4	3
1965	11	.	9	12	13	29	6	6
1972	10	.	5	19	15	20	6	5

Source: ECLA, *Economic Survey, Monthly Bulletin of Statistics, Survey of Current Business* (various issues).

extent and increasing trend of power concentration as well. The bulk of the national product, industrial output and exports falls to the share of the three biggest countries, i.e. Argentina, Brazil and Mexico. Even more noteworthy is the breakdown according to some qualitative indicators of progress. The significance of these three countries determines by sheer statistical computation the main patterns of development within the region.

Due to the disparate rate of development, *the differences in the level of development* of the various countries has also markedly changed. The volume of *per capita* gross national product (exceeding three times the average of the other developing countries) increased from an annual 415 dollars in 1960 to 533 dollars by 1970, and to 620 dollars by 1973 (calculated at 1970 prices). The level of *per capita* GNP in the individual countries is illustrated in Table 15.

Table 15

Per capita national product in various Latin American countries (calculated at 1970 US $ prices)

	1955	1965	1970	1973
Argentina	936	1,261	1,395	1,448
Bolivia	174	161	182	187
Brazil	330	388	481	599
Chile	690	786	876	930
Colombia	296	327	378	417
Mexico	536	676	795	891
Peru	250	317	323	348
Uruguay	870	895	910	890
Venezuela	970	1,251	1,269	1,388

Source: ECLA, *Economic Survey of Latin America*, 1969.

As we can see from the above data, the slow economic growth in the formerly more developed three southern countries and/or the accelerated progress of other countries resulted in a development level in Latin America which is less polarized and more balanced than it used to be ten years ago, although historical differences in the degree of development have not at all faded away. The dispersion of the indices of the individual countries as compared to regional mean values has decreased.

Ranking the countries on the basis of *per capita* national product, Argentina and oil producing Venezuela have indicators of members of the highest category of the medium developed countries. Mexico, Panama, Costa Rica, Chile, Uruguay and Brazil have threshold values characteristic of medium developed countries. On the opposite pole there is Haiti with a *per capita* income falling short of $100, thus being one of the most backward countries not only of the continent but of the whole Third World. The rest of the Latin American countries is to be found in the bracket of $200–400 *per capita* incomes, i.e. in the top category of the underdeveloped world. Out of these, Colombia and Ecuador will presumably enter the group of medium developed countries in the late seventies.

Summarizing the almost 25-year-old development trends of the region, we get a highly diversified picture. Slowest development in the region is to be encountered obviously in the one-time more developed countries which based their economic policies lastingly and very intensively on the socio-economic system of protectionism, and tried to adopt both in theory and practice the institutional systems of the welfare state of the developed capitalist countries. The wide economic gap—which came into existence from the middle of the 19th until the middle of the 20th century—between the historically more centralized or liberal models of the region, has narrowed down quite considerably, whereas the progress of the former "Indian" countries (e.g. Mexico, Peru, the Central American republics, Ecuador) drawing on the historic heritage of the centralized growth model has been stepped up.

The growth dynamism of the individual countries can only be partly accounted for (though by far not to a small extent) by different development strategies and/or evolutionary starting points. It would be wrong not to mention the problems of *natural endowments* having so great a part in the structure and rate of traditional growth, or *the size of the national economy* which exerts an increasing influence on up-to-date growth. The exploration of new natural resources in the past quarter century has had a much smaller effect on Latin American development than in the previous hundred years, or in other developing countries. The most important exceptions to this are the events in Venezuela in the forties and fifties, and those in Ecuador in the seventies. The development of the oil riches of these two countries has radically changed the rate, proportions of growth and the income ratios of these two countries. (In the case of Ecuador this process is still going on.) To a smaller extent Peru received a similar impulse in the period 1950/70 from the recovery of the mining industry which had been dormant ever since the colonial rule, and also from fishing which had forged ahead in the sixties and soon achieved a leading role in the international market. However, in the rest of the Latin American countries, the role of natural resources in growth remained essentially unchanged, and the share of extractive industries in the national product remained by and large the same or even decreased.

As indicated by the development of sectoral progress, industrialization has been the driving force of economic growth in the past quarter century. Industrialization, and especially the development of capital-intensive up-to-date industries demanding extensive markets and mass production depend to a large extent on the economic power of the individual countries. Bigger countries afford more favourable conditions for the unfolding of industrialization, first of all for the development of the heavy industries. Thus, in the group of developed capitalist countries the share of the chemical industry, metallurgy and engineering industry in the production of the manufacturing industry amounted to 58 per cent, whereas in the small countries it amounted only to 43 per cent.

As regards the main characteristics, the economic growth of the Latin American countries took place in a similar way. Brazil and Mexico, the two big countries have been the most dynamic states of the region being able to expand their regional economic power to a maximum extent. Even more tangible are the advantages in structural development originating from the economic potential of these countries. More than half of the output of the manufacturing industry in these two countries is turned out by the branches of the heavy industry mentioned. The indicators of their structural development are more favourable than those of the minor European capitalist countries, and are approaching the structures of the economic great powers. At the same time, in the small countries of Latin America heavy industry is undeveloped or its share in total industrial output is only 10–20 per cent, and even this comes mainly from assembly enterprises.

From the point of view of structural development, Argentina is a unique case. Due to historic developmental standards and an exorbitant protectionist policy, her indices were much more favourable than those of Brazil and Mexico the socio-economic development of which took place in a more heterogeneous environment, and whose general living standards were much lower. Nevertheless, the structural advantages of Argentina have gradually faded away during the past two decades, the price of the prematurely developed structures having been a low rate of development for more than fifteen years and the loss of positions within the region.

The shifts in the main development trends as well as in the balance of economic power of the Latin American region indicate that the world-wide economic and technological progress has affected in a different way the individual countries with dissimilar development backgrounds, economic potentials and patterns of economic growth.

CHAPTER 3

PROBLEMS OF THE EXTERNAL ECONOMIC SECTOR

3.1. THE MAIN DEVELOPMENT TRENDS IN LATIN AMERICAN FOREIGN TRADE

Foreign trade is one of the most critical fields of economic growth and external economic dependence in the developing countries. Until the start of the World Depression, the economic development of Latin America was rooted in the increasingly intensive participation in the international division of labour. A considerable part of the national income, and within that the major part of the output of up-to-date, efficient sectors was realized through exports, whereas the overwhelming part of investment goods being of decisive importance from the point of view of economic progress, was covered by imports. Parallel to the emergence of the so-called "inward-looking" development started by the World Depression and the events of the post-war years, the foreign trade vulnerability of the Latin American countries has decreased, and an ever greater part of formerly imported goods has been substituted by domestic production. Table 16 illustrates the development of the ratio of imports and the national product in some of the Latin American countries.

Table 16

The share of imports in the national product

	1929	1947	1957	1965	1970	1973
Argentina	18	12	6	6	7	11
Brazil	11	9	6	5	8	12
Chile	31	13	10	10	10	—
Colombia	18	14	9	8	12	12
Mexico	14	11	8	8	8	10

Source: C. Furtado: *Economic Development of Latin America*. Cambridge University Press, 1970.

Undoubtedly it was the decreasing dependence on foreign trade which rendered possible a development more independent of world economic trends. The rapid slackening of participation in the international division of labour and the extremely low level of such participation, however, constitutes a serious threat to growth. With such a dramatic decrease in the international division of labour which is an important factor of growth we may have serious doubts about long-term economic results. (Brazil with her huge territory and to a certain extent Mexico are exceptions to this as both countries are in a more favourable position as regards the extensive domestic division of labour.)

At the same time, the rapid and large-scale decrease of the role of foreign trade in the national economy reduced considerably that sphere of the national economy which would be able to make use of the results of world-wide post-war technological and scientific progress and accelerated growth. Assuming

Problems of the External Economic Sector **75**

a similar degree of development and size of national economy, nowhere else do we find such low shares of foreign trade in GNP as in the Latin American countries pioneering in import substitution. It is only in the case of some big countries, such as the Soviet Union, the United States, China or India that this rate is lower than in some Latin American countries regarded as small or medium economies as compared to international standards. As mentioned before, import substitution has been a factor of cost-raising and inflation, depending on the economic size of the individual countries and on the intensity of the process. Besides increasing internal imbalance, however, the stagnation or slow rate of growth of foreign trade had also a considerable effect on the rate of economic growth. (See Table 17.)

Data in Table 17 indicate a number of problems in the external economy of the Latin American countries. The gradual decline of the role of Latin America in world trade is explained to a certain extent by the slow growth rate of foreign trade. This basic trend has been but damped by the new export orientation of recent years. As compared to the pre-war period, only the world economic role of Brazil, Mexico and Venezuela has shown an upward trend.

Comparing the dynamics of foreign trade turnover with the growth rate of the national product, it turns out that, concerning the average of Latin America as a whole, we can by no means speak of a foreign trade which is in harmony with growth. As an unparallelled phenomenon in the world economy *the rate of growth of foreign trade in the period 1950/70 was slower than that of the national product*, moreover in some of the Latin American countries, such as Venezuela, Uruguay or Haiti, foreign trade turnover has virtually come to standstill for a long time.

In the course of the past two decades there rapidly developed all over the world the so-called inverse substitution process, i.e. an ever growing percentage of the less economical domestic production is substituted by imported goods representing a more up-to-date technology or a more competitive quality. An increasing part of the production capacities is switched over to satisfy world economic demands. The resulting intensification of the participation in the international division of labour is clearly illustrated by the fact that between 1950 and 1970 the growth rate of world trade was more than twice that of the gross national product, and within that in the smaller, medium developed countries this coefficient was even higher. At the same time, the coefficient in Latin America for the period 1950/65, i.e. at the time of import substitution was 0.7, for 1965/70 it was 0.8, and it increased only between 1970 and 1973 to 1.8. Thus, despite the enhanced foreign trade sensitivity of growth, foreign trade was able to exert an accelerating effect, moderate as compared to international standards, even as late as the early seventies.

In spite of moderate foreign trade sensitivity, not quite unambiguous though generally clearly discernible connections can be detected between the growth process and the dynamics of foreign trade. With the sole exception of Venezuela, all the countries with a stagnant foreign trade are countries falling behind the general trend of development. Contrary to the experiences of the forties and fifties, no country with stagnant foreign trade relations could achieve a dynamic growth.

Naturally, the dissimilar development trends of foreign trade have also influenced the significance of the individual countries in world trade. Compared to the situation twenty-five years ago, the importance of Brazil, Mexico and the small Central American states in Latin American foreign trade has consid-

Table 17

The development of the foreign trade turnover of Latin America

	Latin America	Argentina	Brazil	Chile	Colombia	Mexico	Peru	Uruguay	Venezuela
Exports ($ million)									
1938	1,720	438	296	139	81	159	76	62	181
1950	6,490	1,178	1,758	372	396	466	189	254	1,147
1965	12,660	1,493	1,596	688	539	1,120	666	191	2,744
1970	17,160	1,773	2,739	1,274	724	1,402	1,044	233	2,691
1973	26,160	3,050	6,203	1,231	1,210	2,446	1,048	322	4,740
Imports ($ million)									
1938	1,450	443	296	103	89	110	60	62	18
1950	5,470	964	1,098	248	365	509	176	201	596
1965	12,030	1,199	1,096	604	454	1,493	719	150	1,297
1970	19,000	1,685	2,849	931	844	2,461	619	233	1,713
1973	30,000	2,100	6,150	1,098	1,100	4,034	1,619	285	2,524
Balance ($ million)									
1938	270	—5	0	36	—8	49	16	0	163
1950	1,020	214	660	124	31	—43	13	53	551
1965	630	294	500	84	85	—373	—53	41	1,447
1970	—1,840	88	—110	343	—120	—1,059	425	0	978
1973	—3,840	950	53	133	110	—1,588	29	37	2,216
Annual average increase of exports (per cent)									
1938–50	7.5	7.5	16.0	8.6	14.1	9.4	7.9	12.5	16.6
1950–65	4.6	1.7	—0.6	4.1	2.0	5.9	8.8	—1.4	6.0
1965–70	6.6	3.4	11.3	9.8	6.2	4.4	9.6	4.1	—0.4
1970–73	13.0	19.6	32.0	—1.1	18.6	20.5	0.1	11.3	20.5

Source: U. N. Statistical Yearbook, Monthly Bulletin of Statistics, June, 1974.

erably increased, as these countries realize almost a fourth of the regional trade turnover. A set-back is experienced, however, by the one-time developed southern and by some of the Caribbean countries, as well as by Haiti and Dominica. The polarizing effect of world trade relations furthers the shifts of the centres of gravity of growth within the region.

The interrelations of accelerated economic growth and an increased participation in the international division of labour are characteristically illustrated by the development of *foreign trade prices* and the *terms of trade*. This indicates

Table 18

World-market price indices of some important Latin American products
(1963 = 100)

	1965	1968	1969	1970	1971	1972	1973
Food-products	100	90	101	118	107	121	181
Agricultural raw materials	101	96	98	98	95	108	149
Metals	129	148	174	172	145	145	210
Petroleum	93	91	90	92	112	126	201
Total	104	99	106	114	114	127	190

Source: *Yearbook of International Trade*, 1963, 1966, 1968, 1973.

on the one hand the role of accelerated world economic price increases and the spread of international inflation to the field of foreign trade relations in the foreign trade expansion of the individual countries, and on the other hand the influence exerted by the changes in the terms of trade on the rate of growth (see Table 18).

Despite considerable fluctuations in the prices of certain products, the world market price level of the main export products of the region has not experienced major changes, and only about one-fifth of increased export turnover can be attributed to the rise in prices. In the seventies, however, the situation has altered. Between 1970 and 1973 almost half of the 52 per cent export increase originated from the rise in prices. Thus, taking the average of the region, the increase in export capacity was by far not of an astonishing speed in the period 1970/73.

However, the fact that the combined statistical data of the region are regarded as a single entity, blurs the differences between the countries dependent on certain export products. The relative stability of oil prices in the period 1965/70 undoubtedly had a part in the slow increase of Venezuelan exports, a country of relatively high production costs. Taking the average of the period, the price level of agricultural produce of the temperate zone, with the exception of meat prices, has remained unchanged. Owing to the lack of adequate stocks, rising meat prices could not be made use of by the greatest meat exporters, i.e. Argentina and Uruguay. The world market price level of sugar and cotton, too, has developed unfavourably for a long time. The rising price level of coffee, however, has undeniably given an impulse to the economies of Brazil, the Central American Republics and Colombia. The world market positions of ore exporting countries have developed even more advantageously: in the second part of the sixties the price level of copper has exceeded the 1963 level by an average of 80 per cent. In those years the steep rise of the exports of Chile and Peru was essentially connected with price fluctuations of the world market, and no actual increase in export capacities could be demonstrated.

The unfolding of inflation and the raw material crisis on a world scale has primarily improved the positions of those countries which export oil and mining products, since speculative demands, related to inflation, are keenest for such products.

The rise in prices of raw materials spread over to agricultural products, too. Their price level increased by 13 per cent in 1972, and taking Latin America as a whole, by an average of 79 per cent in 1973. Thus, the price of cacao had been redoubled in 1973, the export price of Peruvian fish-meal increased by 120 per cent, that of Mexican cotton by 72 per cent, of Argentine wool by 71 per cent, of wheat by 98 per cent, and of corn by 89 per cent. The change in import prices was much more balanced: taking the average of the past decade, the annual average increase in the price level amounted to a mere one per cent. In 1970/72 however, the average price level of imports exceeded the average of the period 1960/63 by 19 per cent, and in 1973 by 41 per cent. The combined terms of trade of the region fell short of the average of the period 1951/55 by 11 per cent in the period 1956/60, by 19 per cent in 1961/63, and following the rise in prices between 1963 and 1966, have virtually stabilized. According to UN surveys, the terms of trade for 1970/73 were higher by 5 per cent than the average for 1960/63. Thus, half of the approximately two per cent acceleration of growth rates in the seventies originated from the improvement of the environmental conditions of the world economy, whereas the other half may be attributed to the increased efficiency of certain national economies as a consequence of joining in the international division of labour.

In addition to the overall proportions mentioned, the tendency of improved terms of trade did not make itself felt evenly in the economies of the individual Latin American countries. According to calculations by UN experts, on the average of the period 1960/70 the terms of trade of Chile were more favourable by 13 per cent, those of Peru by 4 per cent and those of Bolivia by 41 per cent than in the previous decade, whereas in the case of all other countries there ensued a general deterioration of the terms of trade. Since no close relationship can be detected between improving terms of trade and the increased dynamism of the economy, it seems rather probable that the stepped up rate of growth of the individual countries and their expansion in the field of external economy reflect only to a minor extent international price fluctuations and rather show the results of structural development and improving efficiency.

The combined *trade balance* of the region had a significant export surplus amounting to 10–20 per cent of imports up to the end of the sixties. Within regional data, however, there occurred considerable dispersions, whereas the surplus originated essentially from the Venezuelan active balance amounting to a steady annual one billion dollar. In case of leaving oil producing Venezuela out of consideration, surplus diminished to 3–4 per cent of imports, and from 1967 on it turned into an increasing deficit. The share of imports covered by exports amounted only to 78 per cent on the average of the period 1970/73, Venezuela excluded, moreover if Venezuela were included it still would amount only to 87 per cent.

Disregarding certain fluctuations in the business cycle, increased deficits could be explained mainly by structural reasons. There are three countries in the Latin American region with structural trade deficits: Mexico, Panama and Costa Rica. Due to the export of services, the order of magnitude of trade deficits has increased by leaps and bounds in the sixties. The growing deficit of the trade balance was covered by the export of services in the three countries mentioned above. The development of deficits confirms the assumption that there is no close relationship between the dynamism of the economy and the balance of trade. The three countries with prolonged deficits are among the most dynamic ones of the region, while, e.g. slowly developing Argentina and Uruguay claim time and again highly active foreign trade balances.

The rapid development of the export of industrial products reflects the reassessment of political objectives. The international developments mentioned above favouring the participation in the industrial division of labour have a decisive part in the extended introduction of the system of export incentives. The following factors should be mentioned when enumerating successful incentives: the abolishment of overvaluation of currencies, a routine in any protectionist environment; the adoption of mini-devaluations, a system which had been successful in many a country; tax exemptions; the assistance rendered by the governments in matters of export credit techniques and market research; finally, the development of regional agreements affording mutual advantages (Latin American Free Trade Association, Central American Common Market, Andean Group, Caribbean Free Trade Association, etc.). While import substitution had been almost exclusively the most important incentive in the former policy of industrialization, *beginning with the mid-sixties export orientation became increasingly the main objective of incentives in the majority of the Latin American countries.*

The fact that increasing export orientation of the industry is not confined to the big multinational corporations is connected with the enhanced economic role of certain governments. It would be an insoluble statistical task to separate the industrial export of the multinationals on the one hand, and that of the nationalized or private sectors on the other. According to various calculations and estimations it seems that 30–35 per cent of the total export of Latin American industrial products was realized by multinational corporations at the end of the sixties.[1] According to data by Agarwal, the share of multinational corporations in Asian and African total exports rose from 19 per cent to 27 per cent in the period 1966–1970. At the same time, their share in Latin American exports diminished from 40 per cent to 36 per cent, and made up but 17 per cent of the increase of total exports. (In the case of Mexico this ratio amounted to 45 per cent.) Thus, it would be wrong to explain the rapid development of industrial exports merely by the export creating role of foreign capital, this being a frequent explanation given by orthodox bourgeois economists and also by some representatives of the New Left.

3.2. TRENDS IN THE COMMODITY STRUCTURE OF FOREIGN TRADE

Let us now examine how the *commodity structure* of foreign trade relations was affected by accelerated economic growth. A backward export structure based on the export of raw materials and food products is a main characteristic of underdevelopment and semi-colonial division of labour. Within this system, export is concentrated on only one or two products bringing about an economic

---

[1] *Economic Survey of Latin America*, 1970.

structure of monoculture and the dependence of the national economy on the world market positions of one or two products and/or on the main buyers of these products. Up to this date, economists have not quite agreed whether a structurally developed foreign trade unambiguously equalled the national economic requirements of growth stimulation, macroeconomic profitability and technological-organizational efficiency. Comparative advantages of certain countries and/or the trends of the world market often bring about situations (e.g. World War II, the Korean boom, the present shortage of raw materials) when the advantages of the international division of labour are not related to structural development. In the case of balanced world economic trends the surplus value to be realized in foreign trade, the advantages of the international division of labour are naturally closely related to the degree of processing and up-to-dateness. In the age of a world economic environment influenced by established world economic rules of the game, in the period of increasing international cooperation, structural development is an indicator of success.

The termination of dependence on the export of one or two products, the enhancement of the manifold character and *diversification* of the economy represent always a more unequivocal requirement of the economy and national security. Therefore, it seems expedient to concentrate the examination of export commodity structures partly on the one-crop character, and partly on the indicators of structural development.

Latin American exports were of a highly *monocultural* character both at the time of semi-colonial "outward-looking" and in the period of import-substituting growth when the development of the export sector had been neglected. With the exceptions of Mexico, a country striving resolutely for the diversification of her economy, and Peru which has favourable natural endowments, the bulk of exports of all the other countries was made up of one or two products.

The data in Table 19 clearly indicate the trend that the diversification of

Table 19

The share of monocultural products in exports (per cent)

	1950	1965	1970	1973
Argentina				
Animal products	52	40	36	42
Corn	29	39	30	43
Brazil				
Coffee	61	45	41	26
Chile				
Copper	46	69	79	82
Colombia				
Coffee	78	64	63	47
Uruguay				
Animal products	73	73	65	72
Venezuela				
Petroleum	96	92	90	93

Source: Yearbook of International Trade 1953, 1966, 1972; QER, 1974.

exports makes faster headway in countries which have chosen the road of export orientation. Contrary to earlier estimates, *export orientation has not brought about the narrowing down of specialization* or an increase in the monocultural character of the economy, but has explored new sources of growth and enhanced the international bargaining capacities of countries with a dynamic foreign trade.

In the less developed countries diversification generally began with launching the export of some new raw material or agrarian produce. Thus, in the case of Ecuador diversification was represented by petroleum in addition to the traditional banana export, whereas in the Central American republics it was coffee and cotton besides the traditional banana export. This did not result in any structural development, nevertheless it reduced the extent of external vulnerability and instability of foreign exchange situation. In some of the more developed countries, first of all in Brazil, Colombia and recently also in Argentina, diversification mainly derives from the increasing export orientation of industry, and the abolishment of external economic monoculture is coupled with the unfolding of industrial exports. Diversification is closely related in these countries to structural development, more precisely to industrialization.

The traditional foreign exchange earning branch of Latin American economy was agriculture. On the average of the last decade more than two-fifths of the exports of the continent were constituted by food-products. The share of energy-carriers varied between 20 and 25 per cent, that of all other raw materials between 15 and 20 per cent. Towards the end of the era of import-substituting growth, exactly nine-tenths of exports originated from the so-called primary sector. The trends of structural development and/or the starting of industrial exports are illustrated by Table 20.

Table 20

Development of Latin American industrial exports

	Export of industrial commodities (million $)				Export of industrial commodities in percentage of total exports			
	1965	1970	1973	1974	1965	1970	1973	1974
Argentina	84	247	370	600	6	9	12	15
Brazil	125	369	2,000	3,200	8	13	33	40
Colombia	35	62	300	500	6	9	30	37
Mexico	244	486	1,278	1,600	24	40	52	45
Latin America	1,284	3,310	5,800	8,500	11	19	19	20

Source: *Yearbook of International Trade*, 1966, 1972, 1973, *QER*, 1974.

The dynamics of the expansion of the export of industrial commodities is astonishing. Within less than a decade the Latin American countries succeeded in increasing their formerly insignificant industrial exports five-fold and in turning them into a considerable source of foreign exchange revenue. In 1960 industrial exports made up only 16 per cent of industrial imports; in 1970 this ratio increased to 25 per cent, and by 1973 to about 40 per cent. In some of the smaller Central American countries participating in integration groupings, industry is approaching the role of a foreign exchange balancing sector.

Certain shifts are discernible also within the structure of industrial exports. The modest industrial exports of the mid-fifties consisted almost entirely of smelting ore, food-products and light-industrial commodities, whereas the share of machinery in total exports amounted to only one or two *per mille*. At the end of the sixties, however, export lists contained for the first time also products of electrical engineering, precision mechanics and the vehicle industry. The share of machinery increased to three per cent in the exports of 1973.

The export of industrial commodities, however, is not divided evenly among the individual countries, and neither is it divided according to general or structural development. As late as the mid-sixties it was still processed raw materials such as refined oil from Venezuela and copper from Chile which made up the bulk of Latin American industrial exports. Export orientation which has made headway since that time is much more vigorous and efficient, and on this basis the leading part has been taken by Brazil, her share in 1973 industrial exports being 33 per cent. Mexico ranked second. The four countries enumerated in Table 20 realize about two-thirds of the region's industrial exports.

The development of the geographical direction of Latin American industrial exports is also of considerable interest. In 1955, 50 per cent of exported Latin American industrial commodities was taken up by the United States, 20 per cent by the Common Market countries and 10 per cent by intra-regional trade. In 1972 the share of the United States fell to 38 per cent, whereas that of the Common Market countries increased to 23 per cent, and that of the intra-regional trade to 24 per cent. This latter figure reflects quite a considerable division of labour in the case of medium developed countries of rather competitive structures.

The characteristic feature of the industrial exports of the Asian and African developing countries is their increasing concentration on the markets of the developed capitalist states. In Latin America, however, a different orientation of industrial foreign relations of belated industrialization is to be observed, rather similar to that of the CMEA (Council for Mutual Economic Assistance) countries in Europe. The rise of the export of industrial commodities is fastest within the intra-regional trade, its main driving force being regional cooperation. This is manifest first of all in the category of highly sophisticated products, primarily as regards machines and equipment, in the case of which 40 per cent of exports was realized within the region itself in 1955, 55 per cent in 1968, and 44 per cent in 1971.

Certain shifts in the centres of gravity should be noted. Since the beginning of the seventies, the share of intra-regional trade turnover has started to decline in the category of machines and equipment, and an ever growing percentage of exported machines has been absorbed by the markets of the developed capitalist countries. For the time being, it is still not clear whether the shift in recent years has been due to restrictions on market expansion in intra-regional trade and to some financial, organizational, and technical problems, or rather to the effects of the strengthening process of inflation in the developed capitalist countries or perhaps to the growing competitiveness and experiences of the exporters of industrial commodities. The latter would render it possible to leave regional markets of lower requirements, and enter the world market facing thereby conditions of much sharper competition.

Economic growth is reflected also by the *structure of imports*. Owing to the abundant natural resources of Latin America, the ratio of raw materials, foodstuffs and energy carriers in imports is extremely low as compared to international standards, though the share of these items is on the increase. The share of the above three items in imports was 20 per cent in 1955, 25 per cent in 1965, and 30 per cent in 1972, whereas the adequate ratio in the average of the capitalist world amounted to 38–45 per cent. The share of consumer's articles was similarly low: in 1955 it was 21 per cent, in 1970 16 per cent. Within that, however, it amounted to 5–8 per cent in Argentina, Brazil and Colombia, whereas in the smaller Central American countries it reached 30 to 35 per cent. This fact in itself provides greater possibilities for the expansion of the import of machinery and other industrial goods of pivotal importance for the acceleration of economic development.

Latin American import of machinery exceeds by far the world average; the share of machine import in total imports was 33 per cent in 1955, 37 per cent in 1965, and 41 per cent in 1972. Due to the relative development of the vehicle producing industries, the bulk of the considerable machine imports is made up of capital equipment. Thus it is understandable why there exists rather a close interconnection between rising machine imports and accelerated economic growth. Despite the rapid progress of the chemical industry, the share of imported chemical goods in total imports is likewise high; it amounted to 9 per cent in 1955, and to 11 per cent in 1971. The fact should be taken into account that by the end of the past decade the share of machines in the total imports of the highly developed big capitalist countries has been 17 per cent, and that of chemical products only 6 per cent.

To sum up, the commodity structure of foreign trade presents an interesting picture. As regards exports, the commodity structure of the more advanced Latin American countries is similar to that of the belatedly industrialized, though developed small European capitalist countries, while as regards imports, the commodity structure is much more developed than that of the highly developed big or small capitalist countries, and the share of up-to-date engineering products is much higher than in world trade. On the other hand, owing to the fact that the range of joining in the international division of labour has been rather limited, the growth-accelerating effects of the modern structure of imports are not as vigorous as for instance in the developed small countries. Similarly to the structure of industry, the indicators of the structure of foreign trade point to the fact that Latin America is characterized by the structural indicators of the more developed countries and not by those of the backward regions.

3.3. PROBLEMS OF GEOGRAPHICAL DEPENDENCE

Geographical dependence is one of the characteristics of backwardness. A high degree of dependence on the deliveries or buyer's markets of some country may result in pressure by means of trade and also in the restriction of political and economic independence, especially in case of disproportionate economic power relations. Geographical relations which have evolved on the basis of historic, geographical, political and economic factors generally cannot be changed within a short time by the weaker partner, or this may happen only at the price of considerable losses.

In addition to the well-known general consequences, the problem of geographical dependence has manifested itself in Latin America under rather unfavourable historic conditions. The aggravating historic circumstance is due to the nature of the dominant great power of the region. The metropolitan country of Spanish America, i.e. Spain just having entered Modern Times, was the European power which held aloof more than any other power of the principal development factors of the age, using her social energies not as a means of development and economic progress but in the interest of the expansion and later the stubborn defense of the decaying empire. After a short interruption the place of the decaying Spanish colonial empire was taken by Britain in the southern part and by the United States of America in the northern part of the region.

In the second part of the 19th century, following half a century of independence struggles and civil wars, most of the countries of Spanish America became parts of the economic system of British imperialism. The subordination to Britain deepened in the 20th century, i.e. at a time when the decline of the British Empire had already begun and later accelerated, and Great Britain had become ever more the "sick man" of the capitalist world. The slowing down of the development of Latin American countries highly dependent on Britain thus unambiguously reflects also the decline of the dominating power. During World War II, and in the post-war period the subordination to the US, the new dominating power, took on an increasingly unequivocal character. What kinds of additional brakes does the change in the role of the dominant power put on the Latin American countries?

Until the beginning of this century the international economic and political potential of the United States proved too weak as to stimulate in any sense whatsoever the development of the neighbouring Central American republics coming gradually under its influence. As regards the countries under British and/or American influence, the progress of the first group has been the fastest. Large-scale American economic penetration actually began only in the forties, but its impetus soon came to a halt, since the sudden emergence of the United States as a world power, its global commitment, did not leave enough expansion potential for the modernizing of Latin America. Thus, even the minimal stimulating effect in the framework of the system of dependence was strangled, and within the declared American policy of "the carrot and the stick" it was the stick which gained the upperhand, and the dominant power in the region, the United States more and more identified itself with the views of the one-time Spanish colonial policy striving less for the economic exploitation of Latin America than for the exclusion of the rival great powers from the continent. The United States therefore regarded Latin America for a long time just as a security problem.

Another problem followed from the fact that as a producer as well as an exporter of raw materials and food-products the Unites States was by far less dependent on the export commodities of the Latin American countries than formerly Great Britain or Western Europe, and this hampered first and foremost the expansion of the external sector of the countries exporting produce of the temperate zone. At the same time, as a supplier of capital goods the United States introduced highly capital-intensive and labour-saving techniques and organization principles, i.e. those which were least suited to the needs of Latin America.

Finally, it should be kept in mind that the peculiar liberal dogmatism of American great power policy displayed an almost unbelievable *lack of understanding* towards all institutions, scales of value, and socio-economic patterns seemingly contradicting North American evolutionary experiences. In the historically brief post-war period of twenty years, the policy of the United States pursued in Latin America displayed in a condensed form all the symptoms of great power dilettantism.

Owing to their historic heritage and the vestiges of backwardness, the Latin American states, similarly to all other small and underdeveloped countries, could not evade the problems of subordination either. These problems, however, were on the one hand considerably intensified by extreme one-sidedness, by the absolute military and political superiority of the dominant power excluding from the very outset any kind of reciprocity, and on the other hand by the constant historical association with economically less interested power systems of dwindling energy. Except for the half century of expansionist policy of Britain, the attitudes of the dominant great powers could be characterized best by the example of Lope de Vega's "The Gardener's Dog": seclusion from the outside world and at the same time indifference as regards any closer partnership relations.

Due to the negative aspects of the relations between the United States and the individual Latin American countries which became increasingly and rapidly overt, the relationship of American big business and the Latin American ruling classes lacked any kinds of "honey-moons". Not only the left-wing forces but also the so-called populist coalitions emerging as early as in the forties assumed an increasingly anti-US character. By the end of the fifties the Latin American national bourgeoisie increasingly sought for possibilities of easing subordination to the USA. True, the developments of the Cuban revolution have dampened the reformist spirit of the bourgeoisie in the other Latin American countries, nevertheless anti-US sentiments kept on gaining ground and became an almost comprehensive process spreading among conservative and liberal circles as well. Thus, it is understandable that the most important feature of the external economic orientation of Latin America has become the gradual reducing of the large-scale foreign trade dependence on the United States, and at the same time the widening relationships with countries which could represent an alternative to the external sector as against the United States. Under conditions of foreign exchange convertibility it is much more easier to change the geographic structure of imports than that of exports. As a result, the dependence on exports raises more serious problems (see Table 21).

A substantial change in the geographical structure of exports, i.e. the considerable weakening of North American positions in the external sector of the

Table 21

Geographical distribution of Latin American exports*
(in percentage of turnover)

	1955	1965	1970	1973
United States	44	36	34	34
Common Market	15	20	21	24
FRG	4	7	8	10
EFTA	11	10	9	9
Great Britain	9	6	6	6
Japan	3	4	6	8
CMEA countries	2	6	6	5
Developing countries	20	19	19	19
Latin America	10	11	16	16

* Cuba included

Source: Various issues of *U.N. Monthly Bulletin of Statistics* and *QER*.

Latin American countries was realized by stepping up development and parallelly with this by the increase in export orientation. The diminishing share of the United States in Latin American foreign trade is a general process, comprising the whole of the continent with the exception of Haiti, Dominica and Argentina. Nevertheless, the United States still ranks first in the foreign trade of the Latin American countries, with the sole exceptions of Argentina and Uruguay. In the seventies, however, the leading position of the USA has become by far not equally stable in all the Latin American countries. Thus, in the southern states of the continent, i.e. in Argentina, Uruguay, Chile and Paraguay, the share of the USA in foreign trade amounted to about or less than 20 per cent, in Brazil and El Salvador to about 30 per cent, and in Colombia, Ecuador, Venezuela and Peru, i.e. in the Andean countries, and the Central American republics to about 40 per cent. Only Haiti, Dominica, Panama and Mexico represent the extreme cases of one-sided geographic dependence. The efforts aimed at the reduction of the share of the leading partner in the sphere of foreign trade reflect, in addition to economic reasons mentioned above, the national security aspects in the foreign trade relations of the Latin American countries.

Latin America was less affected than other developing regions by the weakening and, later on, the devaluation of the dollar in the early seventies, and by the consequences of the so-called Smithsonian agreement. With the exception of Brazil and Venezuela, the currencies of the region's countries were devaluated to the same extent as the dollar. Theoretically, devaluation had a beneficial effect on the competitiveness of US commodities as against the other highly developed rival powers, nevertheless, no trace of such a trend could be found in the geographical orientation of Latin American foreign trade. At the same time, the overwhelming part, i.e. 60 to 90 per cent, of the foreign exchange debts of the Latin American countries are dollar liabilities, and thus devaluation has relatively eased the burdens of paying-off. Similarly, a number of Latin American countries succeeded in taking advantage of the weakness of the dollar in order to resort to financing by the Eurodollar. But the foreign exchange weakness of the dominant regional power could be exploited only by countries in stronger bargaining positions, as a consequence of which intraregional differentiation was further stimulated by the foreign currency situation.

Within the trend of decreasing general dependence, however, there has manifested itself an unfavourable phenomenon: in the case of machine imports highly significant for development, the diminishing of the share of the United States is much slower. The main body of the machine stock in Latin America is of US origin; spare parts supply, too, is effected according to North American technical norms, and this is hampering the rapid termination of technological dependence. On the other hand, it is rather advantageous from the point of view of industrial exports that the main source of economic growth, the export-oriented industry is linked to a lesser extent than the average to the regional great power and, besides regional trade turnover, it is forging ahead most dynamically on the buyer's markets of the West European countries and Japan.

The diminishing importance of the United States in the sphere of foreign trade is due to a considerable extent to the gaining ground of the Common Market countries, first of all that of the Federal Republic of Germany. The Common Market has achieved leading positions in the foreign trade of Argentina, Chile, Uruguay and Brazil. As a consequence of agreements concluded

Problems of the External Economic Sector **87**

with the Common Market, the European orientation of some countries has got a new impetus in the early seventies. It should be emphasized that in 1972/1973 the Federal Republic of Germany has succeeded in building up positions in the foreign trade of Brazil and Argentina similar to those of the United States, and presently ranks as the strongest direct competitor of the USA on these markets. The rapid development of these relations was obviously influenced by the fact that the Common Market countries, and primarily the FRG, adopt most flexibly new-type forms of economic cooperation going beyond simple foreign trade operations, and also the most up-to-date forms of industrial cooperation. In 1955 the value of Latin American exports to the countries of the Common Market amounted only to 156 million dollars (12 per cent of the total trade turnover), this was raised to 282 million in 1965 (13 per cent of the total trade turnover), and by 1971 it reached 600 million dollars (18 per cent of the total trade turnover). The above development trends throw a light on the 15-year-old conflicts of trade policy between Latin America and the Common Market. Owing to its agrarian policy, and the favours granted to the associated African and Mediterranean countries, the establishment of the Common Market was attacked most vehemently by the Latin American countries. Despite the discrimination easily to be detected theoretically, Latin American export directed to the Common Market countries developed much more rapidly than its export to the EFTA or the North American countries. The differences in the rate of market expansion are most conspicuous in the case of agrarian products which were the very targets of discrimination by the agrarian procedures of the Common Market. 19 per cent of Latin American food-product exports in 1956–1958 went to the Common Market countries, whereas in 1970–1972 the Common Market's share rose to 26 per cent. The overall growth rate of exports to the Common Market was nearly twice that of the average. Concern because of the competition by the associated African countries proved to be unfounded. The increase in import originating from African countries turned out to be more rapid only to a very small extent, and this, too, could be explained by a sudden rise in cocoa and copper imports. The volume of food-product and raw material imports from Latin America is twice of those from Africa. Thus, foreign trade trends prove the fact that decisions on economic policy made by the Common Market have only slightly affected the development of market relations. The development of trade turnover was determined by the fluctuations of demand and supply, and the relative lagging behind of the Latin American countries was due less to the economic and political environment than to limited export capacities.

In recent years, relations with the Common Market have been characterized by becoming more orderly and settled. Three-year trade agreements were concluded with the Common Market by Argentina in 1972, and by Brazil in 1973. The establishment of a joint committee was collectively proposed by the Andean Group for the development of trade relations.

The dynamic development of the relations between the Latin American countries on the one hand, and Japan on the other, reflect first and foremost strivings by Japan to broaden her raw material basis. The Common Market countries display extraordinary activity and readiness to bring about long-term cooperation, whereas Japanese endeavours are primarily aimed at the exploitation of the natural resources of Latin America. To avoid dependence on the United States as regards raw material supplies (as in the pre-war years), the global strategy of Japan attempts to build up a relatively well diversified independent raw material basis which would not be subordinated in any respect to the United States and the multinational corporations. Consequently, Japan consolidates her ties primarily with Brazil, Bolivia and Peru, i.e. Latin American countries richly endowed with natural resources, and in general with countries of the Pacific region. The permanent and considerable adverse trade balance of Japan in her foreign trade relations with Latin America also indicates the significance of Latin America as a primary raw material basis of Japan and not as her important buyer's market.

The relations between the socialist countries and Latin America have undergone a fast development in the fifties. The share of the socialist countries was highest in the early sixties, and after the stagnation in the period 1965–1970, in the seventies again the trend of expansion gained the upperhand. The connections between the two country groups are affected unfavourably by the considerable fluctuation in the centres of gravity as regards geographical and sectoral structures, as well as by the almost permanent adverse structural trade balances of the socialist countries.

Despite certain political measures and strivings of the economic policies of these countries, the share of the developing nations in Latin American foreign trade has remained stagnant. Obviously, the organizational, financial and technical requirements of furthering connections with the developing countries cannot be fulfilled by the Latin American countries which as a rule dispose of but limited resources. The relatively slight decrease in the share of this country group can be explained by the fact that while the rate of the growth of intra-regional turnover in Latin America was relatively quite conspicuous, the weight of the relations with the African, and Asian countries decreased even faster than the discernible international trends.

At the time of import substitution, in the period 1950–1965, intra-regional trade came virtually to a standstill, moreover, it even diminished between 1955 and 1960. Since parallel capacities were created by import-substituting industrialization in the individual countries, the competitive character of the national economies was further enhanced. The establishing of the institutional framework of regional cooperation in foreign trade, i.e. the setting up of the Central American Common Market, the Latin American Free Trade Association and the Andean Group, have created favourable conditions for increasing intra-regional trade turnover, and especially for intensifying the division of labour in industry. Following the rock-bottom of 1961, the expansion of intra-regional relations has had an ever growing part in stimulating economic "outward-looking", and particularly export-oriented industrialization.

3.4. TRENDS IN THE FLOW OF FOREIGN CAPITAL

The joining of the Latin American countries—which have become independent in the first quarter of the last century—in the international division of labour rested on two foundations: one has been the expansion of external markets, the other, large-scale capital imports. Due to the peculiarities of the former colonial pattern, commodity and capital markets could not develop, saving capacities remained negligible and thus, foreign capital succeeded in getting a major part in the development of the new export capacities and also in the introduction of novel technological knowledge and organizational forms necessary for exploiting export potentials. In Argentina for instance, 48 per cent of all investments between 1900 and 1914 were supplied by foreign funds. In the 19th century Latin America became the major region of capital investments of Western Europe. On the other hand, in none of the other developing country groupings did capital imports have such a decisive role than in Latin America. At the time of the outbreak of World War I, the volume of foreign capital investments in Latin America amounted to about 8.5–9 billion dollars, and within that, the volume of direct capital investments to 5.7 billion dollars. Nearly 60 per cent of foreign capital investments, i.e. about 5 billion dollars, were of British origin. 37 per cent of British capital investments fell to Argentina, 23 per cent to Brazil, 16 per cent to Mexico and 7 per cent to Chile. The volume of French capital investments in Latin America was estimated at 1.7 billion dollars, whereas that of North American capital in the first decade of the 20th century at 1.6 billion dollars. More than half of working capital investments served the development of the railway network and the remaining part was allotted to the construction of public works, mining and trade networks.[2]

The role and flow of foreign capital were affected in a complex way by the outbreak of the First and the Second World Wars, and also by "inward-looking" economic growth. In the thirties, the shrinking of the capital market of the developed countries, later on the capital requirements of war preparations, as well as the reserved attitudes regarding the uncertain economic prospects of the raw material producing countries exhausted the sources of capital imports. Thus the economic growth of the Latin American countries was predominantly financed by internal savings. Moreover, during World War II and in the post-war years the Latin American countries spent part of their accumulated foreign exchange reserves on buying up foreign capital interests on a modest scale and/or paying off former debts. And finally, the nationalization of the oil industry in Mexico in 1938 on the one hand directly reduced foreign capital investments, and on the other hand it started in the other Latin American states an initially modest and later on increasingly vigorous chain-reaction aiming at the weakening of foreign control over national resources. As a consequence of the trends mentioned above, the volume of foreign capital investments in 1950 amounted only to eight billion dollars, thus falling short of the 1913 investment level. According to the changes in the world economic balance of power, the United States replaced Britain as the leading capital investor and, from the end of the forties, began to occupy a monopolistic position in the capital market of Latin America.

Parallel with the unfolding of the post-war reconstruction period of the capitalist world economy and/or the intensification of the "inward-looking" period, the capital import of Latin America has increased considerably beginning with the early fifties. Direct net capital imports amounted to an annual average of 430 million dollars in the period 1951–1955, and to 1,057 million dollars in the period 1956–1960.

The suddenly increasing capital imports in the fifties may be regarded mainly as a reaction to the conditions of the import-substituting growth period. A series of public measures in the interest of industrialization, the unprecedented high level of customs and administrative protectionism have isolated the markets of the more developed Latin American countries from foreign competition on the one hand, and have guaranteed fabulous profits to all economic activities enjoying protection on the other. Under given circumstances, foreign

---

[2] W. P. Glade: *The Latin American Economies.* American Book, New York 1969. p. 216.

capital reconciled the viewpoints of the setting ceilings on profits, the circumvention of protectionist barriers and defence of markets by means of taking part in the process of industrialization and by locating subsidiary companies in Latin American countries. Thus, the growth period inspired by the ideology of economic nationalism resulted in a peculiar distortion as a consequence of which the most capital intensive industries of pivotal importance came to a considerable extent under foreign control.

On the other hand, external imbalances and foreign exchange difficulties originating from the gradual exhaustion of "inward-looking" growth necessitated the raising of ever increasing long- and short-term loans for investments. The effects connected with the exhaustion of the "inward-looking" economic growth process were suddenly intensified by the influence of the revolution in Cuba, the spread of anti-capital and anti-US sentiments in all Latin American countries, and by the capital restricting measures of Venezuela as well as other countries. Direct capital investments diminished by almost 40 per cent between 1961 and 1965, while the annual average of total capital imports decreased by 30 per cent. The pattern of capital imports, too, changed considerably. While direct investments had been formerly the bulk of capital imports, in this later period private capital fearing the political risk of expropriation and the decreasing rate of economic expansion reduced its direct investment activity, and capital exports and loans were managed more and more by means of agreements guaranteed by international institutions, public authorities or the state itself.

Owing to the changeover from one growth phase to another in most of the countries of the continent, as well as to the increased joining in the international division of labour from the mid-sixties, capital import undoubtedly ran high again in Latin America. Capital imports, however, reached the level of the fifties only at the end of the sixties. The significance of Latin America as an international buyer's market of capital lessened considerably as compared to the situation prior to the depression or even that of the fifties.

Let us now examine the correlations of the growth periods of Latin American economy and the trends of direct capital investments reflecting most illustratively the macroeconomic processes. The main trends are to be indicated by using the data of US capital representing about 65 per cent of total direct long-term investments in Latin America.

It is clear from the data of 1953 that the explanation of the branch *structure* of US capital investments is rooted, in addition to the points of view of returns, primarily in the efforts aiming at the control of Latin American raw material resources. Within total capital investments the share of the petroleum sector was 28 per cent, of mining 17 per cent, whereas that of the manufacturing industries only 19 per cent. Though the percentage of investments connected with the infrastructure has been quite considerable, it is much lower than at the time of the hegemony of British capital. Import substituting industrialization has directed the attention of North American capital towards the manufacturing industry. By the end of the switchover period, i.e. by 1965, 30 per cent of capital investments came to the manufacturing industry. It should be noted that 50 per cent of the increment of all capital invested in the period 1953–1965 originated already from the manufacturing industry. It is a very interesting feature of development that changes which were believed earlier to be secular characteristics of capital export came about at the climax of "inward-looking" growth.

Thus, while in the heydays of one-time semi-colonial division of labour, i.e.

Table 22

Trend of direct US capital investments (million dollars)

	Total capital invested			Oil industry			Mining			Manufacturing industry		
	1953	1965	1972	1953	1965	1972	1953	1965	1972	1953	1965	1972
Global US investments	16,329	49,328	94,031	4,935	15,298	26,399	1,933	3,785	7,131	5,226	19,339	39,478
Latin America	6,034	9,391	13,528	1,684	3,034	3,245	999	1,114	1,300	1,149	2,745	5,265
Mexico	514	1,182	1,993	10	48	32	144	104	124	214	756	1,385
Argentina	406	992	1,393	.	.	.	.	.	.	200	617	1,836
Brazil	1,017	1,074	2,490	206	57	169	.	51	136	483	723	1,745
Colombia	235	526	739	117	269	327	.	.	.	- 41	160	262
Venezuela	1,308	2,705	2,683	1,006	2,024	1,546	.	.	.	37	246	539
Chile	657	829	621	.	.	.	445	509	359	34	39	47
Peru	268	515	714	209	60	.	170	262	416	17	79	90
Panama	407	724	1,423	.	130	265	.	19	19	4	24	162
Other	278	218	828	17	89	424	11	8	39	12	38	89

Source: Survey of Current Business Aug. 1955, Sept. 1967, Sept. 1973.

at the time of traditional "outward-looking" economic development, the target of foreign capital investments was the *raw material producing* sector as the pioneer of national economic dynamics, in the period of import-substituting growth the main attraction for foreign capital became *the hold on the domestic markets defended by protectionist measures*. This is clearly shown by the trend of capital investments in the manufacturing industries of the individual countries. Capital investments in the manufacturing industries experienced a sudden rise mainly in the bigger and more developed Latin American countries which had broader domestic markets, whereas the markets of the smaller countries which were of a limited capacity did not attract foreign investors. Between 1950 and 1965 total capital investments in the Latin American manufacturing industries grew by 251 per cent, and within that the increase amounted to 933 per cent in Venezuela, 466 per cent in Mexico, 540 per cent in Colombia, and to 280 per cent in Argentina. Accordingly, the size of the respective economies decisively determined the patterns and distribution of capital investments during the period of "inward-looking" growth.

Let us emphasize, however, a remarkable phenomenon: the new growth period which has taken shape since the mid-sixties, furthermore reliance upon the international division of labour, and increased export orientation have, according to all indications, decreased the determining role of the economic potential of a country. When production is oriented towards external markets, the direction of industrial capital investments is not determined mainly by the scale of domestic markets, but it is increasingly influenced by the price and quality of labour, the flexibility of the system of overall economic management, and the character of development policy. This is illustrated by the fact that in the period 1965–1972 Latin American industrial capital investments rose by 95 per cent, and within that—contrary to the experiences of the previous 15 years—the aggregate dynamics of the major countries, i.e. Argentina, Brazil, Mexico, Venezuela and Colombia (90 per cent) did not surpass the average of the region. Moreover, the growth rate of direct capital imports of some smaller countries was considerably higher than the average (575 per cent in the case of Panama, and 137 per cent as regards other countries). It should be noted that in Mexico and Venezuela, two countries energetically striving for national control of natural resources and enforcing a policy of capital restriction in the primary sector, capital which has been ousted from raw material production is switching over to the manufacturing industry.

The modification of the sectoral targets of capital investments, brought about in the wake of the changeover from one growth period to another, resulted in significant changes and in the rapid increase of the weight of the manufacturing industry in the structure of direct investments. As a consequence of these changes, the structure of foreign capital investments in Latin America casts off the criteria characteristic of developing countries and reflects a higher state of development as well as its specific conditions. Thus, in 1972, taking the average of developing countries (Latin America excluded) only 12 per cent of US capital investments went to the manufacturing industry, and if the countries of Southeastern Asia were excluded, too, the percentage amounted to a mere 3 per cent. In Latin America, however, the share of the manufacturing industry rose to 41 per cent (on the average of industrially developed countries to 61 per cent), and within that to 70 per cent in Brazil, 69 per cent in Mexico, 60 per cent in Argentina, i.e. it became equal to the percentage values of the developed capitalist countries.

It is useful to make an international comparison of *the returns of capital investments* in the manufacturing industries. In the period of import-substituting industrialization the cost-raising effects of the underdeveloped infrastructure, the economic environment, domestic markets, etc. could be counterbalanced only by ensuring extremely high-level, artificially constructed profits. Even so, the returns of capital investments in the developing countries have been by and large equal to the average level of the developed capitalist countries. This levelling tendency of the industrial rate of profits was also reflected by the low level of capital investments to the developing countries.

Analyzing the trends of the *utilization of profits*, the picture becomes even more clear. As a rule, the trends of the reinvested part of profits realized clearly illustrate the expectations of the private capital as regards economic environment and development prospects. Rates are generally low when the economic and political outlook is regarded as unstable. As a consequence of this, 52 per cent of profits realized in the developed capitalist countries, and only 31 per cent of profits realized in the developing countries was reinvested in the early sixties. According to data relating to 1972, the rate of profit in the manufacturing industry of Latin America was 12 per cent, while that of reinvestments 56 per cent, i.e. it was raised to the equal of average level of the developed countries. Within the average of Latin America, however, the percentage of reinvested profits amounted to 77 per cent in Brazil, to 52 per cent in Mexico, whereas on the average of all other countries only to 34 per cent.[3] Thus, direct private capital identified the economic environment and development prospects of Brazil and Mexico with those of the developed capitalist countries, while in the other countries it pursued the policy of rapid realization of profits.

The *types* of capital import, too, show certain modifications. Prior to the sixties, the bulk of capital exports reached Latin America in the form of direct long-term capital. Due to the strengthening of capital restrictions and/or nationalizing tendencies in the sixties, the importance of capital import realized in the form of loans became a decisive factor.

It may be regarded as a paradox that the period of the highest rate of indebtedness came with the heydays of "inward-looking" growth. At that time, the highest rate of indebtedness was experienced by protectionist Brazil and Argentina as well as Colombia. The maximum utilization of foreign exchange resources and the narrowing down of external manoeuvreability was also marked by a decrease in foreign exchange reserves. It stands to reason that the indebtedness of the smaller Latin American countries which have relatively smaller resources is always faster and more serious than that of the bigger states.

Owing to the upswing of export orientation beginning with the late sixties, indebtedness has again increased, however, its 10 per cent growth rate in the period 1966–1971 has been less than the average of the preceding fifteen years. It should be noted that the rate of indebtedness in recent years was highest in the major countries acting as pioneers of the switchover to an export-oriented policy, at the same time, however, foreign exchange reserves, too, are most favourable in these states. Thus, in spite of a rapidly rising rate of indebtedness, the foreign exchange reserves of Brazil, Mexico and Colombia amount to 30 to 60 per cent of all debts, and cover a much higher percentage of debts than in the preceding decade. The foreign exchange vulnerability of these countries

---

[3] According to the issues of *Survey of Current Business*, August 1955, September 1967, September 1973.

has diminished, and their possibilities for availing themselves of a selective credit policy have considerably increased. In all these countries economic and external economic expansion pushes forward foreign debts like an avalanche. Though debts are growing in absolute terms, but compared to broadening export capacities, the trend of indebtedness shows a tendency of decrease.

The leading financing countries, too, seem to take into account in their practical policies the increase in demands on financing in the case of a vigorous external economic expansion. They are well aware of the fact that the termination of this peculiar race would result in a sudden standstill of economic expansion, in a temporary or lasting insolvency in the debtor countries, and a freezing of assets in the creditor countries. If the paying off of debts does not surpass considerably 20 per cent of goods and services export capacities, these burdens are not regarded in international financing practice as serious limitations in the financing of countries of a dynamic external sector. In the case of the smaller Central American republics and Venezuela highly dependent on foreign trade, the level of debt servicing is between 3 and 10 per cent; here the restrictive role of debt servicing capacity is not at all taken into consideration. In the seventies it has been mainly the foreign exchange manoeuvrability of Argentina, Chile, Uruguay and Peru which has been affected adversely by the rise of capital and debt servicing burdens over 20 per cent and by the unfavourable development of the ratio of foreign exchange reserves to indebtedness.

The modification of the ratio between loan capital and direct long-term capital coincided with the modification of the financing sources and of *the significance of public and private capital in financing.* (See Table 23.) The importance of public capital export has increased for two decades and is connected with large-scale socio-political movements. Contrary to other developing regions, this trend was discontinued in Latin America in the late sixties. The share of public capital flow through bilateral and multilateral channels within total capital imports amounted to 60 per cent on the average of the period 1960–1965, whereas in the period 1966–1969 to 47 per cent, and in 1970–1971 only to 42 per cent. What are the reasons for the irregularities of Latin American capital import trends?

The diminishing significance of public capital export reflects in a concise way the weakening foreign exchange position of the United States, the dominant power of the region, and also the efforts of some Latin American countries to do away with excessive dependence on public US capital export. In the fifties and at the beginning of the sixties the overwhelming part of public

Table 23

Distribution of public capital export to Latin America (percentage of total public capital exports)

	Multilateral	Bilateral	
		USA	Other countries
1960–1965	22	62	16
1966–1969	40	58	2
1970–1971	50	45	5

Source: Data taken from *IADB Annual Report*, 1972, *QER, Monthly Bulletin of Statistics.*

capital export had been still of US origin, since that time, however, US public capital export dropped back in absolute terms, too.

From the above data it is quite obvious that the United States is not in the position to uphold the former degree of direct credit dependence of the Latin American countries, and since the public financing sources and interests of the other capital exporting countries are limited, the place of US public capital export is increasingly filled by multilateral capital exports. The role of US capital is undoubtedly considerable also as regards the international financing institutions, however, the changeover to indirect presence in financing transactions illustrates the growing strength of Latin American nationalism.

In addition to the weakening foreign exchange position of the USA and its retreat in face of an economic nationalism on the upgrade, the diminishing importance of public capital export is also connected with an increasing interest on the part of private capital in the more dynamic Latin American countries, first and foremost in Brazil and Mexico. Direct capital investments have considerably increased since 1967, and have reduced from the aspect of foreign exchange management the extent of dependence on public capital exports which are *par excellence* suited for exerting pressure on the part of a great power.

3.5. THE ROLE OF FOREIGN CAPITAL IN ECONOMIC GROWTH

Having analysed the main trends of capital movements, let us now examine *the role of foreign capital in the economic development of Latin America*. In the case of countries of smaller economic potential developing under the conditions of permanent dependence, which therefore have rather limited possibilities for accumulation, it is justified to assume that they are obliged to rely largely on external financing funds. In certain developmental stages, especially in the period of industrialization, in the contemporary highly developed capitalist countries, particularly the smaller ones, capital import has had an outstanding part in financing economic growth. The highly developed smaller capitalist states effectuate a capital import of considerable volume, amounting to 1–3 per cent of their gross national product, in order to finance their economic growth, structural transformation, their balances of payments equilibrium or the convertibility of their currencies. In the case of capitalist countries regarded as smaller units from the point of view of the world economy, or those which have entered the stage of up-to-date economic development belatedly, the recourse to external financing is a necessary concomitant of growth and is practically independent of the level of development.

The capital import of the Latin American countries between the middle of the last century and the World Depression equalled 3 to 10 per cent of their national products which actually furthered the development and stabilization of the "outward-looking" model.

Contrary to the general practice, the industrialization of the more developed countries of the region has started essentially on the basis of their own financial resources in the thirties and forties, decades characterized by a shortage of capital in the world economy. Let us now examine the significance of capital imports in economic growth in the past two decades. (See Table 24.)

A number of conclusions may be drawn from the above data. Most conspicuous is the fact that foreign capital—even compared to international facts and

Table 24

The development of capital imports in Latin America, on the basis of macro-economic comparison
(annual averages, million dollar)

	1951–55	1956–60	1961–65	1966–69	1970–72
Net capital imports	756	1,455	1,088	1,786	1,882
Gross national product	56,000	76,000	103,000	126,000	151,000
Gross domestic investments	10,100	13,700	16,300	22,700	28,900
Export	7,033	7,990	8,800	11,200	18,430
Capital and debt service	915	1,217	1,460	2,350	2,970
Capital imports as a percentage of the GNP	1.4	1.9	1.1	1.4	1.2
Capital imports as a percentage of investments	7.5	10.6	6.4	7.9	6.3
Capital and debt service as a percentage of the GNP	1.6	1.6	1.4	1.9	1.9
Capital and debt service as a percentage of exports	13	15	17	21	16

Source: Based on R. F. Davies: La inversion extranjera en America Latina, CEPLAN Documento No. 14 Santiago 1971, *Monthly Bulletin of Statistics* and *IADB Annual Report*, 1972.

figures—has played a relatively modest role in the modernization of the region. Taken the average of the twenty years of the period 1950–1970, capital imports have not reached 1.5 per cent of the national product. This seems an extremely low percentage compared not only to the level of economic development but also to the former role of capital imports as well as to the adequate values of the developing countries averaging that time 5 per cent, and also to the capital imports of the highly developed small countries. It might be of specific interest that the significance of foreign capital had been greatest at the very heydays of import-substituting growth, while the export-oriented development of recent years did not deviate from the long-term average value of dependence on capital import or even fell short of it.

The same picture is gained by comparing capital import with investments. 7.5 to 10.5 per cent of total investments were financed by foreign funds. It was only the socialist countries and Japan which accomplishing industrialization, relied to a smaller extent on foreign resources in the period under review. Without attempting to anticipate a later analysis, let us venture to point out that the dependence of Latin America on foreign capital has been brought about not by the increased demands of modernization on external financing but rather by other factors.

Contrary to the examples of the socialist countries and Japan, the relatively lower level of Latin American capital import may be explained mainly by exogenous factors. (The most important exception being the set-back of capital import after 1957 as a result of Venezuelan oil policy.) In the period 1930–1955, the international shortage of capital may be regarded as the direct reason, whereas the increased capital import following 1955 reflected already the political considerations of the leading capitalist countries (expansion of direct or indirect public capital export). Especially the countries in political key positions were able to rely to a considerable extent on increasing volumes of foreign resources. However, the flow of politically motivated funds was curbed or even retarded in certain cases by the relaxation of the cold-war atmosphere and also by the diminishing political importance of the Latin American region due to various political and strategic factors. As regards economic life, this process coincided with the emergence of international payments problems in the USA, the increased capacity of Western Europe to attract capital, and finally with the rapid expansion of the international division of labour. Under these circumstances, capital investments aimed at penetrating the domestic markets of some developing countries of but local importance by circumventing protectionist measures seemed less and less attractive. This is understandable since—compared to the possibilities offered by the rapidly expanding world market—the importance of national markets decreased.

Thus, after the first period of capital relations between the developed capitalist countries and Latin-America based on the exploitation of raw material reserves and natural resources, the second period, too, came to an end in the sixties. The policy during this stage was founded on making use of the possibilities rendered by markets protected and developed by protectionist measures, by the policy of import substitution and by opportunities in service sector related to the latter policy. The emerging multinational corporations took less and less interest in investing in the economically limited space of the respective national economies. Due to the generally adverse effects on the competitiveness of enterprises, national protectionism proved to be incompatible with the categories of global manoeuvring within the framework of world economy. *Thus, the strategies of the leading great powers and international big corporations assumed primary importance in shaping the order of magnitude of capital import.*

By reason of the quantity of the national product, the level of the capital and debt service burdens does not seem exaggerated either. Viewed from the angle of the export capacity of the region, however, the matter assumes an entirely different aspect. Until the end of the past decade, i.e. until the expansion of exports, the level of capital and debt service burdens rose at a much higher rate than export capacities, and presently these burdens absorb more than a fifth of the steadily increasing export capacities. This trend, too, is a heritage of import-substituting growth which channelled external financing funds mainly to the sphere of import substitution and did not pay sufficient attention to the expansion of export capacities. Contrary to ample experiences gained from economic history, foreign capital which has penetrated these economies in order to seize the domestic markets has participated to a much smaller extent than the average in establishing new industrial export capacities and has not even attempted to establish export-links between the parent companies and their subsidiaries. In 1957 4 per cent of the production of enterprises of North American interests in the manufacturing industry was exported, and this rose to 7 per cent in 1965, and to 9 per cent in 1968, meaning that their export orientation fell short of the average of the manufacturing industries of the Latin American countries.[4] According to a survey in 1965, it was only the enterprises of the food industry which manifested considerable export orientation exporting 22 per cent of their production, whereas the chemical industry exported only 10 per cent, the metal industry 4 per cent, and the electrical industry a mere 2 per cent of their output. At the same time, the motor vehicle industry, i.e. the most dynamic growth factor under foreign control did not engage in any kind of export activity.

[4] *Economic Survey of Latin America*, 1969.

Thus, from the point of view of finances and foreign trade the role of foreign capital in growth should be regarded as only of secondary importance. Capital import did not serve as an amplification of financial resources, and did not aim at improving the adverse balance of the external economy but served first and foremost as a means to seize control over production facilities supplying goods to the domestic markets.

The secondary character of financing, however, does not mean by far that the socio-economic role of capital import is insignificant. Technology import coupled with capital import has brought about lasting changes in the socio-economic development of the region. Optimization of growth to import substitution has led to the increased dependence on technology and capital intensity of the economy since the goods imported have been much more capital intensive than the exported ones (with the exception of mining products). In the mid-thirties the Latin American countries (with some notable exceptions) had no industries producing capital goods, thus investment goods necessary for the readjustment and acceleration of growth had to be supplied by import. As a result of historical development, *labour-saving and capital-intensive* technologies have been developed in the advanced capitalist countries, first of all in the United States which had vested interests in Latin America. The high import-ratio of investments in the Latin American countries had a spill-over effect on the economy as a whole as regards capital intensity and labour saving. Thus, these started a basically distorted, low macro-economic efficiency process of industrialization and modernization in countries grappling with the problems of manpower redundancy. This process was highly enhanced by the direct appearance of foreign subsidiaries. In the more up-to-date sectors of the economy foreign-based technology became a general feature. *The shortage in* Latin American *experts* needed for the adoption of a technically more complicated and capital intensive technology resulted on the one hand in the foreign control of the top-managerial echelons of production and economic life, and on the other hand served—like a vicious circle—as an incentive to technologies calling forth further labour-saving measures.

As a consequence of the shortage of capital and experts, as well as the underdevelopment of the national technological and scientific basis, the import of capital and technology was realized in the framework of so-called package deals, i.e. foreign capital financing capital investments was marketing at the same time its own products and technologies as well, the new projects were managed by its own experts, striving in this way to get maximum profits from all possible types of investment. The few papers dealing with this problem sufficiently illustrate the fact that capital import realized through package deals ensured the owners of funds and technologies possibilities for manipulating all the three cost factors to a much greater extent than the average. The realization of such kinds of capital imports brought about not only higher financial burdens by means of an additional burden on the balance of payments but also resulted in permanent *technological subordination*. Foreign subsidiaries thus represented an alien and exclusive entity within the internal economy and were integrated almost solely with the metropolitan country or parent company. The frequently extolled up-to-date industries of Latin American industrialization, such as the car industry, the chemical and electronic industries represent an exclusive empire of entirely or overwhelmingly foreign capital.

A new situation was created in the capital relations of the Latin American countries by the most recent trends in the development of the world economy and capital exports, mainly by the emergence of multinational corporations. The negative role of foreign capital has become more enhanced by the appearance of the multinationals. They are spreading foreign technological processes as well as consumption habits, moreover they have become the supervisors of the key positions of the economy and the founders of the most concentrated and dynamic private enterprises. Capital import does not result in the foundation of small enterprises anymore in the post-war period. The new enterprises operating with the help of foreign capital are all of considerable size and much more stronger than those controlled by domestic capital.

After World War II the multinational corporations penetrating Latin America have gained an almost absolute supremacy over the national private enterprises of the capital importing countries in the overwhelming majority of cases. The subsidiaries of the multinational corporations settling down in the countries of Latin America rely upon the supply of their parent companies financially, technologically and also in respect of management. As a result, their competitiveness is much higher, and they occupy more and more important positions especially in the capital and technology intensive industries. Thus, in the period 1957–1965 the turnover of US controlled enterprises in the manufacturing industry of Argentina increased nearly six-fold, that of Mexico by 50 per cent, of Venezuela by 25 per cent faster than that of the domestic sector. In 1965 the turnover of US controlled enterprises in the manufacturing industry amounted to 11 per cent of the production in Argentina and Brazil, and to 17 per cent in Mexico. According to calculations of ECLA, North American enterprises turned out 43 per cent of the value added of the strategically important chemical, metallurgical, paper and rubber industries.[5]

The role of the US-centered multinational corporations can by no means be determined by figures indicating the value of capital investments. Owing to their greater dynamism, an even increasing part of the internal resources of the host countries is mobilized by the foreign subsidiaries, and the resources of the national economies are channelled more and more into the development funds of the foreign sector. In the mid-sixties for instance, 83 per cent of the capital of US corporations was supplied by local sources, i.e. the multinationals scarcely acted as middlemen in capital import transactions. The subsidiary companies backed by their centres in foreign countries are hardly affected by the frequent fluctuations of the business cycle and the waverings of economic policies; at the time of depressions they easily and cheaply buy up local enterprises in financial difficulties. Due to their management from abroad, the subsidiaries experience no difficulties in evading taxation rules and import and credit restricting policies. Thus, they are not only curtailing national sovereignty as regards both economy and politics, but give rise to an atmosphere of increasing uncertainty concerning economic policy, oddly enough just at the time of strengthening state interventionism.

3.6. THE LATIN AMERICAN ATTITUDE TOWARDS FOREIGN CAPITAL

In spite of a relatively low-rate capital import, the strong positions of foreign capital have been partly brought about by the inadequate economic policies of some Latin American countries. With the exception of Mexico, where

[5] *Ibid.*

measures had been taken to control the operation of foreign capital as early as the late thirties, no comprehensive and coherent strategical concept has been developed right until recently regarding the role of foreign capital in these countries. True, some isolated measures were taken, e.g. most of the region's countries excluded foreign capital from certain sectors of the national economy (railways, oil, electricity generation), and also took steps to control reinvestment rates. After some years of experimentation, however, there ensued a certain reassessment regarding the expediency of controlled reinvestment rates. According to experiences, foreign capital has reacted much more sensitively than usual to any control measures concerning the utilization of profits, and new investments, too, are considered mostly against the background of the possibilities of profit repatriation. If by any chance such control measures had been accepted by the foreign capital, it demanded an excessive price for it in other fields. Thus, the control of reinvestments may reduce capital supply. On account of the loss of taxes levied on profit transfer, certain economic circles, arguing from the point of view of fiscal interests, are opposed to control measures based on higher rates of reinvestment. Left-wing intellectuals stress the point that the present type of capital restrictions *ab ovo* recognize foreign capital investments as useful. Thus, as a result of different considerations, economic policies in the Latin American countries abandoned the practice of using the rate of reinvestments as a means of control.

The policy of channelling foreign capital into more backward regions (e.g. Northeast Brazil), though it promised additional incentives and aid, yielded no decisive results either. All efforts to persuade foreign capital to participate in the work of surmounting inter-regional inequalities remained unsuccessful. The attitude of foreign capital was characterized partly by its disinterest in such incentives, and partly by withdrawal after having availed itself of those. At the same time, in the general atmosphere of import substitution national economic policies disregarded the character of adapted technologies and administrative structure, the inadequate export capacities of foreign enterprises, as well as their activities aiming at siphoning off domestic capital and running into considerable debts.

With the coming into prominence of increased reliance on the international division of labour in the late sixties, attention was drawn once again to the regulation of the operational conditions of foreign capital. In the period of import substitution the key positions of foreign capital were still compatible with the power of national decision-making centres. By means of a comprehensive control system, the state was in the position to influence the decisions of the individual foreign corporations which, in developing the cooperation between the parent companies and their subsidiaries, were obliged to take notice, within limits, of the domestic problems. Under the conditions of the exhaustion of import substitution and the switchover to the policy of export orientation when economic life has opened up to the world economy, the problem of foreign interests assumed vital importance; in the atmosphere of a more indirect guidance, foreign-owned enterprises were to a great extent able to become independent of the national centres of decision-making and, as a rule, followed the international business strategy of the parent companies.

Though the new policy developed for the treatment of foreign capital was rather diversified from the point of view of ideology and covered the vast scale ranging from the "closed doors" policy of the Chilean people's front government to the Brazilian policy of "wide open doors", it had a common feature: it attempted to fit foreign capital into the strategical framework of economic development schemes. On the basis of earlier adverse experiences, Latin America was the first developing region to begin to shape the framework and means of a policy of controlled cooperation with foreign capital. Let us review briefly the most important experiences gained by the countries carrying on the most typical experiments, i.e. mainly those of the Andean countries, Mexico, Brazil and Argentina.

The established practice of the Andean countries as regards capital control (though it has turned out to be rather differently interpreted in actual implementation) is of great interest. The striving for joint control was the result of the realization that the smaller Latin American countries are too weak both politically and in the economic field to face the gigantic multinational corporations furnishing the overwhelming part of capital import. In the case of integration the lack of capital control may even bring about the control of all the countries of the integrated region from a single bridgehead established in any of these countries. Therefore, the Andean countries accepted the view according to which parallelly to the progress of integration, a joint policy eliminating competition among the member countries and adopting a selective practice in the sectors of the economy ought to be elaborated as regards foreign investments.

As a consequence of the Cartagena Agreement, foreign capital investments, the cession of patent rights and trade-marks were granted only on the condition of obtaining the preadmission of the competent national authorities. Further on, regulations were issued on raising credits by foreign enterprises as well as their export limiting conditions, the annual rate of profit transfer was limited at 14 per cent, and all new foreign investments were prohibited in sectors of strategic importance like finances, press, communication. The schedule of nationalizing previous and more recent foreign capital investments was also fixed. This latter divestment regulation categorized enterprises under three headings: national (more than 80 per cent domestic ownership), mixed and foreign-owned enterprises (51 to 80 per cent domestic ownership, this latter stipulation, however, was later changed, and presently 30 per cent public ownership ensures a mixed enterprise status).

The joint resolution stipulates the obligatory and gradual transformation of foreign enterprises into mixed enterprises within 15 to 22 years, otherwise they were not to share in the benefits of liberalization among the member countries. The selectivity of this regulation is manifested also in the fact that in industrial branches deemed significant from the point of view of integration the common norms are to be strictly enforced, while in branches of lesser importance any country may grant special treatment.

In theory the joint system of the Andean countries for foreign capital control is undoubtedly favourable to increasing the bargaining strength and independent decision-making capacity of the individual countries vs. foreign capital interests, and staves off the danger of invasion of the integrated countries from a base established in one of the member states. Presuming that these measures were successfully implemented, they might serve as a guideline to the efforts aimed at controlling foreign capital in all the other Latin American countries. But the defense based on past experiences, also means certain conceptual disadvantages since the percentage of capital ownership, the specification and restrictions of the ownership functions and rights have moved into the centre of interest. Present-day capital movement trends on the other hand prove the fact that capital is much more profit-sensitive than ownership-sensitive, and its regulation reflects progress rather in the sphere of contractual rights than in the range of property problems. Property-centred regulation may prove effective if (disregarding investments of a monopolistic character for the purpose of exploiting natural resources) it is coupled with a high rate of returns, considerable profits and economic expansion. These latter conditions are nonexistent in the Andean region, and in the implementation of the joint program on capital restriction there also appear considerable differences among the individual member countries. Thus, as the resultant of various factors, the volume of foreign capital investment in the five Andean countries stagnated in the period 1968–1972, new capital investment (with the sole exception of the Ecuadorean oil industry) were negligible, therefore the constructive regulating influence of the recent joint measures regarding capital control could not unfold either.

The capital controlling policy of Mexico has some peculiar features. Out of all Latin American countries it was Mexico where the first measures had been taken against the predominance of foreign capital. The oil industry was nationalized in 1938, the energy sector in 1960, and finally, in the course of the sixties the extractive industries were "Mexicanized", too. By the end of the sixties foreign capital has disappeared from the traditional branches. The most important means of the Mexican control system is sectoral limitation: certain industrial branches are reserved for the national companies, and as a rule domestic capital is entrusted with the development of all those fields where the necessary funds are available, and where foreign investments would but increase indebtedness without establishing genuine new export capacities. In order to restrict foreign capital only state companies are permitted to operate in all the basic sectors of the national economy (such as the oil and petrochemical industries, power generation, railways, transport and communication), only Mexican-owned enterprises are permitted to effect agricultural investments and grant credits, and in six priority branches of the manufacturing industry (e.g. steel, cement, aluminium, glass, etc. industries) only enterprises with a Mexican majority ownership are allowed to operate. In the other sectors and branches of the economy foreign investments are not restricted, though the activities of foreign capital are definitely limited by introducing up-to-date technologies, stepping up Mexicanization, and by the policy of preferential customs duties in connection with the eliminating of export prohibitions.

Regulations introduced at the end of the past decade have extended the process of Mexicanization over further branches of the manufacturing industry, and have also stimulated the further vertical development of industrialization, i.e. foreign enterprises engaged in assembling are compelled to use locally produced component parts to an ever increasing extent. As a new measure, import licences were granted according to the export performance of the enterprises, this serving as an incentive for increased export orientation in the case of foreign firms which had earlier neglected export activities.

By stipulating the operational field and conditions of the foreign capital, Mexican policy has imposed rather severe restrictions on foreign corporations. Despite of this fact, and contrary to all preliminary estimates by the representatives of the liberal school, Mexico has not experienced any decline in the supply of foreign capital. In the period 1960–1972 the volume of foreign private capital investments rose from 1.1 billion dollars to 2.5 billion dollars, and within that the bulk of the increment fell to the share of the manufacturing industry and the services.

The primary aim of the Mexican policy of capital control in the fifties and the sixties was the acceleration of structural development and, therefore, the regional and sectorial restriction of investments served as the chief means of the control system. Later on, permission of capital investments was linked up with the vertical development of the industry, the increased utilization of locally produced component parts and, beginning with the late sixties, with achieving certain export results. While on the one hand these state preferences considerably promoted structural and export-oriented development, on the other hand problems deriving from the denationalization of the up-to-date branches of industry were overlooked, moreover, sectoral regulation had the effect of a boomerang. The increasing foreign control of the most advanced branches of industry eventually forced the government to take a resolute stand. The new Foreign Investment Act promulgated in February 1973, gave priority to the regulation of ownership relations instead of the former structural viewpoints. A ministerial-level National Committee of eight members was called into being and authorized with issuing licenses and controlling the influx of foreign capital. Only a 49 per cent ownership share is permitted to foreign capital investors by the new law, and only the National Committee may grant a higher percentage in the case of some special projects. In order to prevent the denationalization of the decision-making structure, the percentage of foreigners in the management of a company may not exceed the foreign share in ownership. 17 new economic-political criteria are stipulated by the law when granting permissions for new investments. The changes in priorities are most illustratively demonstrated by stress laid upon the role of foreign capital as a complementary factor of domestic capital, and by the emphasis on the balance of payments.

Parallel with capital investments, the process of technological transfer, too, was regulated. All agreements relating to the transfer of foreign technologies came under state control. It was made a precondition of such agreements that the adoption of a certain foreign technology should not restrict the exports of the Mexican enterprise, neither its price, investment and import policies or the development of its own research and technological basis. Thus, stipulations regulating technological transfer were aiming at eliminating the general and most unfavourable restrictions imposed upon the developing countries.

It would be difficult to sum up, for the time being, the effects of these recent and rather radical measures of capital control. In spite of its progressive character, the joint policy of the Andean countries for the control of foreign capital has met with no significant success because only very few new investment projects were launched by foreign capital in the fields drawn under control. In the first months following the promulgation of the new law a certain unresponsive attitude could be felt in Mexico: foreign capital decided on a wait-and-see policy pending the implementation of the new system. Import of direct long-term capital was again on the upswing in 1974, implying the trend that the attraction of the inherent dynamism, profit level and favourable prospects of the economy proved to be stronger than the deterring effect of the restricting measures. Foreign capital is always ready to adapt itself to changing conditions, if these prove to bring adequate profits.

The control of foreign capital in Mexico is effected without any complicated mechanisms, merely by realizing the preferences of the national economic policy. Nevertheless, the lessons of this policy are of a but limited value for other developing countries, since the factors which have brought about the cooperation and the requested movement of foreign capital in a certain direction (e.g. stable political power, a big buyer's market nearby, extensive internal financial resources) are non-existent in the case of the overwhelming majority of the developing countries.

Contrary to its reserved policy towards foreign capital in the fifties, Brazilian economic policy in the past ten years aimed at a maximum participation of foreign capital. Of all Latin American countries it is Brazil which best exemplifies the open door policy. According to Brazilian views, capital exporters are extremely sensitive to any regulation effected by means of ownership legislation or by the strict specification of the sphere of activity. State intervention, therefore, concentrates on the function of *channelling* foreign capital to the spheres deemed most useful by the preferences of economic policy.

Capital control has concentrated its efforts earlier on foreign exchange impact and problems of the balance of payments. It has been realized but recently that more attention must be paid to the adaption of capital and technologies, to the need of the recipient countries, to their adaptation to local production factors, and also to their coordination with the targets set by development strategies. In harmony with expansive internal and external policies, capital import and industrial cooperation is to be concentrated on the extension of export capacities and technological development necessary for a more intensive joining in the international division of labour. Capital investments furthering the processing of domestic raw materials for export became the centre of interest of the preferential system, and this is linked with those sectors and branches which are deemed as of crucial importance from the point of view of technical development strategy and for realizing the aim of achieving the position of an industrial great power. By reason of earlier experiences gained by capital import, present-day Brazilian policy attaches a great importance to the standard of management. In order to enhance export orientation and/or to break down former package deals, more allowances were made for the importing of management elements (mainly in the form of contracts) than in any other country of Latin America, and this in spite of a relatively higher level of scientific and technological development. One of the objectives of Brazilian economic strategy is the purchasing of capital and technologies without any kind of compulsory linked buying, under the conditions of the free market. This, however, presupposes strong adaptive capacities and a high standard of management.

In Argentina the economic nationalism of the Perón era has been succeeded by an economic atmosphere favourable to the influx of foreign capital. Owing to the increasingly fast expansion of foreign capital, more determined government measures became necessary. This coincided with the interests of the national bourgeoisie, too. The stepped up controlling measures in the early seventies were still rather of a restricted character, and proved to be by and large inefficient. This policy consisted of financial aid to domestic enterprises in deep water, the administrative prevention of handing over domestic enterprises to foreign interests, the strengthening of the technological resources of national companies, and the improvement of their technological decision-making potential, the development of the scientific and technical infrastructure, joint enterprises based on the cooperation of domestic and foreign capital, as well as the promotion of the signing of agreements on joint production. At this stage, preventive measures against foreign capital were taken by indirect means strengthening the bargaining positions of the domestic capital. A more vigorous, direct stand, following the example of Mexico, was taken, however, only after the return of Perón in 1973.

The Investment Act of November 1973 restricted sectorally foreign investments and banned all foreign investment in the branches of economy considered of vital importance from the point of view of the security of the national economy (public works, insurance, banking, advertising, marketing, the media, as well as spheres reserved for public enterprises). Enterprises with 51 per cent foreign capital interest are categorized as foreign, whereas those with 51 to 80 per cent national share as mixed companies. According to the law, Congress consent is necessary for the formation of foreign enterprises, whereas the formation of mixed companies has to be consented to by the government. The utilization of local natural and human resources ranks first among the criteria of approval to the formation of such firms. Further criteria are export orientation or import substitution, as well as the adoption of technologies deemed useful from the point of view of the national economy. The ratio of Argentine staff in management and the extent of domestic credits available are specified by special regulations. Priority is granted to foreign enterprises which pass into national ownership within ten years by means of gradual participation of domestic capital. The annual rate of repatriated profits is fixed at 12.5 per cent of the capital or, in the case of interest rates of foreign exchange deposits of 180 days at more than 4 per cent. Foreign enterprises in operation may choose whether they accept as obligatory the stipulations of the new law or acknowledge the 20 to 40 per cent special taxes levied on profit repatriation transactions. The implementation and/or the registration and control of foreign investments is the duty of a newly established specialized department in the Ministry for Economy.

The promulgation of the law is indicative of the fact that Argentina has joined the ranks of countries restricting the movements of foreign capital thereby increasing their economic sovereignty. At the same time, despite these regulations, the conditions of capital import are more liberal as for instance those of the Andean countries. It is quite obvious that the Argentine government tries to establish a balance between the extent of the formerly but vaguely outlined economic nationalism and the enticement of foreign capital.

The bargaining positions of the Argentine economy against foreign capital are indirectly strengthened by the merger of public enterprises into a single powerful syndicate. Patterned after the Italian IRI, CEN (Corporación de Empresas Nacionales) has become one of the biggest economic concentrations of Latin America, the unified management of which undoubtedly has increased the efficiency of measures aimed against foreign capital.

The trend in economic policy of taking a stand against foreign capital has become more and more conspicuous in recent years. Presently it is Latin America out of all developing regions where the most varied measures are adopted for the restriction of the movement of foreign capital. The development of manifold means of economic policy is following partly from the fact that increased export orientations and modernization have added to the capital import dependency resulting from the unstable equilibrium of the economic situation, and at the same time, basic national economic and political interests require the curbing of foreign capital which has penetrated the continent in the period of semi-colonial outward-looking, and the protectionist development policy.

The policy of direct measures against foreign capital is realized partly by enforcing the new preferences of economic policy, partly on the institutional level, and partly by developing the means of control.

The most important reassessment of the role of foreign capital in economic policy is the establishment of the link between export orientation and capital import. The reduced extent of protectionism, the emergence of competitive imports represented but a modest incentive for the extension of the market activities of foreign enterprises which had formerly produced almost entirely for the domestic market only. The first steps towards a more effective system of incentives was the elimination of export restrictions from the contracts on capital or technology imports. A further step was the linking of profit repatriation with the earning of foreign exchange from exports, as well as the stipulation of detailed export obligations for recently established foreign enterprises.

The trend to restrict the exploitation of the resources of the domestic capital market by foreign capital, and to compel the parent companies to hand over more of their resources than ever before, has manifested itself in both the Andean countries and Mexico as well as Argentina. The increased utilization of local raw materials and human potential is also ranking high on the priority list of the capital import policy of the majority of Latin American countries.

The reassessment of economic policies has pointed to the necessity of calling into life government organizations well-qualified for a thorough analysis of the role of foreign capital in the national economy, since the channelling of capital import and the realization of an efficient selective policy requires a profound knowledge of capital markets, technologies transferred through capital import, sources of supply, and also the central registration of concrete capital movements. Previously, a great number of un-coordinated organizations established for different purposes had participated in capital import transactions. In the past two or three years the major Latin American countries began to shape a unified infrastructure of their capital import policies.

The means of control formed within the framework of the modified preferential system present a rather mixed picture for the time being. Though ownership regulation is still of considerable significance, control measures are increasingly operating with elements of the law of obligations. To evade problems following from capital ownership in the transfer of technologies, management and various services furthering exports, it is the connections established on the basis of contractual obligations which expand most rapidly. While the elaboration of adequate means of control has not passed the phase of experimenting, the practical policies implemented in the Latin American countries have certainly not been isolated phenomena.

The problem of the so-called investment atmosphere is put into a new light by the trends of Latin American capital import and the tendency of a policy of more energetic capital control. The doctrine of the liberal school of thought according to which the institutional framework of a given country and its liberal economic policy are the main forces of attraction for capital exports, is no longer tenable. Experience proves that it is not liberal economic policy but economic dynamism, development potential, the endowment with natural and human resources which first and foremost attract foreign capital. If all these factors are existent, capital acquiesces in restrictions, their lack, however, can hardly be remedied by political means.

3.7. LATIN AMERICA AND THE WORLD ECONOMY AFTER 1973

By now it can hardly be doubted that with the so-called energy crisis of 1973 an era of post-war international economic development, lasting for more than 25 years, has come to an end. The worsening imbalance of the world economy, the intensification of its cyclical problems and the so-called energy crisis often divert attention from the main elements of the changeover period in the world economy. Such component forces are: the accelerated rate of scientific and technological progress, the stepped up rate of international political development and the practical strengthening of peaceful coexistence between the two systems, the transformation of the world economic balance of power, the accelerated structural transformation of the developed capitalist countries, the modification of the institutional order of the world economy, and the gaining momentum of the developing countries' struggle for economic independence. In connection with the above-mentioned objectively effective sociopolitical, technological and scientific as well as economic processes there begin to take shape the outlines of a new world economic environment which would render it possible to bring about the consolidation of peaceful coexistence by economic means, to eliminate the vestiges of colonialism, and to step up the socio-economic progress of the developing countries.

At the beginning of the seventies it became ever more obvious that economic growth founded on wasteful consumption is going to collide with the limits set by the finite character of natural resources. It also hampers the mobilization of intellectual and material energies which are necessary to a much greater extent than before for entering the higher (let us term it post-industrial) stage of development, and are indispensable to establishing more up-to-date structures. The international financial system relying on the positions of the United States as a global power had to face the fact that it became increasingly unfit for shouldering the burdens of wasteful consumption and monetary expansion connected, among others, with the Vietnam war. The limits of growth have been marked conspicuously by the breaking up of the equilibrium both in the internal and external economies, and the disintegration of the so-called Bretton Woods institutions. Under these circumstances, the modernization and rationalization of raw material and energy consumption, the curbing of wasteful consumption, as well as the bringing about of *structural development* going hand in hand with an improved growth efficiency—becoming the source of new trends in the international division of labour, too—have become an imperative requirement for the capitalist world.

The development of the productive forces going far beyond the dimensions of the national economy have for a considerable time brought pressure to the *institutional* framework of the world economy. The process shoves along through both micro- and macroeconomic channels. Development in the macroeconomic sphere is marked by the establishment and progress of integrational groupings, whereas in the microeconomic sphere the process is reflected by the expansion of the multinational corporations. Both trends tend to reduce the significance of national sovereignty and of the processes within the national economy. They also bring about new instances of coercive adaptation, and introduce elements into the growth process and decision-making system which often prove to be too confused or uncontrollable for the national governments.

How does the changeover from one period to another in the world economy affect Latin America?

The coincidence of the increased external economic orientation of some major Latin American countries with the switchover from one stage to another in the world economy had a growth-accelerating effect. The increasing inflation in the developed capitalist countries, the incipient runaway of labour-intensive industrial branches based on simpler technologies from the higher wage level areas, the growing part of the international division of labour in the transmission of the achievements of scientific and technical progress, the differentiation of the world economic balance of power have diminished the risks of export orientation as early as the end of the sixties, and have speeded up the development of the countries which have most decidedly chosen the path of export orientation. The rise in the price-level of raw materials conspicuous ever since 1969, which culminated in the raw material and energy crisis of 1973/74, the economic switchover in the wake of the modification of the terms of trade of finished products and raw materials, the development of new international relations both as regards the balance of power and interests have all come about in a relatively fortunate cycle in the countries breaking with protectionist development policy. On the other hand, due to the price explosion in 1973, an unusually serious recession took place in the developed capitalist countries, and, after some time lag, in the Latin American countries as well.

As a resultant of various factors, the economic growth process has gathered considerable momentum in the Latin American region. In the period 1970–1973 the average growth rate of the national product amounted to 6.8 per cent, in 1974 to 8 per cent, and in 1975 to 4 per cent. Although increasingly serious problems are reflected by the set-back in the growth rate, it indicates the increasing weight of the Latin American region in the world economy at a time of a general depression. While the growth rate of the developed capitalist world made no headway in 1974/75, the production of Latin America rose by 12 per cent.

In the case of Brazil, Mexico and Colombia the difficulties experienced in the marketing of Latin American goods on external buyer' s markets, especially the difficulties in the sales of industrial goods had undoubtedly a great part in the decline of growth rates in 1975. As regards Argentina, the reasons for the economic set-back were primarily of a political nature.

The achievements of the economic progress of the Latin American region, however, did not make possible any far-reaching easing of the basic economic problems, moreover, the distribution of economic results was extremely disproportionate among the countries of the region. Even the 1975 report of CEPAL mentions the fact that "... disregarding those seven or eight countries which are in the position to exploit the present situation in their own interest, the overwhelming majority of the Latin American countries has to face the problems of a diminishing growth rate, an increasingly serious employment situation, and the menace of a large-scale deficit in the balance of payments..." In the 15 countries representing a third of the region' s population and production the average growth rate of the national product was 4.2 per cent on the average for the period 1971–1974, whereas *per capita* income rose by only 1.3 per cent. According to CEPAL statistics, in 1974 alone, four countries made additional foreign-exchange earnings of about 10 billion dollars from the rise in the price of oil, and at the same time, the additional burdens of higher oil prices for 19 Latin American countries made up 2.7 billion dollars.[6]

The relatively favourable average rate of growth hides, however, increasing imbalances as well. Imbalances in the external sector and foreign exchange indebtedness increased at a high rate in most of the Latin American countries. Imbalances in the external sector were caused, in addition to deteriorating export possibilities in 1975 and to the downward tendency of prices, also by the belated reaction of the economic policies of these countries to the recession. The favourable tendencies on foreign markets in 1973 and 1974 brought about considerable import programmes, budget expansions, armament spending in most of the Latin American countries, and this resulted in a worsening trade balance already in 1975.

Some adverse symptoms manifested themselves in the sphere of external financing, too. The volume of government aid and advantageous state credits have been diminishing for years, and thus, external financing requirements had to be met increasingly by private sources. Debt service burdens, amounting in most of these countries to 20–30 per cent of foreign-exchange incomes, have considerably increased as a result of deteriorating conditions of raising credits, higher interest rates on loans and the increasing withdrawal of the profits of the capital dreading nationalization.

The process of inflation gained new momentum and internal imbalances were again enhanced in 1974/75 by the spill-over effect of the difficulties of the capitalist world economy, the rapidly rising price-level of imported goods as well as by expansive budgetary policies.

The results of accelerated growth in the seventies were exploited primarily by the most efficient sectors of the economy, whereas the burdens of world economic recession and increased inflation had to be shouldered by the more backward sectors of the economy and society as whole. At the same time, the growing income inequalities increased the socio-political tensions.

International changes affected also the internal political and power relations in the Latin American countries. In countries which achieved certain results in their rates of growth, these results were concentrated first and foremost on the most efficient and up-to-date sectors and on elements and social strata in charge of the power positions of the economy, and strengthened decisively the economic basis of the bourgeoisie. The bourgeoisie strived to make use of its suddenly consolidated bargaining positions for strengthening its political status and, as a rule turned a deaf ear to all suggestions aimed at stimulating social transformation which could have threatened its power positions. Moreover it attempted to adopt a policy of "the mailed fist" against the working classes and/or all forces fighting for social progress. Though the Chilean tragedy of 1973, marking the beginning of the process of a political drift to the right, had been independent of changes in the world economy, the lesser or greater shifts which occurred since that time in the power structures of Peru, Ecuador, Argentina, Mexico and Panama were to a certain extent related to the consequences of international events.

The Chilean tragedy and the drift to the right, however, do not indicate any stagnation in the socio-political progress of Latin America. Owing to an accelerated economic development or to advantageous world economic changes, the bourgeoisie succeeded in consolidating its economic power positions, and

---

[6] *Economic Bulletin for Latin America*, 1974/1–2. 1975.

in a number of Latin American countries, especially in Mexico and Venezuela, obtained additional growth resources which enabled it to build up the material basis necessary for taking a stand against international big capital. At the same time, a part of the forces fighting for a more rapid socio-political development also arrived at the conclusion that, as a consequence of the changes in the internal and external conditions of progress, it is not the socialist revolution or some other radical transformation which is the order of the day, but rather a policy of making the most of the sharpening contradictions between the interests of the national and the international bourgeoisie, as well as the stepping up of the fight against international capital and economic dependence.

The shifting of the political struggle to the international arena is reflected partly by the individual, national aspirations of the single countries, and partly by the joint stand taken by the countries of the region as well as their activation in the field of international politics. The most important problems of the struggle waged in the sphere of international affairs are the following: the development of the fight for the recognition of the economic sovereignty of nations and for the control over their own resources and decision-making centres, measures striving to control and later on increasingly restrict the activity of foreign capital, the nationalization of the internationally significant Venezuelan oil and iron ore industries, reducing dependence on the United States, and finally, the trends pointing towards rapprochement with the West European countries, Japan and the socialist states. The further development of such attempts were curbed to a certain extent by the economic difficulties of Western Europe and Japan in 1974/75, since their main interest in the field of foreign relations shifted towards establishing relations with the countries of the Middle East, and partly by the relative strengthening of the positions of the USA in the world economy after 1973.

Thus, besides the fight waged within national bounds for the reducing of economic dependence, joint regional actions and, beyond that, the common stand taken with the African and Asian countries became increasingly important. In addition to putting an end to the blockade against Cuba, the most important step of the fight waged within regional limits was the establishment of the Latin American Economic System (SELA) in 1975. Without the participation of the United States but inviting socialist Cuba, SELA endeavours to step up the economic and social progress of the member countries, to organize a joint defence against international monopolies, to make better use of the natural, human and financial resources of the region, as well as to establish multinational Latin American mixed companies. As a result of a common stand of the Latin American countries, under the auspices of the UN Economic Commission for Latin America (CEPAL), the Committee for the Cooperation and Development of the Countries of Caribbean Region was set up, creating thereby an organized framework for cooperation in various fields (such as maritime navigation, telecommunication, exploration of ocean resources, etc.).

The joint international actions with the other developing countries marked the emergence of a new qualitative element in the struggle for economic independence. The Latin American countries, especially Mexico, Peru, Venezuela, Argentina and Colombia took an active part in the preparation, elaboration and in the discussions of the fundamental documents adopted by the General Assembly of the Unites Nations, about the new international economic order, and the Charter on the economic rights and duties of the states. These documents may be rated as major victories of the developing countries, and the progressive forces supporting them, in their fight for strengthening the principles of peaceful coexistence and equality in international economic relations. They help bring about mutual international dependence furthering a more efficient and rapid progress. In the course of their fight for the realization of the requirements of a new world economic order, the Latin American countries took a leading part both in the work of the Lima conference of the foreign ministers of the non-aligned countries in 1975, and in the proceedings of the UNCTAD conference in 1976; with elaborating their Memorandum of 1975 they began the shaping of a behavioural code for international monopolies.

However, the results achieved by the joint international actions of the developing countries cannot obliterate the fact that a number of highly heterogeneous national economic and class objectives were united in a coalition by the struggle for a new world economic order. It is but natural that the countries with strongest economic positions are in the vanguard of the struggle and are also the leading spokesmen of the radical demands of the Third World. Despite of the highly disadvantageous consequences of the energy crisis, the countries with economies grappling with serious economic problems, with foreign exchange, food and general growth difficulties, are compelled to side with the "upstart" developing countries, and they try to use these in asserting their own claims.

The radical policies of those developing countries which have gained strong bargaining positions, e.g. Saudi Arabia, Iran, Brazil, Venezuela or Mexico should not be regarded, however, merely as a common defense of the interests of the developing world. The international actions of these countries have undoubtedly influenced the rearrangement of the international economic power relations, and by restricting the sphere of great power and multinational pressure have had a beneficial effect on the external economic manoeuvring capacity of countries highly affected by world economic changes. At the same time, it would be mistaken to disregard the fact that attempts to consolidate their own international, quasi-power positions considerably motivate the radical action of these countries. It is easy to predict that in the long run such policies will separate these countries from the so-called Fourth World. They will strive to stimulate rather the decision alternative of the developed market economies which would champion a closer cooperation with the most advanced, most dynamic developing countries, and would direct their efforts in order to integrate them as fast as possible into a developed capitalist centre. Thus these policies might delay or even prevent the solution of the problems of the so-called Fourth World and those connected with the readjustment of the world economy as a whole. Under these circumstances, the organization of a confrontation with the United States, the developed world and the multinational corporations reflects but an attempt to improve the conditions of collaboration with the developed capitalist countries. Moreover, these countries mark themselves off, as far as possible, from radical policies proclaimed by some of the spokesmen of the Third World. The tactical ratio of the elements of collaboration and confrontation with the developed capitalist countries is determined by considerations according to which a liberally oriented world trade is advantageous to these countries. Therefore, after some favourable corrections, the basic structure of the world economic system ought to be preserved. The joint actions with other developing countries are confined to assuming certain mediatory functions as well as to the undisturbed establishment of economic and technological organizational ties with the more backward African states.

In spite of the contradictions, different objectives and interests of the joint actions of the developing countries, it cannot be doubted that the struggle for economic independence waged on an international scale has already yielded some results. It has strengthened the international bargaining positions of a number of Latin American countries, by restricting the operational field of international big capital and US global strategy. Though the struggle is still confined to the limits of economic nationalism, in the long run, it offers more favourable conditions for the acceleration of internal socio-political development.

CHAPTER 4

REGIONAL COOPERATION IN LATIN AMERICA

4.1. SOME THEORETICAL PROBLEMS OF THE INTEGRATION OF THE DEVELOPING COUNTRIES

The scientific and technogloical revolution taking shape after World War II, and the acceleration of the growth process opeded up a new phase in international economic relations, too. The international division of labour has been enhanced on a world-wide scale by the indivisibility of up-to-date technology and by the growth of economic activites going beyond the frame of national economies. It speeded up not only the international flow of goods, services, production factors, scientific and technological knowledge but broadened to an ever increasing degree direct international production relations as well. On a higher level of development of the productive forces the international economic integration fits into the pattern of the development of both the individual national economies and the international division of labour with increasing weight. Marxist and the majority of progressive bourgeois integration theories regard integration together with the social division of labour as a progressive and deepening process, deriving it from a higher developmental stage of the productive forces.

As a result of the international development trends in the past decades, the problem of integration became the order of the day in the developing countries, too, as in these countries which had broken away from the former colonial empires and were kept within the limited size of their national economies only integration could ensure the rapid and efficient development of an up-to-date economy.

The acceleration of development, and at the same time keeping abreast with the requirements of technological and structural development have proved to be problems the solving of which have become more and more difficult. The overwhelming majority of developing countries, lacking an adequate economic potential, are faced with the danger of a deteriorating social equilibrium and security level. Under present conditions of the world economy about nine tenths of the independent national economies are rated as small units, and this holds true not only for developing countries but for developed small capitalist and small socialist countries as well.

Considering experiences gathered around the middle of the sixties, integration seemed to be able to ensure the possibility of technological-structural development and of joining in the expansion of the up-to-date international pioneer-industries to countries with minor economic potential. The deplorable consequences of protectionist economic policies for the smaller national economies in the thirty years following the World Depression demonstrated that an acceleration of the growth process without serious sacrifices, the improvement and modernization of a country's position in the world economy requires, if possible, the coordinated utilization of resources of the individual regions and broader and unified markets for new industries. The process was speeded up by the expansion of the multinational corporations, twice as rapid as the world average, by the spread of the novel-type menacing factors resulting form the confrontation of small nations and mammoth enterprises.

Thus, contrary to the developed countries, *the rapid regional integration of developing countries is by no means necessitated by the highly developed productive forces but rather by their backwardness and/or the need of their development.* Owing to their absolutely different basic conditions, the problems and mechanisms of the regional integration of developing countries reveal characteristics different from those of the developed countries.

Relevant literature well-known also in the developing countries approaches the problems of regional integration from two different aspects. The traditional liberal school of thought regarded integration resulting from free trade and maximum free competition among the individual countries as the key to regional development. The customs theory worked out by J. Viner[1] and his adherents,[2] derived the developmental influence of individual regional groupings from the ratio of the trade-diverting effect stemming from protectionism and considered to be a negative factor and the so-called positive trade-creating effect. As a reaction to Viner's school of thought not reckoning with international income-flow trends, with the extremely narrow spectrum of alternatives of the underdeveloped countries, with the political, cultural, etc. factors and long-range, dynamic effects of regional cooperation, S. Dell[3] expounded the view that the primary aim of the common markets of the developing countries was not to terminate protectionism but to increase its scale and effectiveness. In the following we consider the nature of the specific problems arising from the integration of the developing countries.

The liberal approach cannot be applied to the present-day integrational practice of underdeveloped countries, and not even to that of the developed capitalist countries. Nevertheless, market mechanisms obviously have had a vital role in establishing the high level of development of certain West European national economies and also in the increasing economic linking-up of the countries of this region. It has been the historic integration of market relations which made it possible that the deliberate shaping of commodity and financial relations as well as means of foreign trade, customs, monetary, and foreign exchange policy have achieved decisive roles in breaking down economic barriers which impeded the emergence of a unified West European region. The very fact of market integration, however, presupposes developed productive forces and market relations, a high level of historically evolved division of labour, and a similarity of the degree of development, economic and institutional structures, economic and foreign trade potentials of the member countries.

According to the first basic problem regarding the integration of developing countries, the idea of integration by means of the market *cannot even be proposed rationally if the economic environment is characterized by the general backwardness of market relations.* However, the other preconditions and background facts of market integration represent at least as serious a stumbling-block as the underdevelopment of market relations. Let us consider the most important impeding factors.

---

[1] J. Viner: *The Customs Union Issue.* New York, 1950.
[2] R. Harrod: *International Trade Theory in a Developing World.* London, 1963.
[3] S. Dell: *Trade Blocks and Common Markets.* London, 1963.

The historically developed *higher degree of* regional *division of labour* played a great part in the successful launching of integration in Western Europe. In spite of the historical and political clash of interests, the member countries of the Common Market have realized already at the start about one third of their foreign trade among themselves. As regards the three economic integrations in Latin America, this ratio was initially 9 per cent for the Central-American Common Market, 8 per cent for the Latin American Free Trade Association, and 3 per cent for the Andean Group. Thus, it is hardly presumable that the common interests in the division of labour would break the trail for the automatism of regional cooperation.

One of the pivotal questions of regional cooperation is the *level of development* of the member countries. It is a well-known fact that the level of development of the member countries of the West European integration, the order of magnitude of their *per capita* national product had been by and large identical, Italy being the only exception. Within the regional groupings of the developing countries differences are much greater. Thus, the order of magnitude of the *per capita* national product in Kenya is the double of that in Tanzania, whereas within the regional groupings of the countries of Latin America differences between the most developed and most backward countries are five and sixfold.

One of the basic problems of regional cooperation is the effect it exerts on the less developed countries. It is a well-known fact from history that up till the beginning of the 18th century differences in the development level of the individual countries had been rather insignificant by present-day standards. The appearance of the gap between the developed and underdeveloped countries coincides with the development of the competitive mechanisms of capitalist economy. Gunnar Myrdal[4] pointed out already in his early works that contrary to the view held by the liberal school of thought, no automatic self-regulators whatsoever function in the capitalist social system, and the built-in mechanisms only deepen the gap between wealth and poverty. Thus, if differences of the development levels within a regional grouping are considerable, reliance upon market mechanisms might become disastrous for less developed member countries, and cooperation will unfold first and foremost among the more developed member countries.

The traditional remedy for disparate levels of development would be unilateral benefits and concessions granted to the *less developed member countries*. This remedy, however, in itself raises serious problems as the quantitative distribution of the advantages of regional cooperation requires complex mechanisms even in the case of countries of the same development level. It is just the distribution of the benefits among the member countries, the lack of adequate compensating mechanisms that constitutes the weakest point in the regional agreements in force. It was pointed out by Ferenc Kozma[5] that benefits granted to less developed countries help the above the average growth of the recipient countries, the balanced development of the region, the intensification of the division of labour within the region and the strengthening of regional cohesion, but at the same time might slow down the development of the region's most developed countries, industries and enterprises.

---

[4] G. Myrdal: *Economic Theory and Underdeveloped Regions.* London, 1957.
[5] F. Kozma: *Some Theoretical Problems Regarding Socialist Integration.* Budapest, 1971. (Trends in World Economy, No. 6.)

Over and above that, in a region tense with national contradictions, there is considerable danger of the less developed favoured countries becoming the primary targets of the penetration endeavours of extra-regional foreign powers and big corporations. For instance, if the favoured country is in a position to pursue a more liberal policy as regards foreign capital, this might result in the establishment of a bridge-head, and foreign capital could glut the markets of the whole region with its products using the less developed country as a base.

Thus, the establishment of regional cohesion requires on the one hand the elimination of the historic differences, the faster development of the less developed countries, and growth rates differing from country to country. On the other hand, according to experiences gained from the economic history, an efficient development policy is always coupled with a more rapid development of the efficient sectors and regions and with the lagging behind of less efficient regions, especially so in underdeveloped countries where the relative lack of resources precludes from the very beginning a simultaneous raising of the standards of all regions or sectors.

The distribution of the means of development according to effectiveness, however, can be envisioned only under rather homogeneous political and national economic conditions. It follows from this that countries where differences in standards between regions and sectors necessitate first of all a national integration of the economy, can accept only restricted regional cooperation objectives, since internal socio-economic integration is first and foremost the task of the national state and not of international integration. Another consequence of the above-mentioned thesis is the fact, that the greater the regional differences in standards, the greater sacrifices ought to be made by the less developed member countries for *a more effective* growth, and by the more developed ones in order to achieve a regionally *balanced* growth.

Let us consider an abstract case. The fastest and most effective development of the East African Common Market would have been furthered if the natural resources of the less developed Tanzania and Uganda had been made use of by an expansionist policy of the more industrialized Kenya. However, such a high degree of unity and solidarity could hardly be expected from independent national economies as in such a case independence would come to an end and the economic union would lose its *raison d'être;* a solution could be brought about only by political union. The less developed countries of East Africa which are adopting antagonistic ideologies and pursue hostile foreign policies were obviously not willing to sacrifice their national interests in order to accelerate regional development. The other aspect of the problem is illustrated by the withdrawal of Jamaica from the one-time West Indies Federation in 1961. As the biggest and most developed member state, Jamaica was not ready to burden her own growth potential by accepting certain mechanisms ensuring the faster development of the smaller member states. Thus in case of differences in the level of development, the backward member states are reluctant to make any sacrifices in order to accelerate the growth rate of the developed ones, whereas the more developed countries are loath to grant unilateral concessions. The level of compromise between the two groups depends on the magnitude of the differences in standards as well as on the dimension of the advantages anticipated by cooperation, and not at least on the weight of non-economic factors. Free trade mechanisms, however, are not suitable for the elaboration of the optimal form and intensity of cooperation and the coordination of national interests in case of differences in development levels.

Similar if not graver problems may arise from *the differing sizes of the national economies* of the countries participating in regional groupings. Under capitalist conditions the coexistence of small and big countries is rarely exempt from conflicts, or it is at least a burden on the small countries since their international bargaining positions and the degree of interdependence are rather disparate. It is less the absolute military or political power than the relative weight within the functions of association which is decisive. Thus, the decisive factor within the EEC has been not the nuclear potential of France, but rather the size of the market capacities of the individual member countries. The foreign trade potential of trade-oriented small Belgium or the Netherlands was by no means negligible as compared to France or Italy, countries with a great population. The foreign trade turnover of the big EEC countries surpassed that of the small ones by only 20 to 40 per cent, while both small and big member countries were characterized by considerable export orientation.

In the case of Latin America, owing to historical reasons, no functional symmetry in foreign trade can be detected. The big Latin American countries outstrip many times the small countries of the region not only as regards the size of their national products, military power and industrial capacity but also considering their foreign trade potential. Unequal balance of power, the heterogeneous relations of power in the region have hampered from the very outset the adoption of common automatisms. F. Perroux has pointed out years ago that big countries are able to spend more on research, development, propaganda, marketing, armament, etc., and therefore the structural integration between small and big countries is realized more often than not for the benefit of the big ones. A small country has precious little chances of liquidating its historical backwardness by accomplishing integration with a country of much greater economic potential.

The most essential problems arising from differing economic potentials of member countries are the following:[6]

– Regional cooperation involves the danger of the big country or its powerful enterprises constituting the main driving force; the small countries realize integration separately with the economy of the big country. Thus, their dependence on the big country participating in the regional grouping will increase. Further, the participation of a big country often curbs the integrational endeavours of the small countries involved.

– As a rule, considerable differences are to be found in the external orientation as well as the economic environment of the small and the big countries. Small countries inevitably take a more active part in the international division of labour, rely to a higher degree on impulses from their external sectors, on foreign resources, vertical division of labour, and they also have to develop an economic policy and institutional system warranting fast adjustment. Big countries never rely to a high degree on the sector of foreign economy; they can afford without losses even the luxury of making use of the political means and institutions of an "inward-looking" economic development. These differences in themselves hamper the regional cooperation between small and big countries. Close links with bigger countries less interested in the broadening of regional or world economic relations can also slow down the development rates of the smaller countries. Similar problems may crop up in the course of insti-

---

[6] B. Kádár: "Small-sized Developing Countries in the Regional Arrangements." *Acta Economica Hung.*, No. 10, 1973.

tutional development as well. Owing to the fact that bigger countries are less interested in this process, it is generally possible to progress only in fields regarded as essential from the point of view of the big country.

– Finally, it is, paradoxically, the big country which receives a greater share of capital imports, and not only in the stage of import-substituting growth but also in the case of an export-oriented model. The examples of Kenya, Brazil or Mexico testify to the fact that foreign capital regards the bigger capital absorbing markets as more promising. The favoured economic position of the big country is frequently strengthened by the political support of the world powers which, acknowledging the role of a regional power subcentre make use of its mediation in their relations with the smaller countries.

Apart from one or two exceptions, the experiences gained from import substituting have convincingly proved that the potential of the national economies of the developing countries is not sufficient enough for developing up-todate industries. Due to the high social costs of protectionism on a national economic level, the protectionist school of thought has accepted in the past decade the strategy of *regional protectionism* as opposed to the "inward-looking" national economy. The theory of administrative regional cooperation directed against market cooperation sets out from the premise that the breaking down of barriers between the isolated small national markets and entrenchment behind a common regional protectionism adds new stimulus and higher effectiveness to the exhausting national import-substituting processes.

The conception of regional protection implying but half-truths is assailable right from the very outset. An East African, Central American or Andean common market based on regional protectionism will remain, even in the case of full integration, a dwarf unit from the point of view of world economy. At the time of the launching of the above theory, the aggregate imports, the industrial buyers' market of the countries of the Latin American Free Trade Association did not reach the capacity of the Netherlands, not to mention the far-reaching division of labour in the small Netherlands which would be highly improbable in the case of these nine Latin American countries comprising an area of 17 million square kilometres. Protectionism developed within regional bounds on account of restricted economic potential could lengthen the life-time of import substitution for some years at best, whereupon growth would necessarily peter out on a regional level. On the other hand, the example of the developed small European capitalist countries illustrates the point that growth based on efficiency and international division of labour can bring about up-to-date structures even in countries which play a minor role in the world economy, like Finland, Norway or Denmark.

The theoretical posing of the problem takes no notice of practical difficulties following from the differing levels of protectionism of the individual national economies. Even the intensity of nominal protective tariffs reflects immense differences, not to mention the differentiated nature of an effective protective tariff system. According to calculations by Balassa,[7] average nominal tariff protection of durable consumer goods was 108 per cent in Brazil (1966), 84 per cent in Chile (1961), and 29 per cent in Mexico, while in the case of capital equipment the respective percentages were 87, 92 and 29 per cent. If there are such considerable differences in the level of customs, the breaking down of

---

[7] B. Balassa: *The Structure of Protection in Developing Countries*. Baltimore, 1971.

customs barriers between the member countries without adopting a common external tariff system will result in increasing out of all proportion the competitiveness of the less protectionist country with a lower degree of tariff protection. It is also highly probable that in such a case not the actual comparative advantages but the differences in customs duties will become the driving force of developing specialization. As a rule, free trade mechanisms function with a more satisfactory degree of efficiency, when the level of tariff protection is low (e.g. EFTA). In countries with high or differing tariff protection the breaking down of barriers impeding trade among themselves, radically upsets the price and cost structures of the member countries. In such cases it is advisable to initiate the breaking down of customs barriers by introducing a common external tariff system. However, owing to heterogeneous customs levels and the necessity of transforming the whole foreign trade system, the actual adoption of common tariffs is a very difficult matter.

Thus, on the one hand free trade mechanism does not offer any solution because of the backwardness and the difference of the development level of the member countries, and on the other hand, regional protectionism offers no way out either on account of the small scale of the national economies and the character of present-day international development trends. Consequently, regional cooperation in the developing countries—as opposed to that in the developed ones—ought to be based, in addition to the maximum exploitation of the advantages offered by the division of labour, *on the coordination of development strategies and investment policies* and on the harmonization of national and regional planning.

Even a brief survey of the relevant theoretical problems demonstrates that the developing countries can neither rely on experiences gained from the regional cooperation of the developed capitalist countries, nor on their theoretical advice. Thus, their cooperation ought to set off on a new path.

4.2. THE SOCIAL PROBLEMS OF INTEGRATION

The most important result of integration has been the cutting of the social expenses of development, and especially of the costs of industrialization. Under world economic conditions when—as a result of *a priori* endowments—foreign trade relations are not favourable to growth and there is no possibility of their being reduced *in merito*, the significance of the optimation of the internal division of labour from the point of view of costs is growing considerably. It is a well-known fact from the history of import substituting industrialization that in countries with narrow domestic markets the forced developing of up-to-date industries demanded extraordinarily high social costs. The establishment of optimum-size plants, the full utilization of up-to-date capacities, the implementation of economies of scale could not be achieved in any of these countries. To illustrate the serious nature of problems resulting from the above situation, let us mention the countries of Latin America. In the motorcar industry, regarded as one of the decisive pioneering industries in countries implementing an import-substituting industrialization, in spite of low wage costs the international price level has been surpassed by 200 to 250 per cent in Brazil and Argentina, by 60 per cent in Mexico and by 300 to 400 per cent in Chile; about half of the difference derived from the additional costs of scattered, small series production.[8] The order of magnitude of differences in the specific cost level of suboptimal serial production within the narrow bounds of the national economy also indicates what possibilities are inherent in the specialization and integration of the individual countries.

The problem of social costs deriving from the limited size of the national economies, however, goes beyond the questions following from a lower effectiveness of the microsphere. The experiences gained in the course of the economic growth process by both developing and developed small countries confirm the theory that the lower the effectiveness of investments, the higher the specific capital intensity and in connection with this, the forced pace of ever increasing capital imports and local investments are direct consequences of the suboptimal scale of the national economy. The dependence on capital imports in itself can restrict a country's international manoeuvrability, not to mention the almost inevitable problems of local nationalism following as a reaction to the above process.

The incessant pressure for increase as regards the accumulation capacity of the national economy—which is not restricted to a certain growth period— resulting from small capacity and the hardly avoidable lasting pressure for investments necessarily become a source of the extension of the administrative sphere not only in economic but in social life, too. Its reactions are to be felt both in the lower specific level of consumption and in the development of mechanisms suitable for the controlling of tensions connected with it. Thus, for countries with a smaller economic potential which are striving to keep abreast more or less with the pace and direction of international development, integration renders it possible not only to utilize the potential energies inherent in economic and technological development, but also *to rely less on administrative mechanisms* and to build up the political equilibrium on a sounder basis.

Integration can achieve an increasing significance in the process of the improvement of the *socio-economic environment*. The fact that the scientific and technological revolution extends the dimension of economic life and/or that contradictions are sharpening within the framework of the national economy results in the speeding up of the concentration of economic power and the establishment of monopolistic positions in countries with a smaller economic potential. The strengthening tendency of the developing of monopolistic positions is hardly affected by the open or protectionist character of the economy, and, at the most, can diminish the negative consequences resulting from the monopolistic positions.

As a matter of fact, under protectionist conditions the isolation of the national economy from external competition means essentially the elimination of competition, since narrower national economic dimensions make it possible to establish only some enterprises which—in certain branches of industry, especially in the heavy industries—practically occupy monopolistic positions. In backward countries industrial monopolistic positions are imposed on the historically developed system of feudal privileges and values, and on a social scale impede technological progress and quality improvement; they also result in rising costs, diminishing effectiveness and decreased competitiveness.

In small countries open to the world economy, and especially in small developed capitalist countries, the drive for competitiveness and the necessity

---

[8] I.N. Behrman: *The Role of International Companies in Latin American Integration*. Lexington–Toronto–London, 1972, p. 185.

of doing away with the disadvantages following from a lack of internal resources and markets require the acceleration of the concentration process. It is a fairly well-known fact that in Sweden, in the Netherlands and Switzerland the six biggest corporations contribute a much greater percentage to the national product than for instance in the case of the great capitalist powers. In countries open to the world economy the negative aspects following from monopolistic positions and the inequalities of economic power relations are undoubtedly restricted by the participation in international competition. However, even an open economy cannot do away with them.

Both in protectionist and open economies the phenomenon of "multinationalization" of the monopolistic local big corporations, their association with the international big capital becomes evident (the interests in choosing partners links the multinational enterprises unequivocally to the relatively bigger corporations). Parallel with the unfolding of this process *it is often just the big corporations which play the part of the trailblazers of foreign great power penetration and they are responsible for the undermining of the sovereignty of the national economies.*

The influence of the economic monopoly positions on power relations and *social equilibrium* should not be underrated either. In the course of historical development the trend of correlation of economic and political power is clearly discernible. In the underdeveloped and small countries the concentration of economic power leading to monopoly positions has resulted in a shift of the political power relations and in the weakening of the foundations of democratic institutions. Lacking any other countervailing forces, a straight and very short way leads from the economic and political monopoly positions to a system which hinders the free unfolding of human abilities, and results in the ever increasing alienation and the deterioration of the quality of life.

Our conclusion is similar to the first one: due to the social consequences of modern development, smaller countries are increasingly interested in integration. Regional cooperation offers the opportunity of the formation of an economic region necessary for the development of competition, the linking up of the requirements of optimum economic sizes and competition mechanisms, and the toning down of negative factors resulting from the concentration of economic and political power. Under circumstances of regional cooperation the requirements of the corporate, harmonizing and arbitration functions of the social organizations can also be met.

It is a well-known fact that in contemporary world economy it is integration which can guarantee the possibility of *structural* development for a growing number of countries. According to experiences all over the world, the growth and specialization of countries with smaller economic potential made progress or developed in sectors where the demands on the mass character of production were relatively the lowest, where competitiveness could be achieved even in the case of smaller economic sizes or where the advantages of the great powers could not be fully exploited in competition. Whether we examine the structural indices of the leading capitalist countries and the developed small capitalist countries, or e.g. those of Brazil, Mexico and the smaller less developed countries, the conspicuous features will be the correlation between economic dimensions and structural development, the higher GDP ratios of agriculture, the extracting industries, the light industry and services, the relatively low degree of participation in the international industrial division of labour, the slower rate and/or less important role of increased productivity and technological progress in countries of insufficient economic potential. It is just their structural underdevelopment which poses one of the main dangers from the point of view of the long-range development of these countries, since their backwardness is most conspicuous in industries based on up-to-date, pioneering technologies.

Backwardness reflected in the structural indicators of the economy also indicates problems in social development and is a factor directly preventing the development of modern class structures. The higher ratio of employed in agriculture, the services and the light industry or small-scale industry naturally results in a relatively smaller socio-political importance of industrial workers. It is not irrelevant either that the minor significance of branches representing the vanguard of scientific and technological progress in the international field, as well as the scarce resources of research and development do not constitute an adequate socio-economic environment for the development of the various strata of the intelligentsia. Under the conditions of the prevailing structure, the social role of the working class and the intelligentsia, having such an outstanding part in accelerating historical progress, is rather limited; the front of effective social resistance that can be mobilized against the concentration of power is much narrower, and the process of social development is less dynamic.

The correlation of integration and the development of class relations is undeniable, especially in the less developed countries. The possibility of the development of up-to-date branches of industry results in the numerical growth of the working-class and the intelligentsia, and also promotes the transformation of their socially often privileged, isolated condition as well as the strengthening of their social positions. The acceleration of internal social integration, the development of socio-political relationships may receive considerable incentive from regional integration.

The relation of the entrepreneur stratum and/or of the individual sectors to integration is of a specific nature. In the developed capitalist countries of Europe the keeping of development within the framework of the national economy is advantageous in our time mostly but to the relatively weaker, less competitive small and medium capitalist strata, whereas the big bourgeoisie, the big corporations demand regional integration. The backward countries developing within protectionist limits show a rather different and complex picture. Small-scale and medium-scale industries process mostly local raw materials or produce to cover local demands, and as a rule perform repair and servicing activities. Thus, resistance against regional integration is much weaker here than in the countries of Western Europe, and general attitudes on integrational problems are bordering on indifference.

Due to vested monopoly interests, however, big industry built up under the wings of administrative protectionism did not bring about automatically efficiency, and competitiveness. On the contrary: frequently it is just the new, modern big industry based on the conception of import substitution which requires the preservation of the protectionist system. Though in the underdeveloped countries it is the relatively more dynamic and efficient foreign enterprises, the multinational corporations which would enjoy the benefits of integration, they are interested first and foremost not in regional, but in world-wide integration. Thus, due to its interests, big industry is opposed to integration as long as the local consumer is willing to cover the additional expenses of national economic protectionism.

Integration has a considerable influence on the development of *income distribution*. The interdependence of the nature of the growth process, the structural characteristics and the distribution of incomes is a well-known fact. In this respect the example of the developed countries does not fully meet our needs, since there, as a result of the fight of the working-class, certain reforms of income distribution have been implemented much earlier. Besides, in many a country the direct or indirect manifestations of power politics (armaments expenses, incomes from capital export) have also considerably modified the relations of distribution.

The problem is of much greater socio-economic significance in the less developed countries where the lower level of incomes is coupled with markedly extreme relations of income distribution. In Latin America—as mentioned above—80 per cent of the population falls behind of the average *per capita* income level, while five per cent has achieved a living standard seven times as high as that level; thus, the statistical average is derived from extreme values. Even more serious inequalities are pointed out by the fact that the difference between the income of 20 per cent of the population belonging to the lowest income brackets and the five per cent belonging to the highest one is 45-fold (whereas in the developed capitalist countries the difference is 15 to 20-fold). The regulation of income distribution is naturally the task of the national economic policies of the individual countries. The question of the role of integration arises if it is taken into account that in the case of extreme conditions of income distribution the dimensions of the individual national economies, too, have a part in the development of the proportions.

It is not by chance that in the case of Argentina, Mexico or Venezuela, i.e. countries with relatively broader domestic markets, the degree of income concentration is much below the average of Latin America and especially that of the small Central American republics. (Brazil is an exception, partly owing to the nature of the development policy of the military regime, and partly to the extremely great development differences brought about by historical development between the northern and southern states.)

In the *smaller countries* the correlation between the dimensions of the national economy and extremes in income distribution is explained by factors as, e.g. the lower level of industrialization, a higher degree of protection, a higher cost- and profit level of the existing and operating industries, and, resulting from the above, the higher ratio of traditional agriculture, the enhanced significance in the national economy of the export oriented sector exceeding by far average productivity. The above factors result, among others, in lower organizational standards of labour. The consequences are unambiguous: like a vicious circle, the extreme income distribution relations of the national economies of narrow dimensions increasingly hamper the structural development of the domestic market and economic life as a whole. Under such circumstances it is impossible to introduce mass production and to establish mass markets rendering possible the cutting back of the social expenses of development.

Due to the high degree of income concentration, the structure of consumption and the way of life of the strata in the higher income brackets is no longer adjusted to that of the overwhelming majority of the country's population but to the consumption structure of countries with a high income level. In countries where the national economy is of narrower dimensions, the more extreme income relations mould the way of life of the upper classes to a greater degree, and increasingly isolate them from the broad masses of the people. The trend is clearly discernable in these countries: social tension, class differences and antagonisms assume ever sharpening forms. The ruling class, isolated owing to the increased social polarization, gradually relinquishes its position as an independent internal leading force and generally becomes a mere puppet of an external power.

Thus, the broadening of the dimensions of the national economy by the means of integration may forestall the accelerated polarization process in the smaller countries and may even prevent the process of socio-political disintegration.

Integration leads to a further strengthening of *the role of the state in the economic life* of the less developed countries. As it was pointed out, it is the big enterprise and/or the microsphere which are the primary driving forces of integration in the developed countries, whereas in the more backward countries, especially in those ones which are developing within a protectionist environment, momentum gained from the sphere of enterprises has proved insufficient. Owing to the increased social costs of the protectionist development and to the many problems manifesting themselves in the macrosphere, it is just the national state and/or its arsenal of political instruments which become the main incentive for integration. Under certain conditions not only the termination of the negative aspects of development but also the benefits deriving from integration need to be approached from the macrosphere. Taking into account the present power relations in world economy, the national state—when having to face the big powers—requires the force of cohesion resulting from regional integration if it wishes to uphold its sovereignty, bargaining position and international mobility. Only the national state is in the position to take into consideration the significance of the non-economic, i.e. national defence, diplomatic and social factors, and to coordinate its integration policy with these elements.

The viewpoints regarding the nature of integrational "motivation" are clearly demonstrated by the integrational endeavours of the Andean countries. When in the middle of the sixties the policy of import-substituting economic development proved abortive, the more competitive Chilean and Colombian industrial bourgeoisie wanted to make use of integration in order to get hold of the markets of the industrially less developed Andean countries. Thus, the idea of integration came from the microsphere of two more developed member countries, and initial conceptions were similar in form to market integration. By the beginning of the seventies, however, the possibilities of joining in the international division of labour in the sphere of industry have broadened for Colombia, while the positions of the industrial bourgeoisie in Chile were considerably weakened as a result of the election victory of the Popular Front. Consequently, the original motivating force of integration has ceased in both countries, whereas cooperation endeavours have survived, since integration represented a safeguard for people's Chile against being isolated, and in the case of the other countries it enhanced their external bargaining power. Under such circumstances, instead of the enterprises, the role of the motivating force was shouldered by the national state and accordingly, conceptions regarding integration also shifted from the position of market integration in the direction of macroeconomic cooperation.

The increased responsibility of the national state for integration may lead to the revision of the part played by foreign and local private enterprises, and it is the unfolding of the integration process which highlights the necessity of founding multinational public enterprises. Thus, when having reached a certain stage, the process of integration serves as a mutual incentive for the development of the institutional structure.

4.3. THE DEVELOPMENT OF REGIONAL COOPERATION IN LATIN AMERICA

Up-to-date growth processes reflect the favourable or unfavourable traits of the historical heritage. Of all economic processes foreign economic relations are perhaps most affected by historical traditions, specialization developed in the course of times, and by external orientation. It is proved by many an example that a radical and fast change in external orientation or the direction of specialization requires serious sacrifices, moreover, incidental results show only at a much later date. It was the division of labour developed in the course of history as well as the high level of European integration which were reflected in the initial successes of West European integration.

In Latin America, the common historical and cultural past has not resulted in intensifying economic relations. The unity of language, religion, politics and law in the Spanish Empire did not bring about integration or regional cohesion. The economic resources of the individual regions have directly met the requirements of the colonizing power. True, some modest progress had been made in gradually developing some kind of economic cooperation among the centres of the vice-royalties, especially between Peru and the peripheral regions. During the wars of independence, however, not only dependence on the former metropolitan country was abolished, but connections with the one-time centres of the vice-royalties were broken off, too. At the beginning of the armed struggle, the creole landowner-bourgeois stratum heading the independence movement set the aim of unification of the former Spanish colonies, their political, economic and military integration, in order to facilitate a joint stand against the metropolitan country. The aims of Latin American integration were outlined as early as 1815 by Simon Bolivar, the leading personality of the independence movement, in the "Letter from Jamaica". However, parallel to the successes achieved in the struggle against the colonizers, less and less emphasis was laid on the aim of integration, and at the Congress in Panama in 1826 even Bolivar abandoned the idea of an independent and integrated Latin America, and suggested a confederation instead. However, due to the conflicting interests of the member states, the so-called Great Colombian Federation established in 1819 with the participation of present-day Venezuela, Colombia, Ecuador, Peru and Panama disintegrated within a decade.

It is not yet clear, whether the failure of the first integration concept can be explained either by Bolivar's prophetic, though historically untimely posing of the question or by reasons formulated by Bolivar himself: i.e. by the differences in climate and temperament in the Latin American countries and by the serious transport and communication difficulties.[9] On the other hand, England, the dominant power of those times striving to seize the Latin American markets, was not interested in the emergence of a politically and economi-

[9] S. Bolivar: *Contestación de un Americano meridional*. Caracas, 1974 (Obras Completas).

cally unified South America. At the same time, the creole ruling class conducting a successful war of independence, was no longer interested in the joining of forces against external danger.

Thus, with the attainment of independence and with externally expanding economic growth regional cooperation was narrowed down still further, and an ever increasing verticality of foreign relations was established instead of horizontal relations. The place of the former colonial powers was taken by England and other West European countries as well as the United States. Moreover, economic, transport, communications relations among the Latin American countries themselves were more and more realized with the active help of a developed centre, whereas cultural and ideological development was adjusted to centres like London or Paris. Under the given trends of development the awareness of regional affinity and cooperation became entirely obliterated, and the wars between Latin American states sowed the seeds of lasting enmities (e.g. between Chile and Peru–Bolivia; Bolivia and Paraguay; Paraguay, Argentina, Brazil and Uruguay; Peru and Colombia).

The idea of integration was revived in the wake of accelerated socio-political evolution. Isolation at the time of World War I, the revolution in Mexico and later the ideology of early "Aprismo" in Peru linked the demand of social progress and the abolishment of dependence on the imperialist powers to the requirement of the intensification of Latin American cooperation. In the fifty years following the revolution in Mexico it has been mostly anti-imperialist governments and political forces which were the champions of regional cooperation. The efforts of the Peron government (the first "middle-of-the-road" government in Latin America) in 1946–1955 in the La Plata region, the integration policy pursued by the government of Arbenz in Guatemala in 1954, and later on by the Popular Front government in Chile and the military regime in Peru early in the seventies first and foremost sought to find means by which imperialist pressure could be reduced and regional isolation prevented.

Pressure exerted by socio-political factors unfortunately did not extend to the economic sphere as well. The launching of import-substituting economic development in the thirties has brought about within three decades parallel capacities in the industry, too, the competitive nature of the individual national economies has increased comprising in addition to agriculture and mining the bulk of manufacturing industry. While an average of 12 per cent of Latin American foreign trade was realized within the region in the period 1946–1953, at the end of the fifties, i.e. at the time of the exhaustion of import substituting policy, the rate of intraregional realization was only 10 per cent. Intensity of regional economic relations remained modest compared not only to intra-European but even to intra-Asian relations, and right up to the beginning of the sixties the trend was of a rather disadvantageous nature. Although much more propitious preconditions were provided for integration by ethnographic, historic, linguistic and psychic factors than in any other region of the world, economic interests and processes did not encourage these countries to draw benefits from these mutual advantages.[10]

The realization of institutional and comprehensive endeavours aiming at integration was started at the end of the fifties. The Central American Common Market was established at the end of 1960 with the participation of El Salvador,

---

[10] V.L. Urquidi: *Free Trade and Economic Integration in Latin America.* Los Angeles, 1968.

Honduras, Guatemala, Nicaragua and Costa Rica, and in February 1961 Argentina, Brazil, Chile, Mexico, Paraguay, Uruguay concluded an agreement on the Latin American Free Trade Association. Later on, Colombia, Ecuador, Bolivia and Venezuela, too, joined the Association.

Earlier endeavours aiming at regional cooperation were always encouraged for *political* reasons, and regional cooperation started in the early sixties was inspired by rather similar considerations of economic policy. Such considerations of economic policy, however, did not have their roots in internal socio-economic processes but should be regarded as reactions to the sudden worsening of Latin America's international positions and to the external "shock effect" caused by this turn of events.

The decisive role of exogenous factors in the emergence of Latin American cooperation is well illustrated by the historical conditions under which the first measures were taken. Conceptions on integration assumed a more concrete form for the first time when, with the passing of the Korean boom, the world economic positions of the raw material exporting countries kept on worsening. Owing to the stagnation of export incomes and to chronic lack of foreign exchange, the countries of Latin America were compelled to resort with increasing frequency to credits and financial aid granted by the developed capitalist countries. This, in turn made it possible for the financing countries or leading trade partners to exert ever growing political and economic pressure. The situation was rendered even more serious by the establishment of new economic groupings such as the Common Market and the EFTA, which even in the long run jeopardize the foreign economic positions of the countries of Latin America. Particularly the clauses of the Treaty of Rome concerning agrarian protectionism and associated countries made the countries of Latin America sense an imminent danger, and this naturally made it imperative to organize a common Latin American front. The views of the leaders of the Latin American countries are reflected by the opening address of the president of Uruguay at the Montevideo conference of the LAFTA in 1961. He stated that the establishment of the Common Market and the EFTA meant a quasi state of war as regards Latin American exports. Therefore, he said, the countries of Latin America had to match integration against integration, intra-American cooperation against intra-European cooperation.[11]

Contrary to the integration processes of the developed capitalist countries, in Latin-America, regional cooperation is strengthened not because of the high development level of the productive forces coupled with the breaking down of the national frontiers, but because of the strengthening of the negative aspects of the international economic and political environment and the diminishing of Latin American bargaining positions. Accordingly, integration strategy takes on preventive functions and is of a defensive character. It is easy to point out parallel features in the historic circumstances of the establishment of the CMEA. In Eastern Europe, however, the joint surmounting of the consequences of the cold war and embargo policy got beyond the stage of strategic defensive within a relatively short period of time, and aspects connected with the regional development of the socialist socio-economic order have played an increasingly important part ever since.

It is of considerable interest to examine the stand taken by the dominant power of the Latin American region as concerns regional cooperation. It was

---

[11] *The Observer*, July 30, 1961.

mentioned that due to her foreign trade considerations Great Britain had a disintegrating role in the 19th century. The role of the United States, the new dominant power taking over from Britain, has been much more contradictory in the course of time. It was the United States which initiated and became the main organizing force of the so-called pan-American conception, of the pan-American conferences the first of which was convened in 1889, of the Pan-American Union established in 1910, and finally of the Organization of the American States founded in 1948. Officially, pan-Americanism declared the unity of American states against non-American countries, but essentially it was nothing else than a means of the United States initially for the ousting of its European rivals and for facilitating North American political, economic and cultural penetration, whereas later on, after World War II, it served to strengthen the continental hegemony of the United States.

Following the establishment of the supremacy of the United States, the political aspect of the cooperation manifested itself to a lesser degree, and in the two decades after World War II, US attitudes towards the mutual economic relations of the Latin American countries were openly hostile. As a result of the revolution in Cuba, American fears and interest taken in the problems of the continent increased. Initially, at the time of launching the Alliance for Progress programme, it was attempted to put a stop to the socio-political unrest of the region by restricting the relations of the USA and the individual Latin American countries to bilateral agreements.

Due to the strengthening and irresistible independence aspirations of Latin America and the weakening of the international positions of the United States, American foreign policy suddenly changed its tactics in the middle of the sixties, and began to urge on the expansion of mutual relations among the countries of Latin America and on speeding up integration both by economic and political means.

It is easy to point out partly *economic*, partly power-political reasons for this change in tactics. As regards the *economic* aspects of the problem, the interest of North American capital in certain stages of growth in the countries of Latin America is vitally important. Growth founded on import substitution and protectionist environment has offered ideal conditions to the big corporations of the USA to conquer certain national markets, industrial bridgeheads, and also to maintain bilateral dependence, since within the narrow limits of national economies the majority of the Latin American national enterprises remained weak and non-competitive.

At the same time, by means of bilateral dependence, US dominance over the region seemed unshakable. Therefore, the introduction of new methods was not deemed necessary.

By the beginning of the sixties, the coming to an end of import-substituting growth and the limits of domestic markets have decreased the profits of US capital and thus reduced its interest in maintaining the protectionist environment. The emerging multinational big corporations strived to build up their activities on a world economic or at least on a regional basis. In the period of so-called global integration, activity on restricted national markets, especially on those of smaller countries, became of less and less interest. According to the estimation of the United States, the benefits of the developing Latin American regional market were to be reaped by the North American big corporations anyway, and thus, American economy was directly interested in supporting the integrational aspirations.

The changes in the strategy of the US big corporations and in the nature of the growth process in Latin America coincided with the changes in the political scene of Latin America. US intervention in Dominica in 1965 may be regarded as the last attempt to uphold US hegemony and bilateral dependence by direct means. The political price to be paid for overt imperialist intervention, the sudden eruption of anti-US sentiments in Latin America, the involvement in the war in Vietnam as well as the untenableness of the strategy of global presence inspired American power policy to adopt indirect means of control. The new global strategy implemented by the US, officially formulated by the Nixon administration, strived to reduce direct commitments, to establish sub-centres of power, and to elaborate mechanisms of indirect control and influencing.

US support for the intensification of Latin American integration fitted perfectly into US strategy and has rendered it possible to reduce the expenses of the defense of the changing US positions, to control the region through countries of key importance, to make use of the increasing mutual dependence of the member countries in order to bar rival powers from the region and to systematically suppress revolutionary movements. Thus, American attitudes concerning Latin American integration are a classic example of traditional great power tactics. In the period when the region is strictly under control, this strategy wishes to maintain bilateral dependence, and it takes an anti-integration stand, but when rival powers appear, it uses more differentiated and complex means.

Let us now examine the *institutional development* of cooperation efforts. Chronologically, it was the Central-American republics where integration conceptions were first given an institutional form. Following the preparatory debates and the working out of the conceptions, the first agreement aiming at establishing a free trade system among the Central American states came into force in 1959, and was substituted in 1960 by a general integration treaty constituting the foundation of integration ever since.

In Central America, too, the history of endeavours aiming at political and economic unity could be traced back as early as the 19th century, since the region constituted a single administrative unit under Spanish colonial rule. Such plans, however, have miscarried one after another. Integration was put on the agenda of history at the end of the fifties because the economy of the five countries, each a dwarf economic unit, met with growing difficulties owing to the worsening of the world economic situation: within the bounds of extremely small national economies the establishment of economically operated up-to-date branches of industry was out of question. (In 1960 the total population of the five countries amounted to 11 million; their aggregate social product came to about 2 billion dollars.) As regards political conditions, the decisive factors were the new situation in the Caribbean region resulting from the establishment of socialist Cuba and US pressure which followed in its wake.

The main devices for the implementation of the basic agreement of 1960 were the reduction of the obstacles to intraregional trade, the drawing up of a system of common tariffs, and the establishment of the institutional structure of regional cooperation. Right at the beginning, 75 per cent of goods figuring in the lists of mutual trade had been exempted from duties, and five years later 95 per cent of the highly increased intraregional trade was effected virtually on a free trade basis. Only such goods are still liable to duties in the mutual trade of these countries which represent delicate problems both politically and financially. Competitive goods, such as sugar, coffee, tobacco, liquors are also liable to duties.

Joint external tariffs, too, were introduced gradually but within a very short time, and by June 1966 in the case of 95 per cent of customs duties common external duties were fixed.

The organ of monetary integration, the Central American Bank for Economic Integration has an instrumental part in influencing the trend of subregional development. The Bank directs both domestic and foreign, i.e. international capital towards the establishment of a regional infrastructure and/or industrial structure, and grants long-range as well as intermediate-term credits for the improvement of the infrastructure, the development of integrated industries, the establishment and modernization of export industries, the furthering of specialization in agriculture and generally for organizational, planning and advisory activities relating to integration. In the first decade after the basic agreement the Bank has made payable a total of 224 million dollars, i.e. it distributed directly about 8 per cent of all investments within the period mentioned. According to preliminary data on the period 1970–1974, this percentage has increased to 12 per cent,[12] meaning that the rate of investment allocation effected through the regional centre is rather high even on an international scale.

As money of account for intra-zonal turnover the Central American Clearing Bank has established the Central American peso, the value of which is identical with that of the US dollar. A special stabilization funds provides for remedying the short-term balance of payments difficulties of the member countries. Thus, contrary to integration efforts of other developing countries, in addition to trade aspects financial cooperation has had a decisive part in Central America from the very outset. Measures taken in order to establish a common infrastructure and joint centres for research and vocational training have also been important aspects of regional cooperation.

The difference in the level of development of the member countries is of but minor consequence: in the most developed country, Costa Rica, the *per capita* gross national product in 1960 exceeded the regional average by 40 per cent, whereas in Honduras, the least developed member country it fell behind the average by 20 per cent.[13] Thus, the ratio of development difference measured in terms of gross national product between the most developed and least developed member countries amounted to merely 1.75 : 1. Therefore, no special mechanisms have been worked out for a faster development of the more backward member countries, and only Honduras was granted a temporary permission to impose a 20 per cent domestic surchange.

The most significant regional grouping of the developing countries, the *Latin American Free Trade Association*, responsible for 73 per cent of trade turnover within Latin America in 1970,[14] had to face from the very beginning of its existence considerably more problems than the Central American Common Market. At the time of concluding the treaty, the Association comprised a region of about 200 million inhabitants and 17 million square kilometres, the gross national product of which amounted to about 5 per cent of world GNP. However, in spite of the common basic problems, differences among the

---

[12] *IADB, Annual Report,* 1972.
[13] J. Ahmad: "Trade Liberalization and Structural Changes." *Journal of Common Market Studies,* September, 1972.
[14] *Economic Survey of Latin America,* U.N. 1971.

member states in development potential were by far more serious than in the case of any other regional groupings. *Per capita* gross national product in Venezuela was sixfold, in Argentina fivefold that of the Bolivian one, and three of fourfold that of the Ecuadorian, Paraguayan and Peruvian ones. Differences in the potentials of individual national economies were even greater. In 1960 gross national product of giant Brazil was 60 times as much as that of Paraguay, 20 times as much as that of Ecuador, and 14 times as much as that of Uruguay.

The original objective of the Association was the establishment of a free trade zone by means of reducing tariffs and breaking down trade barriers at an annual average of 8 per cent, and abolishing all barriers within 12 years. According to the plan, all customs frontiers among the member countries would have ceased to exist by the beginning of 1973. As regards tariff reductions, the basic treaty distinguished national and common lists. An annual average reduction of eight per cent was envisaged for the tariffs of goods on the national lists, however, concessions could be revoked, by the country which granted them, if such commodities did not figure among the items of the common list compiled every third year. It was intended to draw up the common list of such goods the share of which in the total value of trade turnover among the member countries showed a tendency of increase every three years (25, 50, 70 and finally 100 per cent). The first common list came into force at the beginning of 1964, however, the Association could not go beyond the second stage, and the original schedule has been upset. The so-called Caracas Protocol of 1969 envisaged only an annual 2.9 per cent reduction in the trade turnover among the member countries. Thus, the realization of the free trade zone among the member countries has been postponed until 1980. Instead of automatic tariff reductions the method of talks on separate items was introduced, in the framework of which 11,000 tariff rates were discussed up to 1972, more than 50 per cent of which benefitted industrial products.[15]

In order to bridge over the immense differences in the economic potential and development level of the member countries, the Association promised unilateral concessions to countries with *"inadequate markets"*. Initially, it was only Paraguay which came under this heading but gradually all member countries were classified as such, with the exception of Argentina, Brazil and Mexico. The smaller, less developed countries were granted about 7,000 unilateral concessions.

To further specialization among the member countries and to reduce structural parallelism, *cooperation among branches of industry* was launched from 1964 on, taking the form of so-called complementary agreements. Though efforts aiming at industrial cooperation found great response and even political emphasis, by the beginning of 1968 only eight, and by 1972, 16 such agreements were in force, covering mainly industrially developed Argentina, Brazil and Mexico. The most important fields of industrial cooperation are the chemical industry, petrochemistry and electronics.

Apart from encouraging cooperation among the branches of industry, the Association essentially has not gone beyond the limits of free trade mechanism. The free trade approach has taken shape without doubt mainly as a consequence of external pressure in order to keep up the conformity of the GATT. As

---

[15] J.D. Cochrane–J.W. Sloan: "LAFTA and the CACM." *The Journal of Developing Areas*, October, 1973.

a precondition of friendly relations with the developed capitalist countries and with the IMF, concessions granted within the framework of regional cooperation were not to infringe upon the statutes of the GATT which acknowledged only reduction of tariffs granted within the framework of the customs union or the free trade zone, whereas direct, administrative incentives of cooperation would have been objected to. In 1959 R. Prebisch himself[16] held the view that the free trade zone was a mere technical formality. The agreement which came into force, however, could not develop further on account of well-known difficulties. The establishment of institutional regional structures and of a regional development policy could not be carried into effect. The 1971 budget of the Secretariat of the Association was 1.3 million dollars, while the primary means of achieving the objectives are periodically held talks. In addition to the lack of a common policy on development, monetary, financial and fiscal measures, and to the underdevelopment of a common infrastructure it was the free trade mechanism which proved increasingly inefficient in intensifying regional cooperation and in harmonizing the interests of a great number of countries with different development levels.

As soon as the first serious signs of difficulties on the road towards the common goal made themselves felt, five member countries, viz. Bolivia, Chile, Ecuador, Colombia and Peru signed an agreement in 1969 creating the *Andean Group*. This was meant partly to express their discontent with the Free Trade Association, and partly to shape forms of integration cooperation going beyond the concept of free trade. The joint declaration of the Presidents of the five countries issued in Bogota stated that the reduction of tariffs envisaged by the Montevideo Agreement has fallen short of accelerating a true integration in Latin America and of furthening the economic situation on the continent within a reasonable period of time. It was decided therefore to launch a campaign within the Free Trade Association for the passing of measures which could prove acceptable for smaller countries with a relatively less developed economies. Such steps would be essential for the harmonic and balanced growth of the region.[17]

The Cartagena Agreement which came into force in 1969, and was joined also by Venezuela in 1972, envisaged a customs union, harmonized financial, monetary, fiscal and social security policies, cooperation of the up-to-date branches of industries of the member countries, and a common external tariff system by 1980. Right in the first year, i.e. in 1970, 175 items figuring on the common list were declared duty-free in the turnover of the three most developed countries. A joint external policy was elaborated concerning foreign capital, patent rights and registered trade-marks, and other barriers of a noncustoms nature among the member states were broked down, too. The process of rationalization of the products of the dynamic sectors (chemical, petrochemical, vehicle, steel as well as light engineering industries) was begun with, and also a system of regional distribution of new investments among the member countries was introduced. In 1969 differences in the development, economic and power relations were less significant than in the case of the Free Trade Association: for instance, the gross national product of Chile was three times that of the Bolivian, whereas her economic potential was seven

---

[16] R. Prebisch: "Conferencia sobre el Mercado Comun Latin-Americano." *Banco de Mexico*, September 9, 1959.

[17] W.P. Avery: "Subregional Integration in Latin America." *Journal of Common Market Studies*, December, 1972.

times as much as that of Bolivia. The two smaller and less developed countries, i.e. Bolivia and Ecuador, were granted unilateral concessions as regards the distribution of new branches of industry and development resources, and also a five-year respite was granted for the introduction of the common external tariffs and the breaking down of the barriers obstructing trade among each other.

The *institutional* structure of the Andean Group is more developed than that of the other regional groupings in Latin America. The Commission at the top of the institutional structure consists of the representatives of the member countries, being a control body entitled to make political decisions as well. A decision requires a two-thirds majority, and every member state has the right of veto. Another leading institution of the integration is the Junta made up of three members, the Secretariat of which draws up plans and makes technical decisions on matters relevant to integration. The Andean Development Company was called into being with a capital of 100 million dollars for the financing of subregional development. Its activities, however, have been rather restricted due to a lack of funds.

The Andean integration reflects the highest degree of integration in the developing countries, a closer cooperation going beyond the limits of loose regional cooperation, a more developed institutional system with stricter time-limits. It is for the first time that endeavours by developing countries are discernible, aiming at the linking up of integrational objectives and the growth process, as well as at the realization of a more balanced and proportionate growth.

4.4. MAIN CHARACTERISTICS AND UNDERLYING FACTORS OF REGIONAL ECONOMIC COOPERATION

Let us now examine what the foreign trade trends within the region reflect of the structure of the integration process and the nature of problems cropping up in the individual countries (see Table 25).

The results of the examination of the dynamics of regional foreign trade turnover indicate that after a spectacular boom lasting for about a decade, growth in both the Free Trade Association and the Central American Common Market has come to a stop, and the share of regional turnover has been stagnant since 1970. In the countries of the Andean Group which was established only a few years ago, the verve of the initial times is still to be felt. Owing partly to the fact that growth came to a standstill, and partly to the low level of the division of labour among the Latin American countries which originates from the circumstances of historical development, the process of regional cooperation has—in spite of the efforts in the sixties—not achieved, *the measure of the division of labour and mutual adjustment to each others interests which is demanded by regional integration.* Thus, the separate efforts aiming at integration have resulted only in the increase of economic cooperation for the time being.

As regards the Central American Common Market, the coming to a standstill of growth is quite understandable. The stagnation and temporary regression of regional cooperation is generally explained with the worsening of the political atmosphere following the so-called "football war" between Honduras and El Salvador in 1969. It would be mistaken, however, to disregard the

Table 25

The Geographical Significance and Distribution of Trade within Latin America

	The share of regional turnover in total exports (per cent)			Share within regional exports (per cent)	
	1960/61	1970	1972	1960/61	1972
Argentina	14	21	25	25	30
Bolivia	10	9	12	1	2
Brazil	7	11	10	16	25
Chile	8	11	14	7	10
Ecuador	8	11	17	1	2
Colombia	2	11	14	2	6
Mexico	3	8	9	4	9
Paraguay	31	39	24	1	1
Peru	9	6	8	7	5
Uruguay	3	12	12	1	2
Venezuela	8	4	5	35	10
Total of LAFTA	8	10	12	100	100
Costa Rica	6	20	21	12	17
El Salvador	12	32	33	35	29
Guatemala	6	35	30	18	34
Honduras	18	11	3	30	2
Nicaragua	3	10	23	5	19
Total of Common Market	9	23	24	100	100
Bolivia	1	2	17	1	15
Chile	1	2	3	17	10
Ecuador	5	6	11	12	15
Colombia	1	9	15	12	48
Peru	5	2	3	59	12
Total of Andean Group	3	4	6	100	100

Source: Economic Survey of Latin America, 1972, 1973.

fact that significant economic factors, the running out of primary integrational reserves, are also important background factors of this process. As a consequence of the ten-year cooperation the interdependence of the member countries has increased considerably: in 1972 about a quarter of their total trade turnover was realized through intra-regional trade, and within that the share of industrial products exceeded the 80 per cent mark.

Relationship between the degree of economic development of economy, the scale of national economy and the convenient geographic structure is clearly traceable under normal economic conditions. The more backward and smaller a country and the lesser the possibilities of an up-to-date, diversified industry, the greater its dependence on countries exporting industrial products, investment goods, capital and technologies. As a rule, the mutual trade among backward, small countries is confined to the barter of rather simple complementary consumer goods or raw materials, whereas the share of these items in the imports of the individual countries amounts to a few per cents at the very most. In moderately developed countries the exchange of industrial consumer goods may be added to the above (provided that the individual countries do not pursue a policy of vigorous import-substituting industrialization). Further, the mutual trade of such countries consists of a certain division of labour in the ranges of light engineering and chemical industry, and there are some possibilities for an increase in turnover resulting from a major import of investment materials. The share of all these products in imports can reach only 30 to 40 per cent at the best, since more than 70 per cent of the imports of Latin American and other moderately developed countries is made up mostly of industrial goods transferring technology, machines and semi-finished products, the economical and expedient purchase of which can hardly be effected from other less developed countries. This explains why developed countries have such a pivotal part in the imports of the less developed and smaller countries.

Naturally there exists another aspect of the problem, too. The buyer's markets of the narrowly specialized countries, exporting raw materials or labour-intensive industrial goods are also primarily to be found in countries with a higher wage level, a diversified structure and a broad domestic market. Thus, in the case of moderately or less developed countries the reasonable share in the trade with by and large similarly developed countries cannot exceed 30–40 per cent, even if there is to be found a maximum coincidence of economic and geographic endowments, and advanced mechanisms of cooperation are introduced. The countries of Central America, however, are not moderately developed countries but less developed ones, and the primary branches of the economies are of a competitive nature.

Under such circumstances, the ratio of trade with each other presently already approaches the peak value, i.e. the economic optimum, moreover even the present level reflects political sacrifices made for the sake of a possible future Central American Union. However, statements on the decay of the Central American integration[18] should not be approved of, since it represents merely the exhaustion of one reserve of growth, i.e. that of the subregional integration, and a repeated unfolding of integration on a higher level of development cannot be precluded. It should be added that besides economic factors the viewpoint of the isolation of revolutionary Cuba had also a part in US support to Central American integration. The initial attempts at settling US-Cuban relations, the changes in the Caribbean area decrease the importance of the external political momentum as well.

The coming to a standstill of regional cooperation in the countries of the *Free Trade Association* is to be regarded as much less natural. The intensity of cooperation among the member countries had not even approached that of the Central American ones and had come to deadlock at an extremely low level. In spite of the extremely low initial level, the Association, surprisingly, could raise in more than ten years the share of mutual trade relations in total exports only to the level of the decade following World War II. The development level of the member countries would allow a much more expanded rational division of labour. Having a closer look at the individual countries, it turns out that regional cooperation represents considerable weight only in the foreign trade of the Central American countries (Honduras in the seventies marked an exception), whereas of the countries of the Free Trade Association it is only in Argentina and the tiny Paraguay where it reaches a quarter of total export; in Colombia and Peru, the two countries most interested in the economic aspects of the Andean integration, it represents only ten per cent of total exports. The low level of mutual interests and interdependence is obviously

---

[18] M.S. Wionczek: "The Rise and the Decline of Latin American Economic Integration." *Journal of Common Market Studies*, September, 1970.

a major obstacle in the way of coordinating the targets of economic policies and of accelerating the process of integration. Though the significance of regional connections has increased, with the exception of Venezuela, Peru and Paraguay, the measure of the increase of cooperation was still not sufficient to bring about a break-through in the front of conflicting interests. Intraregional exports have expanded most rapidly first and foremost in countries which proved less dynamic in the field of foreign trade in the period under review (such as Argentina or Uruguay) or in countries which directed their trade mainly towards the developed countries and therefore regarded regional integration but of secondary importance (such as Brazil, Colombia or Mexico).

The trends of the *trade balances* of regional turnover, too, show a rather varied picture. Within the Association, trade relations of Chile, Peru, Bolivia and Uruguay show a significant chronic deficit amounting to 30 to 70 per cent of imports, while Venezuela, Mexico as well as Argentina achieve regularly a considerable export surplus. In the member countries of the Central American Common Market, Honduras, Nicaragua and Costa Rica end the fiscal year regularly with a high deficit, while Guatemala, the "major" country of the grouping, usually has an active balance of trade. The situation as regards balances of foreign trade is still more extreme in the countries of the Andean Group where the cumulative ratio of export covering in the period 1969–1973 has been 45 per cent for Chile and only 40 per cent for Peru. At the same time, Colombian exports were nearly two and a half times the volume of her imports. Due to the slow progress made in monetary integration and/or to the lack of adequate mechanisms of compensation such extreme situations in the balance of trade are seriously impeding the progress of countries with a trade-deficit and grappling with foreign exchange problems. In the smaller countries imports by far exceed exports.

The most unambiguous success in the first decade of regional cooperation has been achieved in *structural* development. At the beginning of regional cooperation in 1960, trade with each other—as regards structural development—was similar to that of total trade turnover: 10 per cent of trade with each other was made up of industrial products, while the barter of foodstuffs against energy bearers, i.e. mainly the exchange of Argentine corn for Venezuelan oil amounted to 40 per cent each of the total intracontinental trade. On the other hand, 9 per cent of the rather modest exports of consumer goods was absorbed by the trade of the region's countries with each other. According to data of 1971,[19] the share of industrial products in the total exports of the FTA countries amounted to 18 per cent, whereas this percentage came to 40 per cent in the trade turnover among these countries; a quarter of Association's total exports of consumer goods was made up of items in the trade with each other. Structural development is even more tangible in the case of machine exports, raising a number of problems for the less developed countries. At the initial stage of integration machines made up but one per cent of exports, and 30 per cent of this went to the countries of Latin America. In 1971 the share of machines in exports amounted to 3 per cent, and their percentage in the turnover of the countries of the Free Trade Association was 11 per cent. Trade of these countries with each other has a significant part in stimulating structural development. Beginning with the seventies, however, a new trend could be observed: the phenomenon of improvement in the

[19] *Monthly Bulletin of Statistics*, July, 1973.

structure of exports extended beyond the limits of regional relations to world economic relations as well. Of the increased volume of machine exports 48 per cent was directed to the countries of the Association in 1970, and 43 per cent in 1971.

The improvement in the structure of exports is even more important in the countries of the Central American Common Market and the Andean Group. Within the mutual trade of the countries of the Central American Common Market, the share of industrial products rose to 75 per cent, and within that the share of machines was 15 per cent, this being an undeniably significant achievement in less developed countries.

The main motivating force of regional cooperation seemed to be structural development and/or the intensification of the division of labour in industry. More than half of the increased intraregional relations were accounted for by the exchange of industrial products. Thus, the dynamics of the development of mutual trade relations shows close coincidence with the exploitation of the possibilities of structural development.

There is still one question left: to what degree did regional cooperation effect *the nature of the growth process* in the member countries. Among the objectives of the groupings the stabilizing, growth-balancing effect of regional cooperation was regarded as a major issue. Concerning the stabilizing effect, the fluctuating character of regional foreign trade, which is by far not less than the average, an unambiguous answer may be received. Despite the stimulation of mutual relations, the coefficients for increased turnover represent hardly less dispersion than in the case of total turnover. As a matter of fact, in critical years such as 1969 or 1971, mutual trade relations of these countries have shown even greater fluctuations and recessions than the total trade turnover. Thus, hopes concerning the stabilizing effect have not materialized.

A wide-ranging scientific and political debate has taken place on the growth-balancing, levelling effect of regional cooperation. It is undeniable that in the Central American states which have made greater progress in integration, economic processes reflect a greater degree of mutual levelling up. In the period 1960–1970, a 6 per cent average annual increase was achieved in the gross national product of the five member countries. The rate of expansion has been fairly steady as the difference between the growth rates of countries achieving the fastest and slowest rate of development amounted to hardly one per cent. A similar levelling effect can be observed in the rate of industrial development. Though the range of dispersion is low in international comparison, too, it cannot be regarded as accidental that Honduras, the smallest and most backward country in the region has developed slower than the average.

Regarding the countries of South America as well as Mexico which are lagging far behind in intensifying mutual relations, debates on the interaction of regional cooperation and growth processes seem rather theoretical and overdone. The proportion of trade among each other to the gross national product exceeds the 2 per cent mark only in the case of Argentina and Paraguay while the average for the region is below 1.5 per cent. The sectors of the national economies of the individual countries which are susceptible to the changes in intraregional trade are rather limited. Therefore, suggestions which try to draw conclusions on the growth process as a whole from effects on one or two per cent of the gross national product seem exaggerated. The rate of the growth process, the various phases of the business cycle in the countries of the Association or even in the member countries of the Andean Group are highly different, and though in the latter countries development is more differentiated, it would be an overstatement to conclude from these phenomena on the effects of regional cooperation. In foreign trade where regional cooperation has been from the very outset of a relatively greater importance, interconnections, too, are more direct. Contrary to requirements, however, here too, it is not interdependence but the positions of the bigger countries, the polarization effects which are strengthening.

Undoubtedly, the scarce result of integration in the countries of Latin America reflect the problems originating from differing development levels and economic potential, which naturally could not be remedied by capitalist free trade mechanisms. The development of trade cooperation was seriously hampered by institutional underdevelopment, the extremely lengthy bargaining on annual tariff reductions, by exceptions granted in an ever increasing number, and by the lack of political stability in the member countries. In the less dynamic countries and those which still maintain rather a protectionist environment both entrepreneurs and trade unions have put up stiff opposition to the liberalization of regional imports since they do not want to accept the principle that efficiency and not protectionism should be the source of profits. Characteristically, only two out of eleven member states of LAFTA have fulfilled their obligations concerning the reduction of customs tariffs.

The different rate of the inflation process in the countries of Latin America had a considerable retarding effect on the harmonization of economic-political interests with those of general economic policy. As may be seen, the rate of price increase is of a chronic inflationary nature in Chile, Uruguay, Argentina and Brazil, whereas in all the other countries it is relatively modest. Lacking the restrictions within the region, the inflationary pressures would bring about a chaotic situation in the countries grappling with increasing internal imbalance.

In the developing countries there is no possibility for accelerating the process of natural selection among the competing enterprises, and this, undoubtedly, hampers the process of integration in less industrialized countries. The breaking down of the protective barriers of the national economy, the termination of the economic activities of inefficient enterprises is not feasible by means of agreements on mutual preferences of liberalization. In a dynamic economic space where no overt or covert form of unemployment exists, sectors becoming non-competitive and of marginal importance are easily absorbed by the greater and more dynamic part of the economy. In case of a considerable overt or covert under-utilization of manpower or socio-political tension, however, the elimination of an inefficient sector is a rather difficult task for the political leadership. Under given circumstances the support of existing work-places is a basic feature of short-term economic policy; this principle cannot be jeopardized even by the perspective of a most expansive long-term growth. Thus, the possibilities of integrational policies and growth are restricted from the very outset by predictable social consequences.

Integration was also slowed down by a total lack of mutual capital-ownership relations. By the mediation of the multinational corporations a part of production has been coordinated in a paradoxical way with that of the production of other countries. This coordination has taken place, however, between developed and underdeveloped countries, and therefore an occasional restriction or even nationalization of multinational enterprises did not result in an increased cooperation among the nationalized sectors.

It would be naturally mistaken to analyse the cause of the coming to a standstill of regional cooperation in Latin America as an isolated case, since the slowing down of the process of integration is a phenomenon not limited to Latin America; it is also a consequence of the modifications of the development trends and environments of world economy in recent years, which has slackened the pace of the progress of regional processes on an international scale as well.

The above is not a current problem as regards the intensification of regional division of labour among the developed capitalist countries of Western Europe or the socialist countries, as in these countries the degree of integration achieved and the broader dimensions of the national economy do not induce such serious demands concerning the rate of progress as in the less developed countries. At the same time, it is absolutely clear that the strengthening of worldwide inflation, the development of a capitalist foreign exchange crisis as well as the energy crisis have impeded the progress of integration in Western Europe, too, while in the case of the CMEA countries the fact that the progress of the integrational process has been slower than expected was caused by the coming to the fore of the demands of catching up in the field of technology and joining in the international division of labour.

The slackening of the efforts aimed at integration could be experienced chiefly in the developing countries which are to the greatest extent in need of integration. The fact is that neither the LAFTA, nor the Central American Common Market or the East African Common Market have made any headway points to a common underlying cause. True, the cooperation among the countries of the Andean Group shows some signs of undeniable dynamism, however, the integrational turnover of the Andean Group is but of marginal importance from the point of view of world trade, and does not weigh much even in the external sector of the member countries. The trend can be clearly seen if reduced to a single fact: *the share of the trade turnover realized among developing countries dropped back from 23 per cent to 19 per cent in the total trade of the developing countries in the period 1955–1972.* Thus, the regional agreements, the policy aimed at an intensification of cooperation of these countries among themselves could not achieve a switchover to "horizontalism" from the traditional vertical character of the external economic relations of the developing countries; as a matter of fact, all this led to an even higher degree of dependence on the developed capitalist centre.

The process of integration of the developing countries was impeded by the development trends of world economy in a peculiar way. As it is well known, cooperation efforts were stimulated by the worsening of world economic terms at the beginning of the sixties. Ten years later, however, a number of developing countries were in the position to improve their development performance and relative positions in the world economy. Thus, the so-called Vietnam boom undoubtedly had a beneficial effect first of all on the developing countries of Southeast Asia; the moving out of "runaway industries" escaping from the increasing cost inflation in the developed countries furthered the industrial progress of the Latin American, Mediterranean and Asian countries with a low wage level and "competitive climate", whereas the soaring prices of oil and other raw materials brought huge profits to the producer countries of these raw materials.

The improvement in the world economic positions of some countries, the intensification of integration with world economy naturally put on a break to interest in regional cooperation and especially in shouldering commitments connected with it. On the other hand, under growth conditions when the role of a dynamic sector is played once again by the extractive industries or labour-intensive manufacturing branches of industry, the objective necessity of integration declines, too, since neither the production of raw materials nor light industry or the engineering branches producing labour-intensive mass products or the export of services require the breaking down of the barriers and the integration of the national economies.

Under present conditions, the coordination of the foreign economic measures of the individual producing countries (e.g. oil, rice, coffee producers) and the traditional means of cooperation of their economic policies ensure temporarily an adequate frame for the protection of the positions and interests of the individual national economies, and therefore the path of economic integration with its complex mechanisms, its effects which are often unforeseeable and its contradicting consequences moved back on the priority lists for the time being.

It is worth-while to analyse the *interest* of the major member countries in integration as against the background of events in regional cooperation up till now and of the changes in the international situation. In most of the member countries the original objective of joint protection against the outside world has undergone a transformation in the past decade and new viewpoints have been included.

The combined effect of the above factors is generally known. Though, at the beginning of the sixties, the principle of regional cooperation was supported enthusiastically by all Latin American governments this principle—similarly to a number of other general principles—proved not strong enough either to overcome the practical national interests. As pointed out by P. Herrera,[20] a leading figure of Latin American integration, despite the solemn statements by the governments of the Latin American countries, in practice they keep on to define their development policies within the bounds of their national markets.

The change of US attitude which suddenly became of a supportive nature in the sixties, exerted a peculiar effect on regional cooperation in Latin America, and created an atmosphere of uncertainty and indecision. At the beginning of the sixties, the national bourgeoisie and nationalistic technocrats regarded integration as a kind of possibility to take a stand against the US. The changes in American tactics, on the other hand, reinforced the wide-spread suspicion of the American challenge because one of the beneficiaries of the expanding economic region were undoubtedly the powerful American corporations. The strongly felt "direct presence" of the dominant power in regional cooperation induced the more nationalistic governments to pursue a cautious policy, and in recent years they rather stressed the world economic approach as against regional cooperation, since an increased participation in the world economy would help them weaken considerably the economic positions of the dominant power.

Examining the positions of the three major countries of the region, i.e. Brazil, Mexico and Argentina, the trend is absolutely unambiguous: national interests are put into the fore as against regional economic cooperation.

From the point of view of *Brazil*, the strongest power on the continent, it is not defense, stimulation of foreign trade or structural development which

---

[20] F. Herrera: *Nacionalismo, regonalismo, internacionalismo.* Buenos Aires, 1970. INTAL–BID.

renders regional cooperation important: the conception is subordinated to the long-range objectives of Brazilian power policy which aims first of all at penetrating the economies of the adjacent member countries. Thus, regional cooperation has become the means of the creating of a regional Brazilian raw material basis. Brazilian regional deficit, too, indicates that Brazil pursuing a policy of external economic expansion regards the neighbouring countries more and more as a kind of her economic hinterland. As regards *Mexico*, regional trade represents a means of the conquering of new markets by the strengthening industrial bourgeoisie and has an important part in improving the chronic deficit of the Mexican trade balance, in increasing the export of industrial products, and in diminishing one-sided dependence on the United States.

Argentina has still the strongest positions in regional trade, however, economic and political aspects are of equal importance in shaping her regional policy. Initially, Argentina sought refuge in an expanded zone pursuing a policy of regional protectionism for her less competitive new industries developed within a protectionist environment and, fearing West European agrarian protectionism, also for her agriculture. The Argentine standpoint has been strongly motivated by the efforts aiming at the changing of the policy of national protectionism for one of regional protectionism. The past decade, and as a consequence the shift in the economic power relations of the region have made it increasingly clear that the regional rivalry between Argentina and Brazil going back to a hundred years has been decided in favour of Brazil. Argentina, falling back to the position of the second, and more recently the third economic power of the continent has decided after a lengthy and painful reassessment of her foreign policy to choose regional cooperation and the association with the smaller countries—based on common interests—to bring about a countervailing force against the emerging Brazilian power. The Argentine endeavours become quite unambiguous when taking into account the efforts made by the government to bring about regional cooperation, the policy of rapprochement to the Andean Group as well as some gestures towards people's Chile forgetting thereby traditional hostility. If no contrary impulses will be received from the social side, it may be anticipated for the near future that political aspects, countermeasures to impede the emergence of a Brazilian sub-centre of power, will even more vigorously influence Argentine foreign policy in strengthening and extending regional cooperation.

Uruguay, Paraguay and Bolivia, the neighbours of Argentina and Brazil, after the failure of their initial hopes for accelerating development have chosen a policy of regional cooperation characterized by entrenchment, the seeking of subterfuges, the manoeuvring between the two powerful rival neighbours or from time to time the support of the interests of one of them. The objective of the special, sub-regional cooperation of the *Andean countries* has been—seeing the coming to a standstill of the economic development of their grouping—the avoidance of the fate of the three countries mentioned. The original driving force of the Andean integration has been the industrial bourgeoisie of Chile and Colombia, the two most industrialized members of the grouping. Thus, at the outset, the market conception of integration was dominant in accordance with the interests and the liberal economic and political institutional system of the two initiative countries. In 1970–1971, however, the political-environmental conditions of regional cooperation were basically changed by the increasingly progressive policy of the military regime in Peru, the victory of the popular forces in Chile, and the gaining power of the Torres administration in Bolivia.

Efforts to decrease foreign dependence and to take a common stand against the multinational enterprises became the decisive issues of regional cooperation. The Andean countries, especially Chile pursuing a progressive policy in the early seventies, regarded integration as a political defense weapon against the American blockade aiming at their isolation. Of the forces interested in regional economic cooperation, the Chilean bourgeoisie lost most of its positions and influence between 1970 and 1973. The Colombian bourgeoisie, witnessing the radicalization of the other Andean countries and taking into account the rapidly ensuing successes of her export-oriented trade, showed less and less willingness to accept forms of cooperation reducing her autonomous freedom of manoeuvring. Thus, the centre of interest in the process of integration has been shifted by history from private economy towards the public sector, from the market sphere towards the sphere of production. The public sector and/or the government policies, however, were not prepared to put through such radical changes at such a fast pace in any of these countries. Moreover the political environment of cooperation itself was once again transformed by the counter-revolutions in Bolivia (1971) and Chile (1973). In an atmosphere of ideological conflicts characteristic for the past years, harmonizing the strategy of integration has become increasingly difficult. In addition to the above events, the development of the petroleum boom in Ecuador, the rise in the prices of raw materials had in all of the Andean countries—with the sole exception of Colombia—the result that the traditional export sector has become again the main factor of external growth. As a consequence, the trade turnover of the Andean countries among themselves amounted to only 7 per cent of their total turnover in 1973. Thus, in the Andean countries, too, regional cooperation represents first and foremost political aspects, whereas concerning the economic aspect of the problem it is the Colombian interests linked to the exports of industrial products which are most important.

4.5. THE ROLE OF SMALL COUNTRIES IN LATIN AMERICAN REGIONAL COOPERATION

The common problem of all regional groupings has been the position of countries with modest economic potential. Both the Free Trade Association and the Andean Group have guaranteed special treatment to countries with so-called insufficient markets. Despite of this, the different degree of dynamics of the development of regional cooperation in each of the countries has resulted in polarization within intraregional trade turnover. The share of Argentina and Brazil, the two most industrialized countries of the Association increased from 41 per cent in 1960/61 to 55 per cent by 1972. Within the Central Amerian Common Market the power relations of the small countries are much more balanced, yet the share of "big" Guatemala in intraregional trade has risen from 18 to 34 per cent in the period 1960–1972, while within the Andean Group the share of Colombia has increased from 29 per cent (1968) to 50 per cent.

The shifts indicate the trend that in the foreign trade of all groupings there are one or two leading countries with highest stakes in trade, whereas the interestedness of the majority of the member countries, their range of common interests is on the decrease.

For the Latin American and other developing countries lacking advantageous natural resources or other factors to be realized in world economy, integration continues to be of vital importance, despite the dominant trend of polarization in development and certain favourable changes in the world economic environment of some countries. It might prove useful to examine the possibilities which could further the safeguarding of the interests of the weaker member countries.

The bargaining power of the smaller countries depends generally on the relatively free choice of *joining or leaving* regional groupings and/or whether their participation or abstention would have an influence on the interest in integration of the leading member country or countries. Such a favourable situation may come about in the following cases:

A. a) The small country is in a strong position as regards its international or regional *political* bargaining power. Thus,

– it may be of strategical importance from the point of view of some great power or the leading country of the grouping;

– in a given historical situation its joining or withdrawal may have a political as well as a psychological effect on the attitudes of the other member countries. Both of these political factors can exert great influence on the small countries' favourized position as regards their possibilities in the ranges of trade, development and credit policy.

Bolivia, and to a certain extent Paraguay, are good examples for both the above cases. Bolivia, less interested in regional trade, did not joint initially the Free Trade Association fearing that more expensive and less up-to-date imports from the member countries might cause economic problems. Thus, in spite of her being a founding member of the LAFTA, she declared her intention of association only in 1962, when her right to unilateral concessions (together with those of Paraguay) was recognized by the Association. At the time of the signing of the LAFTA agreement, the more developed countries could not show disinterest in the joining of two countries, since both of them were important from the point of view of the regional rivalry between Argentina and Brazil. However, the relatively favourized position of Bolivia[21] has grown together with the unfolding of her role in regional strategy. At the beginning of this decade, Brazilian expansion has got beyond the first stage of rivalry with Argentina and, without any pretexts, was aspiring to regional dominance. Bolivia situated right in the heart of the continent, was given increasing economic aid by Brazil and recently even support in her endeavours to regain her former coastal territories, because Brazil strives to establish her positions along the coast of the Pacific, too.

b) The economic bargaining power of a small country improves if

– the country has strong supplier's positions in world trade or regional turnover which can originate from the exports of strategically important products or technologies (e.g. oil exports in the case of Venezuela), or from her major influencing and determinant part on certain commodity markets when both the production and exports of a certain commodity are monopolized by the given country (e.g. fish-meal from Peru);

– the country disposes of a strong import potential. A significant international purchasing power considerably increases the small country's possibilities for manoeuvring. The exploitation of the import potential in a planned way cannot only defend the interests of the exports of the smaller countries as against the major powers, but pressure exerted by them through their purchasing policies may even have an effect on the political decisions of the great

---

[21] *Economic Survey of Latin America* 1970, 1973. U. W.

powers. It is characteristic that at the time of a sudden increase in the regional imports of People's Chile, the hostile attitudes of the Brazilian and Argentine governments were soft-pedalled, and Chile was even granted considerable credits by these two countries. In this sense, the import potential may be regarded as a peculiar kind of "deterrent capacity" compelling the stronger partner incidentally to reconsider economic costs in conflict-situations. It may sound paradoxical, but actually the "strategy of forward escape", i.e. the expansion of the import potential in a planned way, may strengthen the capability of resistance against political and economic external pressure, relieve or put off crisis situations, and it may generally improve the bargaining positions of the small countries. The joining or the withdrawal of a small country with a significant import potential is not at all a matter of secondary importance for the other member countries of the grouping.

B. The above factors may offer a chance to the small countries to exert some pressure on their more powerful partners by joining or withdrawing from the integration. As a rule, the small American nations, however, are of no strategic importance at all, have no political key positions or strong export positions, and thus no rational alternatives tempting them not to join the regional groupings. With the sole exception of Panama exporting services and facilities, all the countries of Latin America are members of regional groupings. It might be of some interest therefore to analyse how the small countries can prevent the deterioration of their positions *within* the regional groupings.

a) Small countries may—in principle—arouse the interest of the big countries because of the former ones' international "usefulness". Exploiting the importance of being unimportant, a small country may act as the agent of a great power, can establish financial, commercial, technological and political connections in situations when the overt participation of the big country is not possible or would not seem desirable. From time to time the great powers recompense such services, whereas the "reliable" small countries which are closely linked up with their powerful partners in questions of major importance enjoy a relatively large amount of tactical independence as regards other problems. Thus, within the new hierarchy of American power politics Brazil played partly such a role in relation to the United States, and the same holds true for Bolivia in her relations to Brazil. Since these relations too obviously limit the realization of any national policy of the individual countries, they are not regarded as attractive by most of the countries of Latin America.

b) Small countries may make diplomatic efforts in order to bring about a greater variety in the geographical power structure of regional cooperation. The participation of two or more countries of a similar power potential can reduce the danger of regional domination, and a more diversified power structure ensures more possibilities for manoeuvring. True, there still exists the danger that the leading countries within a regional grouping may come to an agreement to the detriment of the smaller ones, but a more diversified international stage always renders more favourable possibilities to diplomacy. This could be clearly seen in the endeavours of the small former La Plata countries, aiming at a possible balancing force against Argentine and Brazilian economic penetration as well as at the involvement of far-off Mexico.

c) Small countries strive to intensify their connections primarily with the other small countries of the region. In all probability this would decrease the losses following from the foreseeable political and economic pressure of the leading powers, and call into being some elements of mutual dependence. In this connection, however, the economic and political optimums of the geographical structure of external relations do not coincide, since in our times countries with scarce resources, lacking adequate economic and trade potential are obliged to establish relations of increasing geographical concentration for economic reasons. Consequently, the diversification of the structure of external relations entails additional costs, and it is only a government taking into consideration aspects of national defense and willing to make sacrifices in their interest which is able to shoulder such expenditures.

This trend is clearly demonstrated by the subregional association of the five minor Andean countries striving to withdraw from the Argentine-Brazilian sphere of influence. Similar considerations can be traced in the regional cooperation of the small countries of Central America fearing the adjacent great power—Mexico.

d) In the sphere of intraregional cooperation minor countries try to reach agreements on specialization which would render it possible

– to specialize horizontally in turning out products for which there is a dynamically increasing demand,
– to specialize vertically in the production of component parts essential for the most important final products of the member countries.

It would be naturally desirable in both cases if agreements on specialization covered not only the sphere of production but also those of research and development as well as of marketing, since the lagging behind of the small countries could thus be reduced much faster. Under present international circumstances, however, only the resettlement of production has been begun with, and this, too, is concentrated on the small group of countries where wage levels are extremely low. Of all Latin American countries only Brazil and Argentina dispose of a significant research and development potential, whereas the minor nations have no adequate R & D basis.

In the cooperation of countries of different economic potential and standards the careful choosing of the *objects of specialization* has a great part. Due to the different technical and economic nature of the individual products, the import sensitivity of the big country is of a different degree in the case of each product. There are some products (e.g. energy sources and component parts) where import difficulties give rise to immediate problems even in the big countries, while in the case of others, delivery restrictions make themselves felt only in the long run. In some of the Andean countries under military rule it has been requested already that the direction of specialization, the actual field of vertical cooperation ought to be fixed by—much the same methods as the target-ranking of air raids—taking into account the weak points of the industries and the consumption structure of the more powerful partners, the possibilities of switching over from the regional supplier's market to the world market, as well as the costs of an incidental fast changeover. The bargaining position of the smaller countries undoubtedly improves if they succeed in identifying the strategically weak points of their partners.

e) Small countries are especially interested in such new forms of international cooperation as, e.g. *joint projects* which successfully complement their endeavours at regional cooperation. The joint project package deals broken down to the different countries, or regional projects realized by the participation of two or more countries (about 600 mainly infrastructural joint projects were under way in 1972) represent a sectorally restricted type of integration and are especially fit for developing relatively simple mechanisms of cooperation among countries of different social and economic systems, development levels and potential. The setting up of regional projects could supply even small countries with at least one or two projects making a valuable contribution to the development of their technological structures.

C. The possibilities outlined above, may undoubtedly alleviate the problems of the smaller countries participating in regional cooperation operations though they cannot solve the basic problems originating from and inherent in the very nature of being a small country. Integration in itself is by no means a remedy for all ills and difficulties—other *extraregional* solutions are also needed. Since policy is termed the art of proportions, the proper proportions among the local, the regional- and world-market-based economic branches should be established by the development policies of the individual countries by an adequate amalgamation of protected, less protected and competition-oriented environmental elements. Regional cooperation, too, might be favourably affected by the strengthening of the competitive sector: it can lay or broaden the foundations of cooperation. Let us now review those "extraregional" factors which could strengthen the bargaining positions of the smaller countries within regional cooperation itself.

a) Over the past decade the majority of the Latin American countries has realized the antagonism inherent in the growth model based upon the export of raw materials and import-substituting industrialization. The changeover to an export-oriented development naturally offered a stronger stimulus to growth than regional cooperation, since at the time of a boom, the world economy is in the position to further more effectively the exploitation of the advantages of mass production and scientific and technological progress, and leads to a more effective distribution of resources. The attitudes of governments which tend to develop rather their extra-regional relations (e.g. Costa Rica) in order to avoid the excessive intertwinement with the adjacent countries, are regarded as incompatible with regional cooperation by some schools of thought. In this respect, two viewpoints are of interest. On the one hand, it is a general trend that newly founded and more effective industries are oriented rather by the world market than by the regional market. Contrary to the practice followed in earlier years, nowadays it is not exclusively the national or regional markets which have a demand for high-level industrial products, since the importance of extraregional markets has grown considerably in recent years. On the other hand, the possible conflicts between regional and extraregional trends of orientation reflect the negative aspects of present-day regional cooperation. Under normal conditions the general export orientation of economy is a successful complement to regional cooperation and even increases the growth potential of the region. This is perceptible in the case of all countries setting out on the path of export-oriented industrialization, particularly in the case of Colombia, a small country, where the inevitable cutback of protectionism, the introduction of a more flexible system of overall management, have also paved the way for the intensification of regional cooperation. Probably, it is not by chance that within the intracontinental turnover of the countries of Latin America it has been Colombia and other countries pursuing a vigorous policy of export orientation that expanded their trade most dynamically, whereas—before the raw material boom—the shares of Venezuela and Peru, countries industrializing on the basis of import substitution, have declined.

b) As regards the concentration of scarce human and material resources, the small countries once again outpace the big ones. Market mechanisms in the developing countries are inadequate to effect a rapid and efficient concentration of industrial activity pursued mostly on a suboptimal plant scale. As a rule, direct or indirect state intervention becomes necessary for establishing up-to-date plant sizes. The mergers in the car industries of Peru and Chile were already effected on the basis of this principle.

c) The active participation of the small countries in international life was another contribution to the strengthening of their international bargaining positions. World economic environment or bilateral trade limited by administrative regulations and prohibitions are less auspicious for the interests of the small countries. In the case of bilateral relations between capitalist countries it is usually governments of conflicting interests which face each other. Their relations reflect the prevailing balance of power, and this might become a source of losses for the weaker party if there are considerable differences in power relations. It stands to reason that efforts to re-establish normal world economic relations and to bring about multilateral relationships in international economic life, as well as steps taken in the direction of upgrading the status of the international organizations are enthusiastically supported primarily by the small nations. Quite obviously, international organizations, too, transmit pressures by the great powers, nevertheless, small countries participating in such organizations are less exposed to them than in bilateral connections. Similar considerations led Peru in taking a very active role in the preparation of the so-called Programme of the 77, Chile in shouldering responsibility of the Third World trade conference, or Uruguay in taking a stand for the Free Trade Association.

d) Closely related to the above, the effectiveness of the foreign service in enhancing the international bargaining positions of a small country is getting increasing emphasis. The elaboration of the specialized preferential free trade lists generally requires prolonged negotiations, and the rate of progress as well as the value of compromises depend to a great extent on the professional skills and competence of the corps of diplomats. Small countries which frequently have no competent diplomatic corps are obliged ever more frequently to solve their problems concerning international cooperation—often even matters of detail—on the highest level of political-diplomatic leadership, and therefore, owing to their underdeveloped negotiating and decision-making mechanisms their weight in international life is bound to decrease. It is not in the least by chance that Peru, Colombia or Chile under the people's front government have done their utmost to develop the professional level and institutional foundations of their diplomatic services.

Even by partially making use of the possibilities outlined above for a number of small Latin American countries it has become easier to protect themselves more effectively against the worsening of their regional and international positions. The efficiency of such protective measures could hardly be analysed on the basis of long-range development trends because the majority of the small Latin American countries wage a peculiar sort of two-front war and frequently change their tactical orientation. With the exception of the Central American small countries and those of the West Indies, as well as Paraguay and more recently Chile, all the countries of Latin America show the signs of a cautious or definite endeavour to decrease the dependence on the United States. In an unfavourable world economic environment this stimulates primarily regional cooperation, however, when the world economy experiences a boom, all the historic conflicts and misgivings among the neighbouring countries flare up, relations to more developed and distant countries of other continents are strengthened, and at the same time more and more facts indicate a trend of procrastination as regards a closer cooperation with the neighbouring major powers.

At the same time, the problems of the small countries reflect in a condensed form the inherent contradictions of the integration of the developing countries. One of the main objectives of integration is the termination of backwardness, however, the conservation of the underdevelopment of the national economies is a serious obstacle to a successful integrational policy. From the point of view of big countries regional connections are mostly of a marginal character, often mere islands within the comprehensive socio-economic processes. In the case of small countries, however, successful integration presupposes an incessant modernization of the socio-economic structure. This also proves the fact that integration can by no means be a substitute for socio-economic progress but is an integral part of this process.

PART II

ROADS OF DEVELOPMENT

CHAPTER 1

THE NATIONALIZATION OF THE DECISION-MAKING CENTRES IN PERU

In October 1968 the Army seized power in Peru. Recent years have demonstrated the fact that military take-over in that case has not been but one of the numerous instances of *coup d'états* which have become an everyday matter in developing countries. As a consequence of the take-over, radical structural changes were accomplished, the significance of which went far beyond the borders of Peru. Peru, called "the land of metal and melancholy" by Garcia Lorca, has become a dynamic centre since the turn of events in the late sixties, and also the scene of an exciting historic experiment, important even by international standards. Eight years are obviously too short and limited a period for drawing up a historic balance. Nevertheless, the exciting and paradoxical turns of the Peruvian road of development taking an ever more definite shape, the new events belying earlier notions and dogmas, and exerting ever more influence make it necessary to scrutinize the characteristic features of the Peruvian "model" as early as in the period of the first appearance of its peculiarities.

Beginning with the first moment of its establishment, the new power system astonished both theoretical and practical experts with a host of apparent contradictions and anomalies. The Army, regarded traditionally as a bulwark of deeply rooted structures, assumed the role of the guiding force of socio-economic transformation and progress. Without hesitation and with astonishing verve the largely US-educated military launched an offensive against the dominant US positions highly restricting political independence. The social progress of an impressive scale took place, however, without the participation of the masses. It would be meaningless to enumerate the apparent paradoxical turns of events since the socio-economic processes to be outlined will give a clear picture of development. Let us but mention as a starting point that the caracteristics of the Peruvian road of development are rooted both in the political and the economic spheres. Before outlining, however, the new trends, it may seem expedient to comment briefly on the basic characteristics of the country.

Long periods of the historic development of Peru, both at the time of the Inca Empire and in the era of Spanish colonialism, were marked by the centralized character of power and an economy based on mining. The centralized character of the Empire resulted in the establishment of a sprawling public administration and as a consequence of that, in parasitic consumption, whereas mining monoculture brought about an "enclave"-type development confined to a certain section of the economy. Owing to the nature of both factors, development called forth an unprecedented extent of concentration of incomes, wealth and power. The greater part of the national income was controlled by the Peruvian oligarchy, an outstandingly strong stratum in Latin America, more precisely by 40 families.

Another result of this peculiar kind of concentration was regional disparity, the emergence of two Perus. As regards *per capita* income, the capital, i.e. the historic power centre, even with its slums of a million inhabitants, has achieved a living standard four times higher than that of the Andean provinces where the Indian poupulation which does not speak Spanish or has only a smattering of the language, still lives under technological conditions and within the limits of formations developed at the beginning of the colonial era. The extent of dualism may be characterized by the fact that at the end of the past decade the export-oriented, efficient sector comprising the extractive industry, fishing and sugar-cane production, and employing about 100,000 workers has produced one-seventh of the national income and nearly 90 per cent of commodity exports.

Colonialism which befell Peru quite unexpectedly and was of unprecedented cruel nature, had a traumatic effect lasting many centuries in the psyche of the indigenous population. This was followed by the collapse of the historic power centre, large-scale territorial losses, the heritage of two lost wars, the gradual seizing of the national resources by foreign monopolies—in brief, a sequence of failures in the historic experiences of the nation. Together with the numerous and manifold other factors of backwardness, this has resulted in large-scale social apathy and passivity. Though development was considerably stepped up during World War II and in the post-war years, the rate of progress based to a growing extent on export-oriented growth remained extremely favourable as compared to the average rate of development of the developing countries. In the period 1950–1965 the average annual increase in the national product amounted to 5.6 per cent, while the volume of *per capita* national product—despite of the vigorous population growth—rose to 328 dollars by 1968, the year of political change. Thus, the political change was connected with the demand of eliminating socio-economic backwardness, and was by no means a consequence of some kind of a growth crisis. The economic dynamism of the post-war years did not alleviate considerably the basic problems of backwardness and inequality or the strong dependence on the United States. The multi-party system, the bourgeois reformist governments could not and did not even want to tackle the fundamental problems and as a result, the strongest institution of the Peruvian state, i.e. the Army had to bear the brunt of directing the socio-economic transformation. The main characteristics of the Peruvian road of development are from the institutional aspect, *the leading role of the Army*, and from the political aspect *the objectives of socio-economic transformation going beyond the traditional limits of capitalist growth.*

1.1. THE ARMED FORCES AND THE MODEL OF ECONOMIC GROWTH

Analysing the nature of Bonapartism, Marx and Engels had pointed out that in the course of historic development there are periods when the power balance of the conflicting classes invest the state power with a mediating role and a certain amount of independence. At such times an institution of the state power, possibly the strongest one, may superimpose itself upon those classes it had served earlier; it might take an extremely reactionary stand but may open up the way to revolutionary initiative as well. History has proved the relevance of the analysis of Bonapartism. In the developing countries entering a historic period of transition, and characterized by rather uncertain class power relations, it has become increasingly obvious in recent decades that the Army has a direct role in the wielding of power. The background of the numerous military take-overs in the developing countries has been made up of a great variety of motivations, functions, and power relations. However, the military take-over in Peru in 1968 was the first decisive intervention on the part of the military in Latin America which went beyond the traditional role and attempted to open up new possibilities by means of radical reforms not only for political but for socio-economic progress as well. Since that time, "peruanismo", i.e. a society-transforming intervention of the Peruvian type, has become one of the basic models of the Army's role. Before reviewing the genesis of the Peruvian power model it would seem expedient to analyse the specific features of the evolutionary role of the Latin American armies.

It can be stated without exaggeration that of all continents it is Latin America where the Army has the deepest historic roots. In the Spanish speaking part of the region the armed forces had almost always enjoyed an exceptional position in the power structure. In the Pre-Columbian societies the position of the warriors, and at the time of colonization that of the armed conquistadors and/or the armed forces as a whole was restricted by the central power or other powerful institutions (the church, latifundia). Paradoxically, however, the attainment of independence and the political anarchy of half a century following in its wake in the era of liberalism had also strengthened the positions of the Army. The military leaders, the caudillos and their adherents had gained important economic and administrative functions at the time of the wars of independence, and with a few exceptions *"caudillismo"* became the dominant political system.[1] Under the conditions of the emerging "inward-looking" model following the World Depression, and as a consequence of a more intensive state intervention, the influence of the Army gradually extended from the political to the economic sphere as well. A great number of economic projects came under the direct control of the Army, and here were the roots of the personal intertwinement of the economic and military top-leaderships, the foundations of building up a military-economic complex, especially in Argentina and Brazil. Under the influence of the expanding guerilla movement of the sixties, the power basis of the Army became more and more broader and even began to cover explicitly public administrative and social functions. Presently in some countries the Army takes charge of the direction of public administration, education, public health, public works, propaganda, etc. (Acción Civica-Militar) in areas threatened by the guerillas.[2]

In the majority of the countries of Africa and Asia the underdevelopment of institutional structures as well as the lack of organization of the political mechanism renders it necessary for the armies—being relatively young, though technically and organizationally more developed institutions—to shoulder the responsibility of modernization. In Latin America, however, the role of the army gaining ever more momentum almost parallelly with accelerated economic growth finds an explanation in its being deeply rooted in the prevailing historic socio-political structures. According to Mirski, the significance of the Latin American armies in the growth process should not be identified with the situation which has been brought about by historical development in

---

[1] I.G. García Ponce: *Las armas en la guerra de Independencia.* Caracas, 1965.
[2] E. Lieuwen: *The Latin American Military.* Washington, 1967.

the developed capitalist countries, neither with the conditions of such countries which have attained independence but recently.[3]

Thus, the specifically strong power positions of the Army are by no means of recent origin but are rooted in the historic development of Latin America. The wars of independence and the internal struggles following them have crushed the power of the Crown and/or the State, and have weakened the positions of the Church associated with the colonial power. The positions of the Church were further undermined by the process of laicization and scientific and cultural progress. The political power of the feudal oligarchy has faded away in the course of the economic changeover to industrialization following the World Depression. From the very beginning the independent power positions of the rising industrial bourgeoisie were restricted by dependence on protectionist favours granted by the state power, i.e. by the new stratum's non-comparative nature. Contrary to European development, the place of the feudal class ousted from power has not been occupied by a bourgeoisie efficient in the economic sphere, politically well-organized, consolidating the international positions of the state power and capturing ever new foreign markets. In the period of populist coalitions based on a precarious balance of powers the Army could easily assume the role of the arbiter.[4]

The increasing influence of the Army, however, does not merely base itself on the disintegration of possible rival forces. The civil political institutions and classes are less and less in the positions to complete with the efficiency of the Army as an institution of power. Under the given historic circumstances the Army is that highly organized force which is able to cope with the principles of modern management and organization and through the means of sophisticated armament devices is even more closely connected with the most up-to-date technology than the industry. In addition, its staff is better educated, more disciplined and organized than that of any other institution. Due to their historic role and traditions, a peculiar autonomous spirit and a military scale of values have developed in most of the Latin American armies. This military outlook, though not indifferent to class-conscious approaches or material interests, is much more resistant to the "demonstrative" effects and utilitarian mentality radiating from alien environments, to moral and behavioural looseness or traditional corruption than the political or the industrial-commercial spheres. Thus, the Army concentrating material values and knowledge to a greater extent than any other institution, is frequently in the position to carry out independent actions disregarding class interests, and may successfully present itself as the representative and symbol of national values and interests.

In spite of a largely common historic and institutional background, the historic role of the Army has been rather dissimilar in the individual countries and in different periods. Traditional *"caudilloism"*, i.e. personal power wielded on the basis of support enjoyed by the Army and indifferent or hostile to socio-economic progress, was characteristic of the 19th and the first decades of the 20th centuries, i.e. of the periods marked by a policy of semi-colonial "outward-looking". It has survived in some countries in a somewhat modernized form as demonstrated by the systems of Somoza in Nicaragua, Trujillo in the Dominican Republic, and Stroessner in Paraguay.

---

[3] G. Mirski: *Armiya i politika v stranakh Azii i Afriki*. Moscow, 1970.
[4] J. Cotler: *El populismo militar*. Instituto de Estudios Peruanos, Lima, 1972.

The forging ahead of the forces of socialism on a global scale, the extension of the competition between the two systems and/or the accumulating socio-economic tensions originating from general backwardness pose new requirements as regards the tasks of the central power. Such tasks of a new type are especially conspicuous and acute in those Latin American countries which have reached a somewhat higher stage of development, and where besides the traditional praetorian functions an increasing significance is attached to the tasks connected with the strategical motivation of economic growth, the active influencing and controlling of socio-economic processes. The military *coup d'état* in Brazil in 1964, is regarded as the first of the sequence of lasting and direct military take-overs aimed at the defense and rationalization of the prevailing capitalist social system. Similar objects were pursued by the military take-overs in Bolivia (1971), Uruguay (1972) and Chile (1973).

The extension of the field of activity of the power centres and the Army in itself does not explain the phenomenon that in the case of the Brazilian road of development this was linked with the extension of the sphere of activity of the repressive organizations, and with the establishment of oppressive systems. A number of historic examples demonstrate the fact that in underdeveloped countries the resources necessary for the acceleration of belated capitalist development can be procured mainly by forced accumulation. In underdeveloped or medium developed countries progress within the limits set by the capitalist system is still not able to rely on *economic* incentives on a broad scale, and neither can it operate with moral incentives. Thus, as demonstrated by the development of a number of Latin American and South European countries, over a long historical period coercion is the main driving force of belated capitalist modernization. Oppression is singled out of all the sources of power by the processes of modernization and forced accumulation, and the Latin American armies have considerable "comparative advantages" in exercising the function of repression most effectively.

A specific feature of the Brazilian-type military take-over is the combining of a more intensive intervention with a more open approach to the world economy. Previous theoretical views have usually linked military intervention with the protectionist conceptions of economic development, and at the same time the liberalization and decentralization of political and economic management have been regarded as preconditions of joining in the international division of labour. Presently the acceleration of economic and technological progress in the more developed Latin American countries cannot be realized any more without increasing participation in the international division of labour. This, in turn, poses specific requirements as regards the national economies of the individual countries.

Thus, the extension of the internal oppressive functions and the growth of the repressive organizations as the means of that function are furthered not only by traditional forced accumulation but also by the conditions of external economic orientation. In the Latin American countries, lacking natural resources and capital, the main source of external economic expansion is the extremely low price of *labour* as compared to international standards.

The greater the difference between the wage levels of Latin America and the developed capitalist countries (determining the conditions of international realization), the broader the scale—at least theoretically—where joining in the international division of labour can be effected. Thus, the bringing down of the costs of labour indicates not only a maximization of surplus value and accumulation but, due to the new export orientation, it is also a basic requirement of the belated, present-day capitalist development. It is not by chance that only ten years ago hourly wages in the Latin American manufacturing industries amounted to 10–25 per cent of those in the Federal Republic of Germany.

In addition to low wage standards, the guaranteeing—as far as possible—of the stability of the so-called labour-market climate, the production devoid of strikes had a considerable part in improving the conditions of capital returns.

The repressive organizations also have a part in handling the problems of income distribution of external economic orientation. As a result of internal as well as of international class struggles, the previous polarization processes within the income distribution of the developed capitalist countries have ceased to exist. In the socialist countries and in some countries advancing on the road of non-capitalist development the trend of income equalization has become dominant. In order to increase artificially and to a maximum degree the competitiveness of the economy, the present-day belated capitalist development *does not take into account the social aspects of income distribution but attempts to concentrate on economic effectiveness* (naturally under the conditions of parasitism being present, too). The rapid development of export orientation presupposes the distribution of incomes according to effectiveness, and sets off strong differentiation in favour of sectors, enterprises and individuals more effective than the average. Whereas the economic logic of protectionism demands the levelling of incomes, external economic orientation goes hand in hand with income differentiation.

It should not be regarded, therefore, as a mere chance that under capitalist conditions belated growth acceleration can be effected increasingly only by making use of repressive organizations and semi-fascist systems. It is easy to establish the interdependence between the accelerated role of growth and the extension of the activity of the repressive mechanisms and/or the appearance of semi-fascist systems. At the same time, the process produces its own internal and international contradictions which, at a higher stage of development, destroy the political and institutional framework. Social energies suppressed for a long historic period as well as accumulated socio-political tensions may thus sweep away systems founded on oppressive mechanisms by unexpected radical changes. The time of the disintegration of the institutional frameworks based on oppressive mechanisms depends, however, to a considerable extent on the external conditions of development and on the turn of events changing the international balance of power.

The model of a nationalist-reformist military take-over was created by the action of the Peruvian Army in 1968. Dissatisfied with merely organizing accelerated economic growth, this model aims at carrying through structural reforms, eliminating subordination to foreign powers, and nationalizing the decision-making centres. The activity of the Torres government in Bolivia, the military take-overs in Panama and Ecuador have by and large followed the Peruvian model.

Let us now examine the factors as a consequence of which the military take-over assumed the most progressive form in Peru.

Why were the Peruvian armed forces the first in Latin America to take on the responsibility of renouncing their role as guardians of the political *status quo* and to become rather the driving-force of social change?

The assuming of this new role has been the result of a long developmental and maturing process. The attitude and mentality of the Peruvian officers have undergone a transformation earlier than in any of the other Latin American countries, and this has been undoubtedly of great significance in the successful accomplishment of the take-over as well as in the elaboration of the Peruvian model. Following the bloody events in Trujillo in 1934, a peculiar kind of "vendetta" came into being between the Peruvian Army and APRA, the leading mass party of the country. (APRA had originally been a radical movement but subsequently it gradually shed off its radicalism and became a petit bourgeois, reformist party.) The Army attempted to reduce the chances of APRA's coming into power by improving its contact with the masses and by intensifying its preparedness. Taking advantage of and calling for the aid of outstanding civilian experts, Marxists among them, the institute of staff officers' training-courses (Centro de Altos Estudios Militares—CAEM) founded as early as 1958 with the collaboration of the Army's intelligence service, prepared the commanding staff for performing public administrative duties by teaching them the elements of law, economics, sociology and psychology. Extension courses concentrated on the problems of social and economic development. At this staff officers' academy the officers became conscious of the fact that there was a close connection between national security and socio-economic backwardness as well as subordination to foreign powers. It became clear that internal social tensions are eroding the defensive capacity of the country, whereas an underdeveloped economy was unable to supply the armed forces depended on US deliveries and on the goodwill of the North American military-industrial complex. It also became clear that the updating of the economy could not be achieved without the transformation of the social structure and politically activizing the formerly apolitical Army. The ten years' activity of CAEM in itself has played a considerable role in the transformation of the officers' attitudes, and contributed to the shaping of an ideology for the new role of the Army and also to the building up of the necessary cadre until the time of the take-over.

A significant part in bringing about a transformation of mentality and acceptance of that new role had the change in the social composition of the officers' corps. Owing to the extremely great extent of social polarization, members of the rather narrow upper classes and middle strata were less and less willing—especially under the circumstances of the favourable post-war business cycle—to go into the Army, the prestige of which was very low as a consequence of a run of ill-luck all through its history. Thus, it happened that the officers' corps was filled up, as a unique feature in Latin America, to an ever growing extent with members of the lower middle-class and peasantry, and more frequently with persons of Indian origin. The former President, General Velasco Alvarado himself came of peasant stock. Del Prado[5] emphasizes the fact that the positive effect of the ferment within the Army would have been much weaker if the officers' corps had consisted of representatives of the oligarchy.

The mechanism of military service, too, has furthered the political development of the Army. For the overwhelming majority of the officers the system of garrison rotation, i.e. the frequent changing of garrisons proved to be a

---

[5] J. Del Prado: "Forradalom megy-e végbe Peruban?" (Is a Revolution under Way in Peru?) *Béke és Szocializmus* (Jan., Feb. 1971).

great experience. By means of this system they got acquainted with the problems of the population of the remote highlands and of the eastern regions much more thoroughly than any other strata of the ruling classes. And finally, the guerilla activity of 1965 contributed to the reassessment of the traditional concept of the military establishment. It became more and more obvious that the revolutionary tension in the underdeveloped provinces was not due to the penetration of foreign agents but to the general backwardness in these regions, and therefore could not be solved by military campaigns only.[6]

Worries because of the possible consequences of an anti-imperialist policy were mitigated by the failure of the military intervention against Cuba, and also by the poor military performance of the US armed forces in Vietnam. In the years preceding the military take-over, the Army as an institution demanded the annulment of the oil concessions. And last but not least, military take-over was made inevitable by the requirement of bringing about harmony between the historically established control system and the nature of the institution in power. The centralized character of the decision-making model is rooted in the historic development of Peru, since one of the most peculiar features of the Inca Empire had been a historically unprecedented extent of centralization. The power system established at the time of colonization was rather similar to the Inca model. As a result, the liberal economic and political model alien to the historic framework established in the course of evolution has always been rather inefficient in Peru.

Administrative centralism has received further impulse by the present-day structure of the economy and the national psyche of the population. The extremely high concentration of the effective economic sector, the considerable part of mining in the national economy offer by far more advantageous possibilities for political centralism than for instance an economic structure where agriculture and the light industries are predominating. Owing to ethnical factors, historic development, dietary customs as well as climatic factors, the overwhelming part of the population is characterized by a high degree of passivity and submission. Under such circumstances, an increased centralization of initiatives, decisions, organization and executive power seems necessary. On the other hand, central power has almost unlimited possibilities to manoeuvre freely as regards all problems concerning the population, and has to fear much less the sensitivity of the population and its resistance to power than in any other Latin American country.

The military take-over adjusted itself to the historical traditions of the country, and set off in the direction of a centralized model of control. Under the new regime decisions are made by a very restricted circle of commanding officers, the institutions of public administration have mostly but executive functions, and decision processes are characterized by a downwards trend, i.e. decisions are made by higher authorities and handed down to subordinated authorities. Former political parties and institutions have been excluded from the decision-making process. The government strived to freeze in and immobilize the sphere of politics as far as possible because, according to their view, the effectiveness of political parties was extremely low at the time of social transformations and showdowns. Further on, they profess the view that political pluralism results in the slowing down of the decision-making process,

---

[6] C. Delgado: "The Political and the Social Significance of the Revolutionary Transformation of Peru." *International Affairs*, 1972, pp. 526–527.

brings about a fragmentation of forces, and does not further the channelling of social energies into the economic life. Although a political mass-organization, the National System for Furthering Social Mobilization (SINAMOS) was called into existence in 1971 in order to broaden the institutional mass support of the government, this organ established from above did not represent any alternative as regards power positions, moreover during its short functioning it proved to be not a transmission but rather a political nuisance.

Naturally, the Army should not be regarded as an entirely homogeneous body in Peru either; there are quite a few elements in the officers' corps who loathe the idea of social transformation. Neither is the political profile of the services identical: the most consequent representative of progress is the Army, whereas the Navy, owing to its social composition and markedly elitist traditions, professes much more conservative views. Cooperation of the services has been guaranteed up till now by the recognition that there can be no strong armed forces in a country politically and economically weak.

Since the mid-sixties the Armed Forces are wielding power in most of the Latin American countries. This questions the validity of Napoleon's statement, according to which one can do anything by fixed bayonets, except sit on them. One country, Costa Rica, has dissolved her army, in Mexico the role of the army is negligible, in Colombia and Venezuela—though under the indirect political influence of the Army—civil political mechanisms are still existent, whereas in some Central American republics as well as in Argentina after the death of Peron a dual power of the Army and the political institutions came into being.[7] Despite functional diversity it is quite obvious that the function of external defense is of no great importance. In the case of the Brazilian army some traces of a policy aimed at achieving the status of a great power might be detected (though the means of this are mostly not of a military nature), however, here, too, the internal defense of the system is considered as of primary importance. Thus, the functions of the Army are connected with the internal defense of the system in all the countries of Latin America. This defense is directed against the reactionary forces in the progressive nationalist models, and against the Left in models of the Brazilian type. The changing role of the Army is demonstrated, however, by the fact that even in the case of the Brazilian model the so-called praetorian function has been supplemented by some economic objectives, though the problems of social development are still not tackled.

The changing role of the armed forces in the Latin American countries is due to objective socio-economic processes. This new role has to be taken into account in the long run, and all elements should be kept track of which, as a consequence, might further the struggle of the forces striving to accelerate socio-economic progress on a national, regional as well as international scale.

1.2. THE SOCIO-POLITICAL AIMS OF PERUVIAN DEVELOPMENT

In addition to its "institutional" peculiarity, another highly significant feature of the Peruvian road of development is its interdependent socio-political aims. According to the so-called Velasco Doctrine elaborated by the military

---

[7] M. Solaun: *Sociologia de los golpes de estados latinamericados*. Bogotá 1969.

regime, the most important products and natural resources of the country should be nationalized, economic development ought to serve the welfare of the country as a whole and not represent merely a source of profits for certain individuals or groups. Foreign investments, too, have to harmonize with national interests, and the measures taken by the Government should stabilize the independence of the country and improve the living standards of the population.

The four declared principles marked also a certain sequence in the steps taken by the military regime. The sectoral character of the measures unambiguously demonstrated the fact that the eventual restoration of economic sovereignty, especially the termination of foreign control of natural resources vitally important from the point of view of external economy were regarded as high-priority tasks by the military regime. As early as the first week following the take-over, the oil fields and oil refineries which had been controlled by IPC, were nationalized, and this was followed by the nationalization of foreign trade in traditional export commodities, and finally by introducing foreign exchange control. Later on, in 1973, fish-meal production, one of the basic branches of economic life was nationalized, whereas in 1974 the Cerro de Pasco, one of the giants of the copper sector, supplying more than a third of the total production of mining, came under government control.

The offensive against the landowning oligarchy, the strongest domestic ally of foreign capital was launched almost simultaneously with the above nationalizations. The land reform announced in June 1969 was a milestone not only in the history of economic growth, but also in the political development of Peru. Agriculture which had been the strongest sector of the economic life up till the early sixties and still supports the majority of the population, is not simply an economic factor but a mode of life, and a socio-political factor as well. At the same time, the almost immobile subsistence farming of the highland Indians, because of its considerable influence, has become the most vulnerable spot of the national economy and a source of food supply problems, foreign exchange tensions and rising costs. One of the most important as well as most delicate questions of economic policy is how to tackle the problem of agriculture, since economic and political intervention involves much more elements of conflict than in other Latin American countries carrying out a land reform.

The maximum size of estates was determined at 150 hectares of irrigated land by the compensation-type land reform proclaimed from above, however, as early as the end of 1970 already 4 million hectares were distributed, and an additional 4 million was donated to the tillers in the period 1971–1973. By distributing 11 million hectares, the land reform was essentially accomplished by 1975. The relatively fast accomplishment of the reform is in itself a significant achievement, since the escape of capital—a common symptom on such occasions—and the stagnation of production did not assume disastrous extents, and no unending legal disputes ensued. The other specific feature of the land reform is the fact that expropriated big estates were but partly distributed among the peasants, while part of the latifundia were given to cooperatives which came into the possession not only of the land but also of all ancillary and subsidiary plants as well as processing establishments belonging to it. Thus, the agrarian reform law eliminated the big estates, re-established the landed property of the Indian communities, furthered the spread of medium-sized peasant farms and, at the same time, was instrumental in developing the cooperative sector. Though the stand taken by the peasantry regarding collective property varies from region to region, local customs and a historically evolved scale of values in the highland districts inhabited by Indians are not antagonistic to collective work and common property. It should be noted that these communities had developed certain forms of collective farming *(ayllu)* in the Pre-Inca, Inca and colonial societies, and often maintained them up to this very date. Due to the specific features of national historic evolution, the introduction of collective farming did not meet serious obstacles, and operative efficiency also proved satisfactory.

As regards industry, the turning point came with June 1970, when a new Industry Act was promulgated. This law envisaged the nationalization of the production of the most important capital goods, the planning of the country's industrial development, and participation of the working people in the distribution of incomes and shares, as well as in the management of enterprises. From that time on, new foreign investment permits were granted only under conditions varying according to sectors and strict government control. Active industrial enterprises have to allocate 15 per cent of their annual profits to the buying up of their own shares until 51 per cent of their stock-holdings passes into national ownership. This law is supplemented by a rule aimed at gaining the support of industrial workers, and at the same time resulting in the transformation of ownership relations. According to this rule, 10 per cent of profits have to be distributed among the workers in the form of shares with the objective that 50 per cent of the shares should eventually pass into the hands of the workers' collective.

The next step on the road leading to the transformation of industrial ownership relations was a bill passed in 1974 on establishing the state-owned sector. Thus, Peruvian economy has four fundamental forms of ownership by that time: first, the so-called reformed private ownership (the sector based on the traditional capitalist enterprises); second, small-scale private ownership; third, state ownership; and fourth, social ownership. Social ownership is based already on enterprises owned collectively by the working people, and covers mostly small mining enterprises, plants connected with forestry, agricultural processing plants as well as undertakings in the iron and metallurgic industries. Subsequent to the promulgation of this Act, a new element was added to the political and ideological struggle regarding the long-range role of these ownership forms.

The extent of the changes in the ownership relations is clearly shown by the following data: in the period 1968–1975 the share of the public sector in production increased from 18 to 42 per cent, that of social ownership from zero to 10 per cent, whereas the share of the foreign-controlled sector dropped from 34 per cent to 13 per cent.

Similarly to Mexico, the making use of the Indian communities, has its origins not only in economic considerations, but runs parallel to attempts aimed at the revival of pre-colonial formations and values as well as at the forming and developing of a new nation based on the Indian population in the country. The ideological background of these endeavours is highly variegated, comprising views ranging from 19th-century romanticism to up-to-date concepts of social integration. However, the removal of Pizarro the Conqueror's portrait from the state-room of the Presidential palace in 1972, and installing instead of it the portrait of Tupac Amaru, an Inca descendant rebelling against the colonial rule, has unambiguously revealed the attitudes of the military regime, as to what heritage it acknowledges, and by virtue of what heritage it wishes to bring about national unity.

According to the so-called Industry Act, the main branches of industry, e.g. iron and non-ferrous metallurgy as well as heavy chemical engineering will be taken over by the state. Though the new Industry Act leaves some doubt as to the long-range targets of the new regime, it combines the process of increasing national ownership, i.e. of giving production of Peruvian character, with the developing of a novel-type collective ownership. Internal changes were supplemented by a sequence of increasingly resolute steps demonstrating the country's independence in the field of foreign policy. The establishment of diplomatic relations with the socialist countries, the steps taken to end the isolation of Cuba in Latin America, the radical views emphasized at the meeting of the non-aligned countries, the fact that the conference of the 77 countries was held in Lima, the participation in the UNCTAD conferences of 1972 and 1976, as well as the part assumed in the fights connected with the establishment of the New International Economic Order—these are all integral parts of one and the same strategical concept.

The initial targets set by the new regime, and all that has been done up till now, raise some questions as regards the character of the Peruvian road of development. The ideological contents of the Peruvian road of development is one of the most controversial problems of Latin American political literature. The Peruvian road of development is thought by many to show the signs of some or other political or economic philosophy, of the experiences of certain other countries. Thus, this developmental path is sometimes seen as a combination of the Mexican and Japanese types of development, while it is regarded by others as the manifestation of the philosophy of the Opus Dei of Spain or as a school of reformist thought suggested by the UN Economic Commission for Latin America on the basis of the ideas of Prebisch; and finally, there are some experts who are inclined to see the Peruvian model as the crystallization of Yugoslav experiences. While right-wing opposition considers the new patterns of growth as a consequence of communist infiltration, the military regime is regarded by the forces of the extreme left in Peru as the most sophisticated and shrewdest defender of bourgeois rule, and the model itself as the embodiment of the neo-capitalistic, neo-imperialistic type of development.[8] Still other experts scrutinize the model from the point of view whether it offers a capitalist or socialist alternative of development.

The variegated nature of the problems, explanations and interpretations is not at all astonishing. As a matter of fact, the Peruvian model does not manifest any kind of homogeneous ideology. Former President Velasco Alvarado himself tried to define the middle-of-the-road nature of the model by a dual preclusion: "Although Peru does not identify herself with the socialist model, she does not accept the traditional form of capitalism either". The views of the military government are highly pragmatic; theory is formed by the day-to-day pactice of modern production forces and production relations. The standpoint of the Communist Party of Peru was expressed by the statement of F. Arias, member of the Central Committee: "Presently the most important problem of Peru is not the problem whether the developmental alternative is capitalism or socialism. Peru has to choose between national sovereignty, economic independence, social progress on the one hand, and economic sub-

[8] A. Quíjano: *Nacionalismo, neoimperialismo y militarismo en le Peru.* Buenos Aíres, 1971.

ordination, imperialist oppression, medieval backwardness on the other". The present model is regarded as revolutionary by the Communist Party of Peru.

1.3. THE BALANCE OF ECONOMIC GROWTH

In the sphere of politics the Peruvian road of development has given rise to much surprise considering the hopes cherished as regards military regimes in general. The comparison of the trends of political development and economic growth is very intriguing, too.

Though the synthetic index of economic growth, i.e. the average annual growth rate of the national product has fallen somewhat short of the plans, it still might be regarded as favourable. The rate of growth was 2 per cent in 1969, 7.5 per cent in 1970, 5.2 per cent in 1971, 6 per cent for the period 1972–1974,[9] and 4 per cent in 1975, i.e. it corresponded to the average of the two decades prior to the military take-over. Assessing the dynamism of growth, two major natural disasters should not be left out of consideration. As a consequence of the earthquake as well as floods in May 1970, 70,000 people lost their lives, a number of towns disappeared from the face of the earth, hundreds of thousands lost their homes, and material damage mounted to more than half a billion dollar. In 1972, due to a shift of routes of migration of fish, the income of coast fishery dropped to a fragment of its value in the previous years, and the catch became quite insignificant. Taking into account these natural calamities, the results achieved in economic growth show the viability of the Peruvian economy.

However, the growth rates reflecting dynamic development conceal to a certain extent the uneven, and in some of the productive sectors the below-average progress of certain sectors. Thus, the output of mining, having the most important role in the external economic sector, made no headway between 1967 and 1975. Taken in itself, the increase in agricultural output was not very favourable either. In the first two years of the military regime, i.e. in 1969 and 1970, the yield of agricultural production increased by 5.5 per cent, whereas in the period 1971–1975 the average increase amounted only to 2 per cent, and the production of fishing diminished by 50 per cent. Decreasing agricultural output is quite groundlessly connected with the agrarian reform by some critics of the regime, since the regression of fishing was caused by natural factors, whereas the modest increase in agricultural production at the time of the peasantry' s increased consumption orientation and a certain "slackening" of its labour morals might be regarded as a considerable achievement. As a matter of fact, it is rather astonishing that, contrary to other countries, no serious supply troubles were caused by the distribution of the greater part of commodity producing big estates.

Besides the extended services it was the manufacturing industry which became the main factor of growth. The Peruvian bourgeoisie launched desperate attacks against the Industry Act right after its enactment, nevertheless, following a period when its relations with the government had touched rock-bottom, it gradually put up with that law. After a two-years' big drop in industrial private investments, the rate of investments increased by 6 per cent

[9] Economic Intelligence Unit, 1973–74.

in 1973, but still falls short of the pre-1970 level (1967: 5 billion sol, 1970: 4.1 billion sol).[10] Production of the manufacturing industry grew by 11 per cent in 1970, 8 per cent in 1971, 4 per cent in 1972, and more than 10 per cent taken on the average of the period 1973–1975. Due to the above-the-average growth rate of the industry as a whole, the structure of the national economy has improved; the share of the manufacturing industry increased from 20.3 per cent in 1969 to 23.5 per cent in 1974.

The successful efforts aiming at reducing the disequilibrium of the economy may be evaluated as a major achievement. In the first year following its establishment the government undertook the unpopular task of restoring the balance of the formerly inflationary and over-heated economy. As a consequence of the newly introduced restrictive economic and financial policy, the price level which had formerly increased by an annual 15–20 per cent increased but by 6–8 per cent on the average of the period 1969–1973 —an unprecedented event in the history of Latin American industrialization accustomed to an inflationary climate.

Growth, however, had a number of negative aspects as well. The price to be paid for the stand taken against domestic and foreign private capital was a marked set-back in private investments. Though the state, relying on its broadening economic basis, undertook to make up as far as possible for the losses due to diminishing private investments (the investment budget for 1973/74 exceeding that of the previous year by 47 per cent), the observance of the rules of orthodox financial policy has naturally set limits to the expansion of public investments. In the period 1973–1974 already 44 per cent of all investments were realized by the state, whereas in 1968 the share of the state had amounted to only 25 per cent. Owing to the reserved attitude of private capital which still furnished the bulk of investments, the share of investments in the gross national product fell from 16 per cent in the sixties to below 13 per cent by 1972.

The relatively slow development rate of the productive sectors is also a result of the events in the investment sphere. Private capital annoyed at the measures of the Industry Act, withdrew from productive investments and channelled its surplus capital into fields undisturbed by the new development strategy (e.g. services, construction industry and real estates) as well as parasitic consumption. The above-average rate of development of the servicing branches, the rise of the share of the construction industry from 4.5 to 7 per cent within the gross national product in the period 1967–1973, the nearly 50 per cent increase in the expenditure on cars in 1973 —all reflected the non-productive utilization of surplus capital. According to A. Quijano, the consumption standards of the Peruvian bourgeoisie and middle strata reached an unprecedented peak in the early seventies.[11]

The relative decline in internal savings and investments has resulted in an undesired shift of the structures of production and consumption and has not helped to reduce essentially serious unemployment, overt and covert, afflicting about 28 per cent of working-age population.

Thus, the assessment of the growth performance of the domestic economy cannot be quite unambiguous. The dynamics of economic life achieved amidst

---

[10] *Ibid.*

[11] A. Quijano: *Politica social y economica de la Junta Militar*. Revista Latinoamericana, Bielefeld, 1972.

structural transformation is not unfavourable even compared to international trends. At the same time, the full frontal attack on the positions of the bourgeoisie, the surmounting of the difficulties of the resulting flight or less desirable utilization of capital, as well as a dynamic employment strategy necessitated by the political situation could hardly be brought into accord with a restrictive policy aimed at the restoration of equilibrium. By giving priority to reducing the financial disequilibrium—in itself a correct and necessary measure—the solving of problems taking shape in the wake of political changes has been seriously impeded.

The sector of external economy, more precisely: the problems arising from the struggle for economic independence, represented the most serious and dangerous obstacle to the implementation of the new government policy. The dependence of the external economy developed in a rather uneven way.

In 1967, the year preceding the military take-over, more than 18 per cent of the national product had been realized by the exports of services and commodities, i.e. Peru had been more open to the world economy than the average of the Latin American countries. However, high dependence or foreign trade was not accompanied by an economy of a one-crop character, and exports were evenly divided among the different ores, agricultural and fishery products. Thus, it was not foreign trade dependence in itself which created a grave situation; conditions weakened only when it went hand in hand with foreign capital conquering key-positions. Due to the unfavourable historic preconditions of domestic capital formation, foreign capital succeeded in getting hold of the most valuable and exportable natural resources of the country. Out of all Peruvian export commodities only the major parts of the coffee, cotton and fishing sectors fell under the control of domestic capital. The ratio of the volume of foreign capital investments to the national product did not exceed the average, yet, the foreign control of the external economic sector was singularly high as compared to other countries of Latin America.

The primary target of the strategy of achieving economic independence was quite understandably the nationalization of the sector of the external economy. In addition to nationalization acts already mentioned, the government took measures to skim off a greater part of the surplus value of the extractive industry. While a 20-per cent tax was levied on the net incomes in mining in the sixties and it was made obligatory to reinvest further 15 per cent of it, the Mining Law of 1971 stipulated 30 per cent taxes and 40 per cent reinvestments, and extended to mining the norms of workers' ownership established in the industry.

The volume of US claims owing to nationalization was given as 260 million dollars, i.e. total US capital investments diminished by more than one-third.

Similarly to a number of developing countries, the measures taken against the key positions of foreign capital, though objectively necessary, were accompanied in Peru, too, by an erroneous strategical conception, at least in the first two years of the military regime. The steps taken against foreign capital were linked—as in many a developing country—to restricting the export-oriented sector, the main representative and symbol of dependence. Early development conceptions were by no means free of a frequent error of the military cast of mind: independence was identified with autarky and therefore, import substitution represented the pivotal point of developmental strategy. Peru was the last of the Latin American countries to launch—rather belatedly—an import-substituting industrialization which quite naturally drained off energy and diverted attention from the export-oriented sector of the economy. Due to the neglect of development priorities and to the measures taken against foreign capital, the production of mining, i.e. the backbone of the export sector, has virtually stagnated ever since 1968. Thus, exportable commodity supplies became scarce just at a time characterized by the expansion of world trade and, later on, by an extremely fast rise in the prices of raw materials. The set-back in the yield of fishing in 1972/73 was the direct result of a natural phenomenon, i.e. the shift of the coastal migration routes. This problem, however, would have obviously not assumed serious dimensions if the fishing sector had been updated, and up-to-date cold-storage and processing ships had been built or bought. Under such circumstances it seems highly questionable whether the decrease of the ratio of foreign trade dependence to 16 per cent is a favourable development, since exports, despite of favourable world economic trends, scarcely increased since 1968, whereas imports, after a considerable drop lasting for many years, surpassed the level of 1967 only by 1973. The fact that the value of foreign trade turnover in 1974/75 increased by 50 per cent may be attributed largely to price fluctuations. As a result of the favourable price trends in the world economy, economic development had not, however, to face external economic restrictions. At the same time, it might be argued that—in spite of the ousting of foreign capital—the external economic positions as well as the internal strength of the country could have been considerably strengthened by a strategy concentrating more on the export-oriented sector of the economy.

In addition to the initial neglect of export orientation, the policy of building up foreign exchange reserves—a mentality reflecting once again the military cast of mind, and, in itself, a correct policy aiming at maximum safety—also put a brake on the rate of economic development. At the end of 1968 the foreign currency reserves of the country amounted to $111 million, whereas by the end of 1974 reserves reached the sum $968 million, which meant a nearly three times more favourable rate of import coverage than that of the average of the developing countries. In the case of favourable world market chances and urgent internal development needs, however, a policy aimed at import restriction and building up large foreign currency reserves is again a highly questionable venture.

The policy of treating export-oriented productive sectors as mere ancillary fronts was extended to tourism as well. Peru has an abundance of archeological and cultural historic treasures, and this provides her with extremely favourable conditions and comparative advantages as regards Latin American tourism. The use of these possibilities, the development of the infrastructure and organization of tourism, however, are still in the embryonic stage, and incomes from tourism are amounting to a mere fracture of the possible earnings. Thus, the concept behind import-substituting industrialization has disregarded one of the most rapidly and economically realizable growth factor with the help of which the development of the backward regions of the interior of the country could have been stepped up as well.

The balance of capital turnover took an adverse turn, owing to the capital-restrictive economic policy and also to the situation of the external economic sector. Already after the nationalization of IPC the lobby affected by this measure organized a large-scale campaign, and pressure was exerted by the US in order to enforce the Hickenlooper Amendment. Though things had not come to a head on the government level and no sanctions were applied, the North American monopolies tried to bring the Peruvian government to heel by means of embargoing all financing activities. For instance the EXIM Bank, a prominent representative of all financing transactions in Latin American countries, declined to undertake any funding until IPC had not been paid indemnity. US pressure made itself felt also in the policies of international banking houses. In addition to US capital and credit discriminations, private capital has been fleeing from the country in the first five years following the military takeover, even according to official data on the balance of payments. The exodus of capital effected through non-official channels assumed fantastic proportions. The Government had to resort to legal price-regulation in 1974 in order to curb under- and over-invoicing manipulations in the field of exports and import. The net capital import of the country derived essentially from long-range state capital exports the volume of which, however, scarcely exceeded one per cent of the national product on the average of the first five years.

The recent development of new oil fields and/or the raw material crisis have again created new conditions for the relations with foreign capital. In the past two years a number of large-scale agreements have been signed with foreign monopolies to further the exploitation of oil-wells and copper mines. As a result of these agreements copper export is expected to rise conspicuously from 1976 on, and considerable oil export is counted upon beginning with 1980. The shift of emphasis may be characterized by the fact that out of the three billion dollars allocated for investment projects in 1973–1975, one-third will go to the development of the infrastructure, and one-seventh to mining, while it is hoped to arrange for 60 per cent financing of the investment budget from external, i.e. foreign funds. Following a rapid decrease in the dependence on foreign capital, the share of foreign capital in investment funds—though under the conditions of highly improved Peruvian bargaining positions—would reach an all-time peak.

Besides subordination in the sphere of investment financing, the most critical field of external economic dependence is the geographical composition of foreign trade, i.e. its large-scale dependence on the United States. A considerable percentage of Peruvian exports is absorbed by the US, the bulk of the mining products of Peru is processed in North American plants, whereas the overwhelming part of Peru's investment imports and industrial equipment, as well as her system of technological standards originate from the United States. Thus, on the basis of its dominant position, the United States both as a buyer's market and a supplier of commodities of vital importance and up-to-date technologies may be in the position to influence and determine the political situation in Peru at any time. Quite understandably, the new government considered the extent of the market positions of the North American super-power critical from the point of view of political and economic independence, and in order to diminish the external vulnerability of the economy it began, concurrently with measures taken against foreign ownership relations, to curtail the foreign trade positions of the United States. While the share of the USA in the exports of Peru amounted to 43 per cent and in her imports to 40 per cent in 1965, these percentages were reduced to 36 and 29 per cent respectively in 1974. Peru's foreign trade relations are still dominated by the United States, nevertheless the sphere where it could exert pressure according to its interests as a super-power shows a diminishing trend.

The government of Peru did not want to fill the void left by the weakening of US positions by relying unilaterally on another great power but strived to develop a widely spread geographical structure of foreign trade rendering possible a higher degree of manoeuvrability in the sphere of external economy. Against all expectations it was not only Japan, the strongest rival of the United States on the market which occupied the surrendered North American positions, but there emerged another foreign trade alternative vs. the United States: the increased participation in Peruvian foreign trade of the West European countries, other Latin American countries, and to a lesser extent of the socialist states. In 1974 for instance, 14 per cent of Peru's exports were directed to Japan, 8 per cent to the FRG, 3 per cent each to the Netherlands, Argentina, Belgium, Great Britain and Italy. Considering the results of the foreign trade war between the United States and Peru, it can be stated that in the course of the past six years the balance moved from the initial stalemate increasingly in favour of Peru. The attempts of the military government to reach geographical diversification in foreign trade have not been completely successful since in spite of losing some foreign trade positions it is still the United States which dominates the financial scene in Peru. Under the present conditions of the world economy mainly the US monopolies have at their disposal such huge capital which is necessary for the further development of Peruvian mining and oil production. The new regime succeeded in ousting North American capital from the bank network, telecommunications, the sugar industry, fishing and the fish processing industry, and also to a certain extent from copper mining and refining, however, the vested interests of the big commercial enterprises (Sears, Oechsle, Monterrey) remained intact, moreover the new oil concessions were also obtained by North American oil corporations, though under less favourable conditions than before.

On the other hand, the offensive launched by the United States has not achieved its purpose either. Considering the capital-intensive nature of the Peruvian economy, the credit embargo actually resulted in a critical situation, however, it failed to create an economic state of emergency and to squeeze out political concessions. The United States did not enforce the Hickenlooper Amendment, and the freezing-in of credits as well as the initial suspension of capital investments proved insufficient for the strangling of the military government. In the resulting stalemate the events on the international commodity markets and the unfolding raw material crisis strengthened the bargaining positions of Peru. The great power was obliged to acquiesce in the decreasing dependence of the small country, and from 1971 on the US began to normalize its relations with Peru.

It would be mistaken, however, to attribute the relative successes of the military regime entirely to the achievements of the political model. Certain favourable external factors had also a part in the progress of the Peruvian road of development. The humorous truism "Dios es Peruano" (God is Peruvian) reflects the realization of the fact that the exploration of oil fields as well as developments on the world market have stabilized the international bargaining positions of the military regime in the critical situation of the period 1972–1974.

The developments of the world economy in 1973–1975 not only directly stabilized the positions of the Peruvian government but determined indirectly, mainly beginning with 1975, the internal balance of power and gradually the political image of the military regime as well. Owing to the external economic strategy described above, the economy of Peru was able to exploit the possibilities of the price boom only in respect to and to the extent of the rise in prices, whereas it proved incapable of increasing its export supply and of stepping up structural transformation. At the same time, in order to reduce the political imbalances, the government permitted a large-scale (80 to 110 per cent) increase in imported consumer's articles, building materials and military equipment. Thus, as against the policy aimed at a balanced or even active foreign trade in the first six years of the military regime, a considerable disequilibrium manifested itself in 1975/76. Due to the stagnation of agricultural productive forces and purchasing power, internal division of labour satisfied less and less the requirements of an expanding external sector. The problems of the imbalances of the internal economy were sharpened by the guarded attitude of private investors, the lasting stagnation of budget revenues (with the exception of 1974), and at the same time, by the increasing budget expenditure and/or costs of the public sector which were accompanied by the consequences of a world-wide inflation. Thus, it was clearly indicated by the developments of the period 1973–1975 that at a time when private capital is extremely cautious in investing and the local bourgeoisie insists on its high-level consumption, the internal resources of the country are not sufficient for solving the problems of the reconstruction of the socio-economic structure. In order to be able to finance the costs of adapting the economy of Peru successfully to the changes in the world economy, as well as to bridge the gaps in the rather restricted development funds and an adverse trade balance, the opening up of additional external financial resources became an urgent necessity. Tension developed as a result of this situation, and as a consequence, conflicting opinions manifested themselves within the army which seriously jeopardized the operative capacity of the military regime. The question under discussion was whether under given conditions a radical socio-political transformation offered a solution, or the pace of change should be slowed down and relations with both the domestic and the international bourgeoisie should be improved.

The spectacular shift in the internal balance of power within the military regime took place in August 1975, by means of a military coup during the meeting of the non-aligned countries. Though the new president who masterminded the takeover, the former prime minister and minister of finance F. Morales Bermudez had a leading part ever since the takeover in 1968 and was known as a representative of "revolutionary pragmatism", and General Maldonado, the new prime minister declared himself a committed partisan of radical social transformation, the ousting of General Velasco has initiated a new phase in the experiment of the Peruvian armed forces.

The first months of the so-called "second revolutionary phase" announced following the coup, were characterized by conciliatory gestures towards the private sector and by measures aimed at the slowing down the pace of revolutionary transformation. The outstanding steps in the economic policy of the new stage were the following: in February 1976 new regulations were issued favouring private small and medium enterprises, in March the representatives of the Left and extreme Left in key positions of the press were dismissed, restrictive measures were adopted against the trade unions, and finally, in July the fishing fleet nationalized in 1973 and grappling with serious financial difficulties was denationalized, the sol was devaluated and the monopoly of oil prospecting and producing was renounced. Yielding to the pressure of the bourgeoisie and to the increasing right-wing tendencies and even an attempted coup within the army, in July 1976 there resigned Prime Minister Maldonado, one of the representatives of the reformist wing of the Army, as well as Foreign Minister General De la Flor Valle, the outstanding advocate of Peru's radical stand in the grouping of the underdeveloped and non-aligned countries.

As already mentioned, the significance of the Peruvian road of development went far beyond the limits of a single country. The lessons drawn from the achievements of the peculiar Peruvian road of development force us to face and critically re-examine a number of views and concepts derived from history. Peru sets a good example how to put new life and vigour into traditional social institutions and how to put them to the service of modern growth processes. In a historic emergency situation the Army as a political institution can perform the part of the directing force of development at least as well as a one-party system or personal dictatorship (a common phenomenon in the developing countries), or as the combination of these two systems. The military take-over favourably met the requirements of the control system deriving from socio-economic development.

The representatives of the political sciences and systems analysis tend to think in extremes. Problems originating from the centralization of the social guiding systems of certain developing countries were often regarded as permanent efficiency factors. In the second half of the sixties concepts regarding the decentralized decision-making models superior came into vogue again. The achievement of the Peruvian model, highly centralized as regards macro-level decisions (even compared to international standards), indicate, however, that a centralized model may be operated very efficiently if it is rooted in the socio-economic development and if the standard of decision-makers and decisions is high. At the same time, the Peruvian model demonstrates the fact that social transformations are feasible even in the case of the economy becoming increasingly dynamic. Other developing countries, too, tried to effect a radical social change, and in some cases these attempts proved successful. However, the objective political and economic factors have not been taken into account properly and the problems of implementation coupled with the acceleration of social transformation often resulted in growth crises, in the weakening of the economic foundations, or permanent stagnation. The example of Peru has proved that there does exist an alternative to such an economic predestination: socio-economic transformation and—using a military term—redeployment is feasible not only in a stationary situation but also on the march.

In choosing the targets of development strategy, the policies adopted by the military government did not prove infallible. Despite of its uncontested power position the military government was unable to resolve the problems deriving from objective subordination, dependence on the international division of labour, and from the great number of conflicting objectives.

It would be premature to regard the shifts in the military and political power-structure as a defeat. Although the government did not succeed in carrying through the comprehensive social transformation announced after the take-over as speedily, efficiently and radically as it had envisaged, after 1968, army control effected a number of irreversible changes in the formerly archaic economic and social structure of the country, and also eased the extent of unilateral dependence on foreign powers. Even if there may be some truth in the interpretation of the political events of 1975/76 according to which the military regime in Peru—similarly to some Latin American armies —will fall into line with the guardians of the *status quo*, it should be kept in mind that the *status quo* of 1976 is the product of an eight-years's socio-political development extremely fast as compared to other developing countries, and under given international and domestic political conditions the protection of these achievements is not to be dismissed lightly.

The Peruvian road of development, clearly discernible among the other development types in Latin America, obviously demonstrates the variety and manifoldness of developmental roads. One of its most important indicators is the fact that the social costs of progress and increased independence have remained low even as compared to international standards.

CHAPTER 2

THE FALTERING ECONOMIC GROWTH OF ARGENTINA

2.1. DECLINING RATE OF ECONOMIC GROWTH — A CASE STUDY

Twenty-five years ago the analysers of the economic development of Argentina used to search for the specific factors which had rendered it possible for the one-time Spanish colony to become within a century the most developed country of Latin America, moreover to rate sixth on the international list based on the volume of *per capita* national product. As regards economic development, Argentina kept level with the white dominions of Britain, i.e. with Australia, New Zealand and Canada, moreover, with respect to social services and cultural standards even outstripped these countries. This main characteristic of the Argentine development was reversed in the fifties and sixties. Argentina has to face the following serious problem and to seek an answer to this question: why has the country's long-range economic and political progress slackened in a world economic environment expanding with considerable speed, and why has the country, regarded not so long before as rich, slid back into the group of the "developing" countries? In short, where has Argentina taken the wrong turning? We are going to summarize the most significant factors of Argentina's faltering economic growth, a country developing dynamically up to the middle of this century, and to throw light on the causes of difficulties lasting for two decades. We shall also discuss the main characteristics of the latest development stage beginning with the early seventies.

Right up to the middle of this century the economic development of Argentina represented one of the most typical examples of growth caused by external factors. The formerly sparsely populated colonial territory became peopled by large-scale European immigration, and economic growth was shaped mostly by external factors, external buyer's markets and/or by expanding capital imports. Institutional structures, economic, political and cultural standards inherited from European countries took root. Argentina became the "most European" overseas country. Contrary to most of the former colonial regions, growth brought about not only quantitative but also structural changes. In Argentina, rapidly becoming one of the leading agrarian exporters of the world, a significant industry developed at a historically early time as a result of the processing of certain export products as well as of a domestic market expanding concurrently with quickly increasing exports. By the end of the first decade of this century one-fifth of the national product came from the manufacturing industry[1], and the major part of the country's demand in industrial articles was covered by the domestic industry. It is characteristic of the technical level of the industry that not only consumer's goods but agricultural machines, too, were produced locally, moreover in 1930 the first oil-refinery of Argentina was turned out entirely by the domestic industry. In the first half of the century Argentina ranked among the more developed countries not only as regarded

[1] C. Furtado: *Economic Development of Latin America*. Cambridge, University Press, 1970.

the level of incomes but also from the point of view of the *structural* indicators of production and social progress.

The World Depression and later on World War II have naturally badly shaken the Argentine model, too, which was founded on the expansion of the international division of labour. However, import-substituting industrialization and protectionist economic development launched in the thirties and gaining momentum in the forties, were able to uphold the traditional rate of progress until the early fifties. The consequences of the decrease in the growth rate of the national product from 5 to 3 per cent, the standstill in the expansion of exports, the considerable drop in the effectiveness of investments, and the growing of inflation made themselves felt at the beginning of the fifties. Within a short period of time, Argentina, formerly a prosperous land, became the sick man of Latin America.[2]

According to certain views, the falling behind of the Argentine economy is due to the nature of its dependence on and *extremely strong economic ties with Britain*. It is pointed out that in the 19th century when Britain had become the growth centre of the capitalist world this close interrelationship had exerted a dynamic, stimulating effect. However, with the considerable weakening of Britain's world economic positions and the lagging behind of her internal growth, Argentina, similarly to the Commonwealth nations strongly linked to Britain, also fell victim to the disease emaciating Britain. It is an undeniable fact that the patterns and dynamism of growth of the dominant country necessarily influence the development of the dependent countries as well. Thus, the nature of the geographical direction of the external economy is by no means an irrelevant growth factor. A number of examples taken from the events of recent decades, however, clearly demonstrate the fact that even countries with a much weaker international manoeuvring capacity succeeded in transforming the structure of their medium-range external economic relations. Therefore, the strong links with Britain do not furnish a satisfactory explanation.

Explanations setting out from the phenomenon of the unfolding international agrarian protectionism or from the slackening of demand for agricultural products on the world market are likewise erroneous. The growth of the Argentine economy came to a standstill the effects of which made themselves felt as early as at the beginning of the fifties, i.e. at a period of overwhelming global food demand and prior to the rise of agrarian protectionism. On the other hand, the lower agrarian price level between 1957 and 1963 was followed—despite the agrarian protectionism of the developed countries—by favourable and active demand tendencies, and these were thoroughly exploited by other agrarian exporter countries. At the same time, the share of Argentina in agrarian world exports diminished conspicuously indicating the fact that the slowing down of her growth was not due to external factors but rather to the inherent problems of her economy, viz. to the errors and mistakes of the development strategy of Argentina.

The roots of the mistaken *development strategy* go back to at least half a century though the failure of the economic strategy became evident only in the early fifties. By the time of World War I, the economy of Argentina had almost entirely exhausted all the reserves of extensive growth. The stage of her structural development, furthermore the coming to an end of Britain's role as the centre of economic growth would have called for a shift in the

---

[2] A. Ferrer: *La economia argentina.* Fondo de Cultura, Mexico–Buenos Aires, 1963.

emphasis of development. Instead of agrarian exports, industrialization, and especially an export-oriented industrialization founded on the processing of agrarian products, ought to have become gradually the key sector of economic growth. At that time, the relative development level of the infrastructure, servicing agrarian exports, would have rendered it possible to build up step by step an internationally competitive industry. Political power was concentrated in the hands of the liberal landowner class which, however, did not become aware of the requirements of the subsequent stage in the economic growth process. As a result of the shock caused by the World Depression, the step towards structural transformation was taken with a time lag of about twenty years. By that time, however, the conditions of the world economy did not make it possible to realize any alternative to protectionist, import-substituting industrialization. As the sources of British capital exports became exhausted in the period following World War I, the infrastructure of the Argentine economy began to deteriorate just at the beginning of the process of industrialization, this being a factor considerably impairing competitiveness.

Thus, large-scale industrialization has started with a time lag of two decades. Contrary to economic requirements, it assumed a purely import-substituting character and eventually was carried through in an atmosphere of extreme protectionism going far beyond the objective needs created by the development level of the economy and the intensity of external competition. Owing to its potential and technical development, the Argentine industry ranked first in the Latin American region, nevertheless the government afforded a higher extent of protection to the industry than any other country on the continent; occasionally the effective rate of this protection amounted to many hundred per cent.[3] The protection guaranteed in order to step up to a maximum the pace of industrialization as well as excessive industrial profits, however, undermined and sometimes even eliminated the competitive spirit, and resulted in a considerable reduction of the rate of technological development and effectiveness.

To accelerate the change in the pioneer sectors of economic growth, a disproportionately high percentage, i. e. 50—60 per cent, of investments was allocated to the industry, primarily to the heavy industry in the late forties. Perón's five-year plan promulgated in 1947 aimed at transforming Argentina which had no steel and coal mining industries into a "steel empire" within a single plan period. In addition to the disproportionate system of incentives, it was the unfortunate choice of development priorities, as well as the forced development of capital-intensive branches lacking any internal basis, which had a considerable part in the exhaustion of the protectionist development strategy in the early fifties.

Until it had used up its growth reserves, protectionist development forged ahead dynamically. The main financing source in the fifties was agriculture the produce of which were initially bought up for 30–40 per cent of the world market prices by IAPI, the monopolistic state foreign trade organization established in 1946. Agricultural surplus value thus skimmed off, undoubtedly released appreciable means for industrialization. Under the combined influence

---

[3] B. Kádár: *Gazdaságfejlesztés és nemzetközi munkamegosztás a fejlődő országokban* (Economic Development Policies and the International Division of Labour in the Developing Countries). Budapest, 1967.

of agricultural prices, fiscal and investment policies, however, the decrease in agricultural production began already in the late forties.

Another source of protectionist development was the foreign currency reserves amounting to 1.2 billion dollars accumulated at the time of World War II owing to an active trade balance. The greater part of these foreign exchange reserves were used up by Perón's import-substituting industrialization within a few years. Moreover, due to this strategy, the state run into debts totalling over one billion dollars. As a result of the exhaustion of foreign currency sources and other credit-raising possibilities, protectionist development lost all access to foreign exchange financing funds.

Errors committed in the course of economic development, however, throw a light only on the direct causes of the lagging behind as regards growth. It would be expedient to indicate briefly the peculiarities of socio-political development directly affecting growth.

The part of external factors in the growth process is demonstrated not only by Argentina's increasing reliance on the European markets and European technology, by the adoption of European-type institutions, but by a large-scale *immigration* as well. Immigration affected the development of Argentina to a greater extent than any other country of the Latin American continent. It had been unparallelled not only in its dimensions but also in its concentration in time. Immigration practically had taken place within a couple of decades and thus, the assimilation of the immigrants was accomplished much slower and brought about enormous difficulties. At the end of the colonial rule the country had a population of hardly one million. In the first century following the attainment of independence, especially between 1870 and 1914, eight million immigrants, most of them Italians, had arrived in Argentina. The absorption of such a large-scale and rapid immigration inevitably brings about problems from the point of view of national cohesion. A famous saying by an English diplomat Lord Bryce is quoted by John Gunther: "The Argentines are no longer Spaniards, but had not yet turned into anything else either."[4] Immigration reflects a significant concentration in space and sectors. Whereas in the vicinity of the capital the percentage of those born abroad was 70 per cent, in the provinces of the interior it amounted only to 20 per cent. In 1914, 47 per cent of the economically active population, 50 per cent of the industrial workers, 66 per cent of the industrial entrepreneurs, and 74 per cent of the merchants had been foreign-born.[5] Thus, the immigrants occupied the up-to-date sectors of the economy and consequently, it was here that "Europeanization" reached its highest degree.

There developed a highly specific social stratification and *class structure* which came to have a great influence on the political progress of later times. Initially the immigration of a great number of small entrepreneurs ousted by European big industries, the lasting atmosphere of hectic founding activity in business life, and later the sprawling administration in the wake of increased public intervention—all these factors added to the development of an entrepreneur and clerical staff category of one of the highest percentage not only in Latin America but in the whole world. Proprietory, technical and clerical staff strata amounted to 30 per cent of the earning population in 1914, 40 per cent in 1947, and 45 per cent in 1960, and within that, the share of independent

---

[4] J. Gunther: *Inside South America.* New York, 1967.
[5] G. Germani: *Social Modernization and Economic Development.* U.N. Geneva, 1970.

entrepreneurs (mainly craftsmen and retailers) has been 20 per cent ever since 1947.[6] Thus, the bourgeois way of life, mentality, political attitudes are built on a rather broad material foundation.

In addition, the organization of the industrial workers is much higher in Argentina than in most of the developing countries. There were four million trade union members already in 1950, and thus even highly conservative military governments had to reckon with the political influence of the trade unions. The revolutionary potential of the workers and their trade unions, however, could not be channelled in the required direction. The Socialist Party founded at the end of the last century professed the very same end and means as its sister parties in the highly developed capitalist countries of Europe. The fact that at the turn of the century the overwhelming majority of immigrants came from Italy created a special problem as anarchism had great influence there at that time. Thus, due to the ideological inexperience of the working class, national psychological characteristics resulting from the ethnic composition of the population, the extremely high ratio of small owners and clerical staff, and to the general state of the economy, *anarchism* proved much more enduring than in South Europe. Despite the hundred-year-old democratic traditions and a higher degree of economic development, violence, political *violencia*, represented a steadily increasing tendency.

When the population of mainly Latin origin (the majority came from southern Italy) was made to accept the Anglo-Saxon institutional system, this must have resulted in a number of *paradoxical* situations in historical development. In spite of the fact that the institutional system and policy of economic liberalism and the "outward-looking" model have struck roots, in spite of the close relations with Britain and the higher level of economic development, it has been Argentina where—owing to a number of peculiarities of the socio-economic structure—the first and most powerful political mass movement and mass party of Latin America (Perónism and the Justicialist Party founded by Perón) have come into being. The economic boom during and after World War II created favourable conditions to the efforts of the Argentine national bourgeoisie which strove to become independent from weakening Britain, and also to win significant mass-support. In its foreign policy, Perónism stressed the specific position of Argentina, her middle-of-the-road character between socialism and capitalism, whereas its political programme for domestic affairs emphasized the concept of a state rising above the classes and creating social justice *(justicialismo)*. Similarly to all other populist movement of Latin America, its social basis was rather heterogeneous, while initially expansive nationalism and the heritage of Italian fascism were granted a considerable part in its ideology.

Perónist policy relying on the strong economic positions of Argentina exerted an accelerating effect on socio-political progress. It crushed the power of feudal oligarchy, considerably raised the real wages of the workers, extended the scope of social services, and—though by means of social demagogy—activated the masses, granted the Communist Party legality, established diplomatic ties with the socialist countries, etc. The collapse of the Perónist system was due not to political but to economic reasons. The historic resources of growth which had extraordinarily increased during World War II, as well as the political support by the masses were put into the service of an erroneous

[6] *Ibid.*

economic policy. Though the government was strong enough to skim off once the accumulated growth resources, the launching of a new self-inducting, cumulative growth process releasing new energies was made impossible by an erroneous economic policy.

Oddly enough, the military overthrow of the Perón regime which had become increasingly cumbersome to the United States, too, took place at a time when only the national bourgeoisie and part of the intelligentsia began to become estranged from the regime, whereas organized labour still enjoyed the after-effects of the boom of the previous years. Thus, Perón's downfall came at a time—a rare event in history—when the regime still enjoyed the support of the broad masses, and therefore its overthrow did not entail the withering away of its ideology and the connected mass movements. The wavering economic policy, the anti-workers measures of the governments following Perón, the political "banning" of Perónism created favourable conditions to a nostalgic view of the period under Perón's rule.[7] At the same time, as a result of the resistance of the Perónist and other forces, no subsequent government had the mass basis needed for the realization of an efficient policy. As in the years following the World Depression, and later, under the rule of Perónism, the political automatisms of the liberal model had been abolished, the Argentine model operated less efficiently. Decreasing efficiency was coupled with growing centralization and incessant conflicts between the Anglo-Saxon and Mediterranean elements of the overall management system. Moreover, direct coercion gained more and more ground in the operation of the system.

2.2. THE HERITAGE OF PROTECTIONISM

The discarding of the heritage of protectionist economic policy started rather belatedly and under difficult conditions in Argentina. The overthrow of the Perón regime in 1955 restricted the consequences of the take-over virtually to the socio-political sphere. As a matter of fact, the development strategy of protectionist industrialization remained intact; the bourgeois radical reformers strived to carry out the concepts of the Perónist development policy without Perón—and better than the former dictator. In the stabilization of 1963, at the beginning of the presidency of Illia, belonging to the Radical Party, the traditional method of the International Monetary Fund was followed: it was attempted to find a short-range remedy for overcoming the adverse balance of payments as well as inflation. This, naturally, could not ease the problems of long-range development. Following a transitional relaxation, the imbalances of domestic economy and politics once again increased, and finally resulted in another military take over in 1966. The conservative and nationalistic Ongania Government relying on the armed forces, experimented with remedies ranging from traditional means of economic policy, to restrictive measures and government pressure. From the traditional means it retained the expansive money and credit policy; the adverse budget and ample money supply served primarily to finance industrial growth. At the same time, energetic price and wage restrictions were introduced with the aim of preventing the internal imbalance

---

[7] G.A. O'Donell: *Modernization and Bureaucratic-Authoritarianism*. Institute of International Studies, Berkeley, 1973.

to assume exorbitant proportions. According to the French newspaper *Le Monde*, the spending power of wage and salary earners had dropped by 30 per cent between 1966 and 1970.[8]

The initial results of the orthodox stabilization policy of the Ongania regime were marked by a certain buoyancy of the economic growth process, and by the reduction of the rate of inflation to an annual 10 per cent in the period 1968–1970, as against a 25 per cent rate for 1955–1965. However, the stabilization policy pursued since the second half of the sixties had collapsed by 1970, and the inflation controlled temporarily by restrictive measures again manifested itself violently beginning with 1970. As a consequence of the menace of a total chaos in economic life, of growing political tension going hand in hand with a stepped up rate of inflation, of the riots in Rosario, Cordoba and Salta reflecting the increasing discontent of industrial workers and mounting disarray in all spheres of life, another military take-over took place early in 1971, this time ousting the conservative Army elements. This once again cast a light on the close correlation between failures of the growth process and political changeovers in Argentine history.

The slow development or stagnation of the Argentine economy is explained by economists mainly by the mistakenly chosen instruments of economic policy and by incorrect timing. It seems, however, that low-grade efficiency of growth reflects—more than in any other Latin American country—rather the consequence of the hesitation whether to discard or transform the erroneously selected development strategy which had not taken into account the endowments and capacities of the national economy. Political disequilibrium and power vacuum are by no means unrelated to this irresolute attitude. The first difficulties and warning signs in the realization of protectionist developmental strategy had manifested themselves as early as the beginning of the fifties during Perón's first regime. In spite of this (disregarding the feeble attempts in the sixties) no radical change in development strategy was carried through until the beginning of the seventies, i. e. after nearly twenty years. This has been the second strategical vacillation lasting two decades within this century—a staggering international record as regards slow reactions in development policies.

The difficulties of growth of the Argentine economy were aggravated in three fields as a consequence of the belated changeover from the *erroneous development strategy*: the previous disproportionate funds allocation between the industry and agriculture was continued, the problems resulting from structural prematurity grew more and more serious, and the sphere of the national economy which could have exploited the advantages of the increased international division of labour narrowed down incessantly.

The forced industrialization to the detriment of agriculture took on an internationally unprecedented extent in Argentina. In the first postwar decade the policy of deliberately exploiting agriculture could be explained partly by the system of economic targets set under the given international economic and political conditions, as well as by the crushing of the economic and political power of the feudal oligarchy whereas with the cessation of the external emergency situation and the crushing of the economic and political power of the feudal oligarchy the neglect of agriculture could not be accounted for by non-economic factors. Nevertheless, the trend of decapitalization in agricul-

---

[8] *Le Monde*, Apr. 30, 1971.

ture continued all through the fifties and sixties. Only an infinitesimal share of new investments and capital exports were channelled to agrarian development, and measures aimed at curbing the inflation generally tried to prevent the further deterioration of internal imbalances by restricting agricultural prices and taxes levied on agricultural exports. At the same time, compared with previously imported products, the purchase of the products turned out by newly developed industries operating on high production costs entailed significant additional expenditures.

Under the influence of measures following from the economic policy, the rentability of agriculture diminished considerably, and it actually ceased to exist altogether. Emigration from agricultural regions increased, holdings, especially small-peasant farms, were abandoned in great numbers. During the fifteen years following the downfall of Perón, the neglect of agriculture, a negative phenomenon of basically economic nature, had assumed a number of socio-political negative features as well. The killing of hens laying golden eggs took on proportions in Argentina unparallelled in any other country. Between 1945 and 1970 the average annual growth rate of Argentina's agricultural production amounted to a mere one per cent, thus falling short considerably of the yearly rate of population increase. Therefore, it was the stagnating agricultural production which directly brought about the decrease in export commodity stocks and caused the phenomenon of faltering growth. In addition to slowed-down growth, the neglect of the modernization of a potentially high-capacity agriculture, the large-scale dependence on weather conditions resulted in elements of uncertainty and annual fluctuations which proved insufferable in a country with a developed industry.

The extent of the gap between the prices of agricultural and industrial products is demonstrated by the fact that the index of internal terms of trade decreased from 100 in 1935–1939 to an annual average of 75 in the period 1940–1949; it had been 83 between 1950–1955, and reached only in the sixties the pre-war level unfavourable in itself for the development of agriculture.[9]

The development of agriculture was affected adversely, however, not only by the draining off of energies but also by the performance of industry. *Forced industrialization* brought about a great number of relatively big enterprises in all sectors (e.g. 30 such enterprises in the motor vehicle industry). In order to guarantee success, the new industries were protected by much higher customs duties and administrative protectionist measures than would have been justified by domestic competition suited to the size of internal market and/or by the exclusion of imports. Inordinate protectionist measures resulted in the slowing down of technical quality development, as well as in rising costs. Due to exaggerated import substitution, the ratio of imports which still could have been substituted diminished to a minimum and proved insufficient for the stimulation of industrial expansion. Thus, the whole of the growth process was retarded by the breaking of the impetus of industrialization, the pioneer force of progress. The expansion of the internal division of labour came to a standstill, the numerous ineffectively operating industrial enterprises founded under the conditions of almost absolute protection had to face growing difficulties in disposing of their products on the domestic market losing its dynamic verve, whereas owing to its low competitiveness, the export of

---

[9] C.F. Diaz Alejandro: *Essays on the Economic History of Argentina.* New Haven, Yale, 1960

industrial products was out of question. Thus, the Argentine manufacturing industry operated on a 60 per cent capacity basis in the sixties (and within that, the capacity utilization in metallurgy was 54 per cent, in engineering 53 per cent).[10] The consequences of this vicious circle have been unambigous: the growth process is retarded by ineffective production, as a result of which the under-utilization of production capacities brings about a further increase in costs, deteriorating effectiveness and imbalances.

It follows logically from the above that an industrial-technical basis developed within the framework of a protectionist environment should be regarded as significant only according to its quantitative order of magnitude, whereas the effectiveness of its growth rate is highly dubious. The increasing share of the manufacturing industry in the national product was regarded with unquestioning faith as a great success of economic growth by the representatives of an economic policy giving top-priority to the benefits of structural development. The average annual share of the manufacturing industry in the national product was 30 per cent for the period 1950–1955, 33 per cent for 1960–1965, and 37 per cent for 1970–1972, and within that almost three-fifths of the production came from the branches of the heavy industry. Thus, on the strength of the indicators of structural development Argentina, despite her grave economic difficulties, has kept level with the highly developed capitalist countries. Moreover, regarding the structural indicators of the production she ought to be considered one of the most industrialized countries of the capitalist world. As a matter of fact, the development of Argentina is a striking refutation of the absolutistic view of the structuralist theory. In spite of the extraordinary structural development, the productive capacity and export capacity of the up-to-date industries is low, and the level of technology lags behind international technical progress. Thus, instead of an essentially healthy, stepped-up structural advance, there came into being a *premature* structure which does not conform to the general development standards of the productive forces.

The troubles in the *economic cycle* also reflect the symptoms of premature growth. Besides the permanent internal imbalance and the slowed-down rate of the growth process it is the effectiveness of investments which sheds light on the irregularities of the accelerated Argentine structural development. In international comparisons, based on long-range estimates, Argentina rates among the countries with the least effective investments. Although a significant percentage, i.e. about one-fifth of the national product was invested on the average of the period 1950–1970, the rate of increase remained rather low, and the effectiveness of the investments amounted to barely half the average of that of the developing countries and but one-third of that of Brazil. This modest increase was achieved by means of enormous capital expenditures and, consequently, by the forced restriction of consumption. These two factors resulted in growing political and social tensions.

The capital-wasting character of growth aggravated the problems of *employment*, brought about forced employment situations and called into being a great number of fictitious jobs. This, naturally, further diminished the low effectiveness of the economic cycle. The increasing number of impractically employed labour force working with negative effectiveness added to the deterio-

---

[10] D. Schydlowski: "Short Run Policy in Semi-Industrialized Economies." *Economic Development and Cultural Changes*, 1971/3.

ration of social labour discipline and to the worsening of the general atmosphere at work-places. It also stepped up the rate of the emigration of skilled labour and resulted in the appearance of the "Wandering Argentine".

Perhaps the most serious consequence of the surviving protectionist heritage, closely connected with the facts outlined above, has been *the permanent narrowing down of contacts with the world economy* ant the neglect of the sector of external economy. The share of imports in the national product decreased from 18 per cent prior to the World Depression to 7 per cent by 1970. For a medium developed country with a relatively small population such a negligible participation in the international division of labour and such a significant restraint of imports inevitably result in retarded growth even in case of a rational internal economic policy, since they considerably restrict imports necessary for technological progress, the improvement of economic efficiency, and the maintenance of a healthier competitive climate.

During World War II and the cold-war period efforts aiming at increased internal autarky may have seemed justified for reasons of national security. Parallel with stepping up the international division of labour, with the widening sphere of alternative buyer's and purchaser's markets, however, there were less and less reasons for granting priorities to the internal division of labour versus the international division of labour. The developmental policy of the Argentine government strived to alleviate the serious adverse growth balance in the early fifties by a permanent restriction of imports and by diverting an ever growing percentage of export commodity funds—the growth rate of which was rather slow anyway—to the domestic market, thereby limiting import capacities.

Owing to the gap between the prices of agricultural and industrial products developing in the atmosphere of protectionism, to the low price level of agricultural products which were more sensitive to price restriction, as well as to the high prices of the products of the recently established industries, the consumption structure of Argentina rapidly drifted off the growth optimum. While internationally Argentina ranks first as regards *per capita* food and protein consumption and actually consumes more than the biological needs of man would require, it does not dispose of adequate export commodity funds. It became obvious that the forging ahead of inflationary forces was accompanied as a rule by levying new export duties and taxes or increasing the existing ones. Thus, to restrain inflation and to skim off domestic purchasing capacity the development policy made use not of imports but of exportable agricultural commodity funds.[11] Characteristically, the "iron-handed" conservative military government was not an exception to this domestic supply sensitivity either. The weakening of Argentina's position in world trade as a major food exporter should be explained therefore not by the slackening demand of external markets of by some kind of discrimination but by the unyielding vestiges of protectionist development policy. A typical feature of long-term Argentine growth is the fact that the interests of the *internal economy* are granted high priority as opposed to those of the external economy—a policy characteristic generally only of great powers. The quantitative volume of Argentine agrarian exports fell short of the average of the thirties by 30 per cent in the period 1945–1955, and by 18 per cent between 1955 and 1965.

---

[11] G. Maynard: "Argentine 1967–70." Banca Nazionale Del Lavoro, *Quarterly Review*, December, 1972.

In addition to neglecting the production background of the sector of external economy, there prevails a relative indifference towards the problems of external economic manoeuvrability and the progress of independence. Despite the autarchic tendencies represented by the nationalistic political system in the greater part of the past four decades, *the economic positions of foreign capital interests have strengthened*, and the indebtedness of the country has grown considerably. Thus, the volume of North American capital investments increased from 320 million dollars in 1950 to 1500 million dollars in 1974, i.e. it grew by far more rapidly than the national product or the value of exports. As a consequence of foreign capital penetration, 50 per cent of the turnover of the biggest 50 Argentine enterprises fell to the share of foreign firms,[12] whereas in the chemical industry, engineering and in all the so-called dynamic industries the share of foreign corporations was even higher. As a matter of fact, Argentine economic policy greatly relied on foreign investments in order to step up the process of import substitution. Foreign firms, however, were affected much less by the consequences of the slowing down of the rate of import-substituting growth, in fact a great number of Argentine firms grappling with serious difficulties were bought up by foreign enterprises at lower prices than their actual value. As a result, the turnover of foreign controlled firms grew twice as fast than that of the enterprises of the public sector controlled by the Argentine bourgeoisie. The rocketing profits from the increased turnover achieved without any substantial capital import were repatriated to a growing extent from the less dynamic Argentine economy. Furthermore, growing capital indebtedness has come to play an ever increasing role in rendering the problems of the balance of payments and of debt redemption more and more serious. The outstanding paradox of the protectionist, nationalistic and economic growth of Argentina has been *the increasing denationalization of the economic life*.

The permanent political imbalance since the fall of Perónism had a considerable part in the lasting difficulties of the Argentine economy as well as in the failure of the attempts of the economic policy to wind up the heritage of protectionism. It was the political sphere which represented the main restricting condition for the consecutive governments. As a result of the lack of support by the Perónist masses, and owing to increasing socio-political tension, none of these governments had a political basis rendering possible the implementation of a more comprehensive, long-range development policy.

One of the factors giving rise to social unrest and political tension was undoubtedly the inadequate growth rate of the economy. After the rapid growth in the period 1970–1950 which had been quite conspicuous even if compared with international trends, the average annual growth rate of the national product between 1950 and 1975 amounted to 3.4 per cent, i.e. *per capita* national product grew only by an annual average of 1.5 per cent. On the basis of such a slow growth rate it is naturally impossible to raise the incomes of the entrepreneur stratum as well as of workers and employees to roughly the same extent. Owing to the changes in the political life of Argentina in the past twenty years, the advantages deriving from the modest increase in the national income made it possible to improve only the income conditions of the ruling class and the middle strata providing the mass support for the regime. At the same time, the living standards of the working classes have stagnated or even deteriorated ever since 1950—an almost unprecedented phenomenon in

[12] U.N. *Economic Survey of Latin America*, 1972.

international life. The index of the real wages of the Buenos Aires workers rose on the average by 50 per cent between 1914/15 and 1928/29; it was essentially stagnant in the fifteen years following the World Depression, i.e. all over the first phase of import substitution, and it increased again by 22 per cent between 1946 and 1951, as a result of wage policy measures in the wake of the Perónist take over in 1946. Concurrently with the slackening of import-substituting industrialization and/or due to the steps relating to economic policy by the consecutive governments after Perón's fall in 1955, the wage index dropped in the late fifties to the level of the twenties.[13] The extremely modest increase in real wages in the early sixties was eliminated again by the conservative economic policy of the government at the end of the sixties. The sharpening socio-political tension has easily discernible economic causes. There is probably no other country in the world, especially one which had been relatively developed, where the real wages of the workers have been stagnating for more than four decades. The political conditions of the economic growth in Argentina were adversely influenced by the fact that in Buenos Aires, i.e. in the decision-making centre, the living standard of the workers in the industry, the very backbone of economic life has not improved at all in the long run.

The increasing social proliferation of violence is demonstrated by the ever growing political role of the *Army*. By right of its traditions and history, it has been the Argentine army patterned after the Prussian model which served as an example in Latin America for the professional armed forces specialized for meeting the requirements of defense against an external enemy. Until this date the Argentine army—composed of landowner's and middle strata elements—is the most elitist armed force of Latin America. As opposed to the Peruvian army, it was ill-suited for exercising direct control as it had a sparse mass basis, and neither its background nor its preparedness rendered it fit to take over the government. The first intervention of the Army took place in 1930–1932 following the outbreak of the World Depression, whereas the second occurred in 1943–1945. In both cases intervention proved to be merely an emergency measure. However, following the anti-Perón *coup d'état* (1955–1958), and later, following the overthrow of the governments of Frondizi (1962–1963) and Illia (1966–1973) it became more and more clear that the Army was heading towards an overt take over. During the 18 years between 1955 and 1973 there had been a military regime for 12 years. The shortening intervals between interventions by the military, the lengthening of the periods of direct army government point to the fact that, similarly to other Latin American armies, the Argentine armed forces, too, strive to justify their very existence by seizing power. Contrary to other Latin American countries, however, the elitist Argentine army, which is traditionally alienated from the civilian population, is less suited for fulfilling that task. This is reflected also by the slow rate of development in Argentina and by constant disequilibrium and permanent political tension.

When enumerating the factors resulting in the lasting political disequilibrium, closely connected with growth troubles, the so-called *national-psychological shock* having a traumatic impact on broad strata of the society should also be taken into account. Within a decade, i.e. within a historically rather short period, the international and domestic image of Argentina has basically changed. Until the fifties of this century and especially in the pre-war years

---

[13] G. Germani: *Social Modernization and Economic Development*, U.N. Geneva, 1970.

Argentina had been one of the developed countries with a high income level and relatively efficient democratic institutions. For decades she was regarded to be a veritable Land of Promise by the immigrants from Europe. Moreover, public opinion in Argentina regarded the country for a long time as a rival of the United States on the Western Hemisphere. In an outstanding work of Latin American literature written at the turn of the century, Ariel, impersonating Argentina, takes a brave stand against Caliban, symbolizing the United States.

As a consequence of the grave difficulties of the fifties, Argentina became within some years an underdeveloped country grappling with the effects of a permanent crisis of growth, and governed by means of military dictatorship. As concerns the international level of development, Argentina lags behind considerably, and recently her economic potential lost ground in Latin America, too, first to Brazil and then to Mexico as well. The radically changing self-image of the nation—a change which took place within the life-time of a single generation, —an atmosphere of social malaise, and finally, the loosening of the scale of values have shattered the confidence in the future of the country.

The commencement of emigration is an unambiguous proof of an increased national-psychological crisis and of the weakening public spirit. The citizens of the country which had taken in the highest proportion of immigrants began to discover in increasing numbers their foreign origins, and became suddenly aware of the fact that actually Argentina meant very little to them, and emigrated to countries holding out the promise of better living standards and a brighter personal future. The "Wandering Argentine" has become a rather frequent phenomenon in the United States and in other countries of the American continent. The troubles deriving from economic and political imbalances have been further enhanced by the increasing social malaise, the lack of confidence, and the emigration of numerous representatives of the intelligentsia and the skilled workers.

2.3. THE NEW ECONOMIC STRATEGY IN THE EARLY SEVENTIES

Following the fifteen years of crisis after the fall of the Perón regime in 1955, and the hesitant, often antagonistic endeavours to resolve problems, there began to unfold in the early seventies a new development strategy more suited to meet the objective requirements of the situation, and holding out promises of a more permanent and consequent development trend. Chronologically, this is related to the coming into power of the more reformist wing of the Army in March 1971, followed by the restauration of Perónism. After having wielded presidential powers for two years, General Lanusse, the representative of the more reformist Army elements handed over power to the Perónist Campora who won 50 per cent of the votes at the presidential elections in March 1973. Within a quarter of a year Campora prepared and masterminded Perón's return and taking power. The composition of the votes cast in the elections which raised Perón once again to power demonstrated that the working classes, the majority of the petite bourgeoisie and the middle-class which have become weary of the lasting political chaos followed again the lead of Perón. Thus, Argentina which has not known a strong government for more than twenty-five years got a government with a firm mass basis. In spite of political disequilibrium and/or the maintenance of an unstable temporary balance, the period 1971–1974 was characterized by a kind of conceptional continuity.

The causes bringing about this turn of events were partly of a domestic, partly of a foreign political, and partly of an economic political nature, however, all these elements manifested themselves rather as interdependent factors. In the early seventies, the more enlightened strata of the Argentine ruling classes, first and foremost certain Army circles came to the conclusion that the socio-economic problems could no longer be resolved by some kind of economic and political tinkering or by the coercive methods of the past fifteen years. But large-scale reforms, as it turned out, were inconceivable without the support of the Perónists representing the most significant organized force on the political scene. Recovering from the state of stagnation therefore demanded the broadening of the political basis, the including of Perónism in the government system and in connection with this, the restauration of constitutionalism. At the same time, the gradual losing ground of Argentina both on the Latin American continent and in international life, as well as the dominance of the United States and Brazil becoming ever more conspicuous demanded also an active policy in the international sphere which could have furthered the making use of the advantages of the international division of labour for accelerating of economic growth.

The most spectacular initial sign of the gradual turn in external orientation was the Salta Declaration in July 1971 in which the military government of Argentina and the president of the people's Chile determined the basic principles of positive interstate relations (elimination of ideological restrictions, sovereignty, mutual equality, pluralist character of international relations, etc.). The declaration was also a significant step in thwarting the attempts to isolate Chile on the Latin American continent.

The first unambiguous signs of a positive change in Argentina's foreign and external economic policy became manifest already in 1971. The search for alternatives aimed at a cooperation against North American and Brazilian dominance took a more definite shape. The emphasis on cooperation with the Andean countries was also one of the steps of a new opening. Argentina is the most active and most interested member of the Latin American Free Trade Association. The stressing of the cohesive force of the Spanish cultural heritage, the insistence on the economic and foreign political similarity of interests between Spain and the Latin American countries was also a new element in international relations. The concept of Argentine foreign policy was founded on the view that Spain, especially in the post-Franco period, could play an important part in building a bridge between Europe and Latin America as well as in developing a certain counterbalance of forces against the United States.

The pragmatic character of Argentine foreign policy was characterized first and foremost by the change on the country's relations with the socialist states. The former hostile and later non-committed attitude was replaced gradually by a policy of cooperation. Establishing relations with the People's Republic of China, the People's Democratic Republic of Korea, Cuba and the German Democratic Republic, the radical reassessment of the external economic relations with the European socialist countries in 1974, and finally the economic cooperation with Cuba which was unique as regards its scope and volume—all these were integral parts of a changing external orientation. Favourable conditions were created to the exploitation of the possibilities of a long-range cooperation by agreements with the CMEA countries signed in 1974. Argentina also signed an agreement with the Soviet Union envisaging an annual trade turnover of 200 million dollars. To promote exports, the Soviet Union offered credits to the amount of 600 million dollars, Romania and Poland 100 million dollars each, while Hungary's credit offers totalled 50 million dollars. Czechoslovakia made an offer of unrestricted credit funds; 30 per cent of the funds used could be paid back by means of untraditional Argentine exports goods. Both the goods exported and the credits granted by the socialist countries were aimed mainly at the development of an independent Argentine energy basis. It was a noteworthy episode that in the course of negotiations with the Soviet Union the Argentine delegation showed interest in establishing eventually relations with the CMEA.

The quantitative and structural development of Argentine export was furthered mostly by the agreement signed with Cuba in 1973, which created a buyer's market for the Argentine engineering industry (vehicles, diesel engines, investment goods of various sorts). The agreement envisaged an Argentine export of 1.2 billion dollars for the realization of which Argentina granted a 600 million dollar credit. According to Cuban sources, already in the first year Cuba bought goods to the value of 600 million dollars in Argentina, surpassing considerably the sum envisaged. Thus, the unfolding of the Argentine–Cuban relations meant at the same time the breaking up of the embargo against Cuba, moreover after some initial wrangling and yielding to Argentine pressure, even the North American subsidiaries in Argentina joined in the Cuban business by delivering commodities. In addition to the price increase in raw materials, the enlarged relations with Cuba and other socialist countries, too, contributed to the nearly 60 per cent increase in Argentine exports in 1973.

The extent of *dependence* on *external economic* relations is much less in Argentina than in the majority of the developing countries; this is due to the development trends of the Argentine economy in the past twenty-five years. Foreign trade is responsible for an unhealthily small percentage of the national income, and the significance of capital import, too, increased only during the sixties. The relatively small extent of subordination to the United States has an especially favourable influence on the external economic and foreign political manoeuvrability of the country. Nevertheless, more than a third of capital imports came from the United States in the past years, and the North American corporations have penetrated the key industries (although European capital, viz. the Italian Fiat, Olivetti and Pirelli, the West German Hoechst, Siemens, the French Renault and the British-Dutch Shell are still predominant in the Argentine industry). Financial and technological dependence do not entail foreign trade dependence; since the new opening the share of the United States has actually diminished in Argentine foreign trade. The share of the United States in Argentine imports dropped from 24 per cent in 1970 to 21 per cent in 1974, whereas in exports it was reduced from 10 to 9 per cent.

Of all the countries of Latin America it is Argentina where the distribution of foreign trade according to geographical regions is most favourable: taking the average of 1970–1972, Italy got 18 per cent of the exports, the Netherlands 12 per cent, Spain 8 per cent, whereas the share of the Latin American countries was more than one-fifth. Thus, three-fifths of Argentine exports was directed to economically small and medium-size countries without excessive foreign political ambitions; with such nations it is easier to bring about mutual dependence as they are of a similar economic size. Efforts for maintaining traditional

Western European orientation are also considerable. The place of Britain was taken by the Common Market as a whole. After negotiations lasting for several years, Argentina concluded a three-year non-preferential trade agreement with the Common Market and, in return of concessions granted, undertook not to "disturb" the market of the EEC. Owing to the relatively evenly distributed geographical structure, the sphere where pressure can be exerted on the foreign trade of Argentina by the great powers is rather narrow. To obviate the consequences of the energy crisis and/or to avoid becoming dependent on US-controlled oil corporations, Argentina concluded a 550 million dollar agreement with Libya at the beginning of 1974, according to which Libya was to supply three million tons of crude oil annually in 1974 and 1975, whereas Argentina undertook to build an oil refinery, schools, housing developments in Libya in addition to granting her technical assistance. Economic cooperation with the socialist countries as well as other developing nations, e.g. Cuba and Libya, gives Argentina more possibilities for manoeuvring both in the spheres of foreign policy and external economy.

A decisive step of the new economic policy was the Foreign Investment Act of November 1973. Similarly to Mexican economic policy, this law restricted foreign capital investments geographically on the one hand, and prohibited foreign investments in branches regarded as vital from the point of view of national economic security (public works, insurance, banking, advertising, marketing, telecommunications as well as certain fields reserved for public enterprises). According to the stipulations of the law, all enterprises with foreign capital interests exceeding 51 per cent are considered foreign, whereas enterprises with national capital interests of 51 to 80 per cent are classified mixed enterprises. The establishment of foreign enterprises needs congressional approval, whereas for the foundation of a mixed company it is sufficient to obtain government consent. Among the criteria of approval the utilization of local natural and human resources ranks first, followed by export orientation or import substitution as well as applying technologies expedient from the point of view of the economy. Special regulations prescribe the percentage of Argentine staff in the management of such enterprises as well as the extent of credits to be granted *in loco*. Foreign ventures which pass into national ownership within ten years by gradually including local capital are given priority. By the law, the annual level of profits to be repatriated is set at, 12.5 per cent of the capital, whereas in the case of foreign exchange deposits the rate of interest is exceeding 4 per cent. Operating foreign enterprises have the choice of submitting to the new law or acquiescing in the 20–40 per cent surtaxes levied on profit repatriation. The enforcement of the law and/or the recording and control of foreign investments have been entrusted to the newly established specialized agencies of the Ministry of Economy.

The passing of the law indicated the fact that Argentina has joined the ranks of the countries controlling the movement of foreign capital, thereby strengthening the economic sovereignty of these nations. Despite of such control measures, however, the conditions for capital imports are more liberal than in the Andean countries, and it has become quite obvious that efforts are made to set the limits of economic nationalism in a way which holds out favourable conditions to foreign capital.

The bargaining positions of the Argentine economy as against foreign capital are indirectly strengthened by the establishment of a powerful syndicate of all public enterprises. CEN (Corporación de Empresas Nacionales), which has been called into life patterned after the Italian IRI, has become one of the strongest economic concentrations of Latin America, the unified management of which has undoubtedly enhanced the chances of a successful stand against foreign capital.

The first step of the world economic opening both chronologically and as regards significance was the change effected in Argentina's diplomacy in the field of *external economic relations*. The incentives of the external economic orientation, however, are of considerable interest, too. To increase the profitability of agrarian exports the export taxes on a number of products, such as wool, sunflower, meat were abolished or reduced. Contrary to other Latin American countries, however, direct state intervention had a great part in shaping export orientation. Thus, the National Corn Council, the central public purchasing authority was invested with the power to market the 1973/74 wheat, sorghum, maize and sunflower crop both within the country and abroad. By this measure the state succeeded in establishing strong export positions. Export incentives based on administrative constraint in the car industry are also of great interest. The production and import quotas of the producing enterprises are linked to their export achievements, and enterprises may increase their domestic turnover only when they have fulfilled the export quotas prescribed by the government for each prototype and plant. For instance plants which have exported 15 per cent of their 1974 production and 35 per cent of their 1975 production may increase their annual car sales by 8 per cent. In case of falling below the plan, domestic marketing activities are restricted proportionately. By this system of forced incentives the government strived to raise export quotas in the car industry gradually to an annual 200 000 units and to establish a basically export-oriented car industry by 1978. On the other hand, by ousting marginal and less efficient producers, the government wants to reduce the diversity of the car industry. Owing to its highly administrative character, this system of incentives may serve as a very instructive experiment not only for Argentina but for all other developing countries as well.

The third new economic factor of Argentine growth is the transformation of the *concepts of development strategy*, i.e. the modification of the traditional allocation of resources needed for economic growth. The changes in the developmental strategy of the new stage are clearly reflected by the medium-term plan elaborated for the period 1974–1977. Perónist strategy was very cautious to avoid the mistakes of earlier years. In the field of politics it tried to revive the one-time populist coalition, and at the same time the government held out the prospect of introducing measures for safeguarding the interests of the small and medium bourgeoisie, raising the share of the working classes in the national income from 33 per cent to 47 per cent by 1977, and even promised to take a definite stand against foreign corporations. In the implementation of the "Programme of National Reconciliation" proclaimed in the name of Perón a prominent part was assigned to the so-called Social Pact *(Pacto Social)* within the framework of which an agreement was reached by the trade union centre, the association of private contractors and the government on freezing both wages and prices. As regards economy, the development strategy broke radically with the conceptual heritage of the period 1946–1955. At that time Perón's economic policy was founded on import-substituting industrialization, whereas after his return to power he primarily intended to explore and to develop the national resources of the country, first and foremost those of agriculture, and put the main stress on export orientation. Plan targets envisaged a 7.5 per cent annual increase in gross national product, an increase of 12 per cent in investments, 9 per cent in agriculture and 20 per cent in exports. The industrialization policy of the Perón regime in 1946–1955 was of a highly capital-intensive character. However, the second development concept definitely emphasizes the development of labour-intensive branches and the maximization of economically effective employment. Of the one million new job opportunities envisaged until 1977, the share of the construction industry (the development rate of which is planned to be the double of the average) will be 30 per cent, and the share of the manufacturing industry, according to the plan, will amount to 24 per cent. Thus, the great majority of the new jobs will go to the productive sectors.

Since 85 per cent of Argentina's oil demands are covered by domestic production, the agreement concluded with Libya, and the stepped up development of the sources of energy (oil, natural gas, nuclear power stations) did not bring about growth problems. The number of cattle which has been stagnant up to that time, increased by 10 million within a single year. The modification in the structure of capital loan imports also reflects the reassessment of the development strategy. Thus, out of the 665 million dollar long-term credits settled for 1974, 300 million were allocated to the financing of the exports of capital goods and 200 million to the development of agricultural production. Earlier the bulk of foreign credits was assigned to import-substituting industrialization or to the development of the infrastructure. Parallel to the shift on emphasis both as regards structure and sectorial balance, considerable efforts are made by the economic policy to bring about coordination among the small and medium-size enterprises in order to increase the international competitiveness of the country. Thus, 100 million dollars were allotted to the financing of a syndicate of small and medium-size enterprises.

It seems too early to take a definite view on the latest phase of economic development in Argentina. The new economic strategy which has been shaped after the return of Perón undoubtedly perceived correctly the main direction to be followed in economic development. Nevertheless, the improvement of the economic situation, the recovery of growth was not only due to the new strategy but rather to exogeneous factors favourable to the realization of this strategy. It was first of all the favourable weather which was the cause of the Argentine boom: both in 1973 and 1974 crops were good. Corn exports which amounted to 5.6 million tons in 1972 rocketed to 10 million tons, and what is more, at a time characterized by an unprecedented high level of agricultural prices. Thus, the facts that exports nearly doubled, that the trade balance showed a surplus of almost 1 billion dollar in 1973, that foreign exchange reserves also rose by 1 billion dollar—all these were due first and foremost to the weather conditions and to increased world market demand. At the same time, the meat industry which is less dependent on weather factors could exploit the favourable market conditions to a much smaller extent. As a consequence of insufficient incentives and organization, meat exports decreased by 100,000 tons in 1973, although foreign exchange earnings increased by 18 per cent, owing to world price increases.[14]

At the same time, the examination of the factors of growth proves the fact that the new development strategy has not resulted in any essential changes in

---

[14] *QER*, 1973, Annual Supplement.

the domestic economy, and actually has not eased the problems deriving from the heritage of protectionism and political disequilibrium. The internal economy kept on to persist in its expectant attitude.

The sudden disappearance of the favourable growth symptoms of 1973/74 was essentially due to exogeneous factors. The short-lived relative political stability was brought to a close by the death of President Perón on July 1, 1974, following which the permanent Argentine power crisis was deepened by an extremely serious government crisis. Mme Perón (Isabelita) who assumed the presidency after the death of her husband, was able to rely from the very beginning only on a very unstable coalition of right-wing perónists, middle class radicals and the armed forces; thus her political basis was extremely narrow. In the course of the fight for succession and interpretation of Perón's ideological heritage the Perónist party broke up into two factions and, at the same time, the Trade Union Federation (CGT) gradually became estranged from the presidential power basis; the development of a government crisis was accelerated and deepened also by certain subjective factors.

Almost simultaneously with the political crisis there manifested themselves the consequences of the 1974 world economic recession. The increasingly export-oriented development policy was paralyzed almost in the stage of upswing by the meat restrictions of the Common Market and by the ever deteriorating chances of the Argentine industrial commodities the competitiveness of which had never been strong. While the surplus of the trade balance amounted to 1 billion dollars in 1973 and to 435 million in 1974, it showed a deficit of 1.1 billion dollars in 1975. As a result of the deteriorating economic and political situation the exodus of capital was stepped up, and it was due to this fact that whereas in 1973 the balance of payments had been active by 921 million dollars, it had a deficit of 51 million dollars in 1974 and of 1.4 billion in 1975. Foreign exchange reserves dwindled from 2,026 million dollars in the middle of 1974 to a mere 620 million dollars. Owing to the fact that Argentina was less and less in the position to pay off the loans and thus indebtedness increased rapidly, her debts abroad ran to 10 billion dollars at the end of 1975.

The disintegration of the basis of the government, the paralyzation of the executive power and the effects of the world economic recession made themselves felt simultaneously, and this brought about a series of inconsistent and overhasty measures in the economic policy resulting soon in an acute crisis. In the crisis situation which has developed in the period following Perón's death the economic leadership of Argentina dared not carry through the overdue structural reforms and rehabilitation programmes envisaged. On the contrary, it tried to blunt the edge of the consequences of political imbalance by the deficit financing of the budget and by fictitious employment. Between 1972 and 1975 the number of those employed in public administration rose from 283,000 to 407,000, in the public sector of the oil industry from 14,000 to 50,000, and at the same time the subsidies of the increasingly unserviceable national railways amounted to 2 million dollars a day. In 1975 public revenue covered only one-fourth of budgetary expenditures, and state debts equalled one-sixth of the annual gross national product. The rate of inflation reached the all-time high of 335 per cent in 1975, investments dropped by 7 per cent, whereas, even according to official data, the gross national product decreased by 2 per cent. Political tension and the extent of the increasing incompetence of the state organs are reflected by the proliferation of political violence:

during the 21 months of the presidency of Mme Perón 1,700 fell victim to political terrorism.

The take over of the armed forces in March 1976 put an end to the rapidly sharpening government crisis. Up to this time the military government headed by General Videla has shaped its political line rather cautiously. Of the elements of Perón's development concept the military government upheld that of agricultural development and also the continuation of the world economic opening, however, the new rulers wish to lay less stress on economic nationalism and etatism.

The present and even the medium-range tendencies of the world economy are undoubtedly favourable to the efforts aiming at the restoration of the international significance of Argentine agriculture. The medium-range changes on the world markets of meat and wheat, i.e. the two main export goods of the country are improving. At the same time, the economic effects of the so-called Pacto Social have not made themselves felt since 1974.

Presently, the most serious problem of the Argentine economy (and this will not change in the near future) originates from the political sphere. The majority of the population has become weary of the problems deriving from the political vacuum. At the same time, with the death of Perón the basic safeguard for the preservation of the policy of "national reconciliation" has disappeared for at least a couple of years. The question whether to have a bourgeois democratic development or a shift to the right has remained unanswered for the time being. Thus, the contradictions of the Argentine growth model may come to the fore at any time as political tensions grow in the country.

CHAPTER 3

THE BRAZILIAN WAY OF ACCELERATING GROWTH

3.1. SCOPE AND PRINCIPAL METHODS OF BRAZIL'S DEVELOPMENT STRATEGY

An extremist group of generals seized power in Brazil more than a decade ago. This military take-over in 1964, however, proved to be not merely one of the numerous *coup d'états* the world has seen in Latin America. Ever since, the new regime has been attempting to direct the socio-economic development of the country to a course advantageous to the interests of the Brazilian big capital. This it tried to achieve—in conformity with the emerging US global strategy of the sixties—by applying political violence and coercive measures on a scale almost unprecedented in the 20th century, and also by a maximum intensification of the exploitation of the working masses.

The development of Brazil in the past ten years has been in the limelight of international interest for several reasons. The Army which had seized power started the stepped up process of eliminating economic backwardness with capitalist methods, making use thereby to an extremely great extent of the repressive organs. At the same time, the stress in the economic growth process was shifted from import substitution to the joining in the international division of labour. As a result of accelerated growth, the significance of Brazil in the world economy has grown. Presently, she ranks first in the group of developing countries both as regards economic potential and industrialization. Moreover, Brazil with a population exceeding 100 million and a gross domestic product of 80 billion dollars has become the 10th on the list of the world's leading producers. This undoubtedly indicates a spectacular increase in the rate of growth, and recently the international press has duly coined the phrase of "Brazilian economic miracle". However, the capitalist formula of accelerated growth brought about a serious set-back in the socio-political progress, as well as the sharpening of socio-political tensions. As a matter of fact, there does exist a Brazilian miracle other than that cited: why couldn't the government—in spite of abundant natural resources and advantageous external economic conditions—find less brutal, more human solutions for overcoming traditional backwardness? Let us now examine the origins of the so-called Brazilian course of development and its peculiar features which may be regarded as typical of the model.

The circumstances of the military coup are well known. The favourable conditions of the development of progressive forces by the Goulart-regime of a pluralistic character were abolished by the Army representing the most powerful groupings of Brazilian big capital. Within this politically extremely centralized model, the Army leaders are invested with all decisions pertaining to power functions, whereas the fields of economic management and public administration have been handed over by the professional politicians—contrary to the recent Peruvian model—to the stratum of technocrats. The effects of the so-called restrictive political conditions were diminished from the point of view of top decisions by the pushing into the background and even temporary elimination of the political opposition. The ceasing of the fluctuations of the political cycle, i.e. the stabilization of the Army's power made it possible to enforce the ideas of big capital concerning development strategy. Though after the first years following the take over characterized by an undisguised reign of terror, the military government would have preferred to rely on a broader political basis, even this rather modest political opening served only the interests of a more active policy of big capital, whereas the restrictive measures controlling the political activity of the broad masses were not eased at all. Thus, the "Brazilian course of development" is exorbitantly growth-oriented, and the rapid updating of the economy was not followed by any progress in the socio-political sphere.

The military regime regarded as its main development strategic tasks the revival of the process of economic growth which had slowed down and partly even come to a stop in the early sixties, the modernization of the country and/or the elimination of economic underdevelopment. However, in the course of the economy oriented development of the past ten years a rather more disguised objective—not at all derived from economic targets—became of vital importance: the establishment of Brazil's great power position as soon as possible. As a result of a peculiar intertwinement of the traditional values and objectives of the Army wielding power and of the ambitions of the Brazilian big capital, an unprecedented extent of energy was concentrated (as compared to other developing countries) on establishing a sub-centre of power in Latin America with a view of raising it to the status of an international force of the first rank in the future.

Owing to its class background, objectives, and its appraisal of the international strategic situation, initially, the military regime did not want to challenge the positions of the United States either in the international field or on the Latin American continent. Moreover, by strengthening the political, economic and technological relations with the US, as well as by assuming the role of a voluntary gendarme on the continent, Brazil tries to obtain the political favour of the dominant power in the region and rely on it in realizing its own "quasi imperialist" aspirations.

The Brazilian regime strives to gain the status of a regional and later of an international power first and foremost by means of economic and external economic expansion. In the new Brazilian strategy acceleration of economic development has been given absolute priority over the elimination of dependence on foreign powers. In spite of this, economic development is subordinated to such an extent to long-range power aspirations, which is without precedent (perhaps only China might be cited as an exception).

A statement by R. Campos, regarded as "the father of the Brazilian economic miracle", casts light on the political ambitions of the military regime: "The United States seems to relax gradually its hold on Latin America, whereas we are ready and willing to shoulder this responsibility".[1] At the time of the pre-election campaign in Uruguay in 1971, it was this power policy which induced the Brazilian leaders to organize naval and land manoeuvres and to threaten Uruguay with military invasion in case of an election victory of the Broad Front. Similar motives were underlying the menacing attitudes towards the People's Front government of Chile and the progressive Peruvian regime.

It would be an incorrect and biassed approach to trace back the Brazilian

[1] *The Financial Times*, January 14, 1972.

strategy to internal, "autonomous" factors of the Brazilian "course" of development. Though the military take over is closely related to the change of development periods and to the growth crisis rooted in the Brazilian evolution, the specific course of development taking shape coincides with the modification of the global strategy of the United States and its new policy towards the countries of the Third World. As a result of the changes in the international balance of power following the collapse of the classical colonial system, the United States has been less and less in the position to fill the vacuum everywhere with the same intensity. As the consequence of the failures in the early sixties and later of the miscarriage of her policy in Vietnam, the US has striven to reduce her international commitments, to shape some kind of a "growth pole" conception, and to develop the countries of pivotal interest as model states. In compliance with this strategy America backs the endeavours of these countries to bring about accelerated growth and also their expansive aspirations. Thus, the stepped up growth rate of Brazil fits in with the new strategic guidelines pursued by the United States in Indonesia, South Korea, Iran, Spain and other countries. By supporting the present-day great power ambitions of Brazil, the United States attempts to employ Brazil as her strategic Trojan horse in the Latin American region.

It is rather instructive to examine what sources and methods the Brazilian military regime used in order to avoid the menacing growth crisis, to carry through the changeover from one period of growth to the next one, and to step up radically the rate of economic development.

The origins of "the Brazilian economic miracle" are traced back by superficial or biassed researchers generally to the radical change of view as regards foreign capital. At the time of import-substituting growth, during the three decades prior to the military take over, and partly due to the historic emergency situations in the pre-war, war and post-war years, and partly to the policies of the so-called populist governments (Vargas, Kubitschek, Quadros, Gulart) representing energetically the interests of the petite bourgeoisie and the middle class, the role of foreign capital had been rather restricted. During these three decades the part of foreign capital in the economic growth of Brazil had been much more modest than at any previous periods of Brazilian economic history or in any other country of Latin America.

After the military take over Brazil opened wide doors to foreign capital, and tried to channel international capital exports towards Brazil by tax exemptions, and by granting favourable repatriation conditions. The liberalization of capital imports resulted in a mass influx of foreign capital, first of all in the capital- and technology-intensive key branches considered of strategic importance. As compared to the average of the previous ten years, the annual average capital import nearly trebled in the period 1965–1972. At the same time direct North American capital investments rose from 1.1 billion to 2.5 billion dollars, and within that investments in the manufacturing industry from 723 million to 1,743 million dollars, while the volume of foreign debts increased from 3.5 billion to 10 billion dollars[2]. The dependence on capital imports will be analysed later on, but now it will suffice to state that as regards the sources of accelerated growth, capital import —despite its rapid increase— represented but 7 per cent of domestic investments on the average of the period

---

[2] *Survey of Current Business*, September 1967, September 1973; *QER* 1973, Annual Supplement.

1964–1971.³ Thus, the sources of the new stage of economic development have been primarily of domestic origin.

The part of the repressive apparatus, reliance on administrative coercive methods ranks first among these sources. The Brazilian trade unions had been greatly manipulated under the previous governments, too, and in the thirties and forties had been granted even "corporation" status; however, from the late fifties on they assumed an increasingly active part in directing the economic as well as political movements of the working classes. The new military regime put an end to the trade union movement and integrated it into its own government mechanism. A number of the most active trade union leaders had been physically liquidated, whereupon, the workers deprived of their unions, became exposed to the price and wage regulations of the government.

Under the circumstances of the galloping inflation, the real wages of the workers were easily cut down by the wage policy of the government which was further enhanced by the effects of stepped up internal migration. The level of minimum real wages dropped by 35 per cent between 1964 and 1970, stagnated between 1970 and 1972, and showed some slight increase only since 1973.⁴ The literature on Brazilian economic growth prefers not to mention the fact, though the above data prove unambiguously, that the oppressive political system shifted the burden of accelerated growth on the working classes.

While the development policy regarding the labour force was founded on the adoption of *coercive measures*, the policy concerning capital operated mainly with *economic incentives* aimed at the increase of domestic savings and the stepping up of accumulation, though certain administrative elements were discernible in this field, too. Thus, as early as 1964/65 the military regime enforced a tax reform as an initial step. The list of tax-payers was enlarged, the tax control system was computerized and severe reprisals were made against defrauders. As a consequence, the formerly—even by Latin American standards—extremely low "tax-morale" improved rapidly, and tax revenue more than doubled within five years. Incentives were held out to private interprises to initiate economy measures. Full tax exemption or partial tax allowances in case of investments on the stock-market or in backward regions of the country, as well as monetary corrections neutralizing the effects of inflation proved effective additional stimulants to capital, the investment inclination of which has increased immensely as a result of the political consolidation and controlled labour market. In the period 1960–1964 investments had made up on the average 17 per cent of the national product, whereas the average for 1971–1973 amounted to 21 per cent,⁵ i.e. the increase was 150 per cent for ten years. The above data unambiguously point at the sources of the "economic miracle".

Let us now examine the methods and preferences of the economic policy which have furthered the unfolding of the process of accelerated growth.

Two chronological stages can be distinguished within the change-over process of economic growth. The four-year period 1964–1967 was characterized by the winding up of the crisis situation which had taken shape in the early sixties, as well as by the creating of favourable conditions to economic expansion. The process of high-speed expansion of the external economic sector was

---

³ *IDB Economic and Social Progress in Latin America.* Annual Report, 1972.
⁴ *The Financial Times,* November 13, 1973. Survey.
⁵ *U.N. Statistical Yearbook, QER,* 1973.

launched in 1968 and has been continued uninterruptedly ever since. It might be illuminating to approach the methods of this process from the angle of political power.

Although the economic role of the state has tended to increase since the thirties, it would be mistaken to speak about some kind of clear-cut trend of interventionism. Interventionist measures of the economic policy were mostly improvised and politically conditioned, and as a rule had but short-term objectives in mind. The individual units of the expanding public sector had a considerable autonomy as regards decisions, there was plenty of elbowroom for independent and counterbalancing measures, while budget discipline was extremely lax. In 1964 a "general staff" was called into being for the coordinated direction of the highly independent units of the public sector; R. Campos, the well-known advocate of the combination of the idea of free enterprise and up-to-date management was appointed head of this Ministry of Planning and Cooperation (MINIPLAN). The newly established authority began to coordinate the plans of the units of the public sector, i.e. of the autonomous economic development agencies, to harmonize them with national economic, regional and sectoral programmes, and finally to work out a long-range strategy for the national economy. The first published plan of the MINIPLAN covered the period 1970–1974, and envisaged an 8–10 per cent rate of growth, and within that set the target for a 10–12 per cent increase of industrial output and a 7–8 per cent rate of agricultural development. The plan also prescribed the transformation of the structure of investments and provided for a diversification programme. It envisaged the diminishing of the annual rate of inflation to 10 per cent by 1974, the increase of the rate of investments to 18 per cent, a three per cent increase in job opportunities and the raising of the value of industrial exports to the level of one billion dollars.

In addition to the operative, i.e. tactical development plan a number of strategical plans for economic development were elaborated; the so-called medium-term plan covered the seventies, whereas the long-term alternative covered the whole period until 2000. The military government did not deem it necessary to publish these plans. Nevertheless, it is well known that the global objective was to reach first the present production potential and later the development level of Britain and France in the eighties and the middle of the nineties. As regards the institutional framework of economic development, let us but mention the reorganization of banking and the establishment of a national stock-market. Out of the numerous institutional factors of economic life the part of the banks in the realization of the developmental strategy has considerably increased.

In the first stage the new regime regarded the reduction of the ever growing internal and external imbalances as the most important task. The rate of inflation, an indicator of the extent of internal disequilibrium, had been 52 per cent on the average of the three years preceding the military take over, whereas in the very years of the *coup d'état* it amounted to 89 per cent. At the same time the balance of payments showed a deficit of $912 million, and the rate of export coverages of imports was 90 per cent.[6] The putting an end to internal disequilibrium, a constant feature of Brazilian history, and the curbing of the excessive rate of inflation seemed an absolute necessity from the point of view of external economic expansion since inflation had a restrictive

[6] *Ibid.*

effect on both capital imports and the trade balance. The changing of an economic and financial atmosphere rooted in historic traditions has naturally proved to be not a result to be achieved overnight. Moreover Latin American economists hold different opinions on the evaluation of inflation. Therefore, Brazilian economic policy has endeavoured only to check the rate of inflation.

The reduction of the increase in the individual cost factors was brought about both by certain economic and political mechanisms. The defencelessness of labour made it possible to adopt subsequent wage corrections which as a rule lagged far behind the rate of inflation. This system prevented the increase in the real wage factors of manpower on the one hand, on the other hand it led to a systematic reduction of real wage, and consequently, to a decrease in purchasing power. Maximization of interest rates resulted in diminishing credit costs, whereas the restrictive budgetary policy of the government in the very first year succeeded in cutting down the deficit formerly amounting to five per cent of the gross national product to one-tenth of it. (In 1973 the budget showed some surplus—the first time for 20 years.)[7] The economy was suddenly shaken by the devaluations which became necessary in the wake of the accelerated inflationary process, and this led to the rise in import costs by leaps and bounds, as well as to dubious speculative ventures which gave further impetus to inflation. Following the lead of other Latin American countries, Brazil too, adopted the method of periodic mini-devaluations (crawling peg). Exchange rate modifications carried out generally every six weeks resulted in reasonable foreign exchange costs, brought about improved balances and succeeded in practically eradicating the black market as well as fluctuation symptoms originating from the foreign exchange economy.

Brazil was one of the first states which —contrary to the expertise of international organizations—adopted monetary corrections in order to reduce distortions caused by inflation. According to the movement of the wholesale price index indicating the rate of inflation, the level of wages, securities, annuities, credits, bank deposits, etc. are occasionally adjusted, and—in addition to the fixing of the ratios of the bearing of inflationary burdens—this also diminishes the psychological effects of inflation. And last but not least, the restoring of economic confidence and the stepped up rate of growth rendered it possible to make better use of the formerly underutilized capacities and/or to increase the volume of mass-produced goods which, in turn led to reduced fixed capital costs and diminished inflationary pressure.

As a resultant of the interventions and/or processes mentioned above, the rate of inflation calmed down, and the average annual rate of increase in wholesale costs dropped to 38 per cent as compared to the average of the period 1965/1967, to 21 per cent compared to the average of 1968–1972, and by 1973 the costs decreased to 16 per cent.[8] It is noteworthy that the reduction of the rate of inflation ensued at a time when world economic inflation took shape and expanded to external economic relations at an ever increasing degree, and when Brazil joined vigorously in the international division of labour becoming thereby increasingly dependent on the fluctuations of the world economy. The gap between Brazilian inflation and world economic inflation became narrower, and inflation could not hamper the unfolding of Brazilian external economic expansion.

[7] *QER*, 1974/1.
[8] *U.N. Statistical Yearbook*, *QER*, 1973.

The reduction of the adverse external balance was started parallelly in several spheres and sections. The policy of capital liberalization and enhanced capital imports of the first stage went hand in hand with a rigorous protectionist import policy. Following the enforcement of these new measures, the protection of the Brazilian economy against external competition has increased during the first years. Between 1965 and 1970 the government succeeded in arresting the process of growing foreign trade deficit which had increased in the period 1960–1964. However, this was not due mainly to increased exports but rather to the restriction of import growth and to the establishing of a healthy trade balance.

Export orientation unfolded in great dimensions only after the laying the foundations for such a policy during the first three years. A considerable part of the direct development resources of the state as well as of imported capital (more than one-third, according to estimates) was channelled by the government to existing branches which were thought to be easily transformed into export industries. The changeover to export orientation was introduced by a number of measures regarding the allocation of funds. These measures aimed partly at expanding the capacities of comparatively advantageous raw materials and agrarian produce (iron ore, sugar, soya beans), and partly at increasing the competitiveness of existing, mainly up-to-date industries (vehicles, means of transport, machine tools, general engineering, etc.) the effectiveness of which, however, was low as a result of the protectionist atmosphere. From the early seventies, a definite development policy got under way. This policy aimed at the establishment of factories with a markedly export-oriented economy based first and foremost on utilizing processed local raw materials. Thus, the main emphasis of industrial expansion has been shifted gradually from the industries producing for internal markets to ones which make use of domestic resources.

Let us now examine the most important means of launching the policy of export orientation. Although the newly introduced policy on the rate of exchange has eliminated speculative fluctuations, it has not fully done away with export hampering influences inherent in such a policy. Between 1964 and 1970 the price level of Brazilian export commodities rose by 392 per cent, while the exchange rate of the cruzeiro increased by 277 per cent. Taking into account the industrial price indices it is evident that the exchange rates resulted in a 16 per cent disincentive to industrial exports on the average of 1967–1969, and in 1970 in a 11 per cent disincentive.[9]

Thus, while the policy of exchange rate had a major part primarily in improving the economic climate, it had no stimulating effect in the field of industrial exports. Incentives were provided mainly by means of a budgetary policy. Such means were principally the exemption of exported industrial goods from production taxes (IPI) and turnover taxes (ICM), the declaring duty-free of items needed for the production of such commodities, and the income tax redemptions granted to exporting firms. According to a survey in 1971, the total of all these different subventions amounted to 36 per cent of the domestic market prices. The orders of magnitude of subsidies within the individual branches of industry are naturally different. The export of cars, bicycles and motor-bicycles, instruments, chemicals were backed by 45 per cent, those of textiles by 41 per cent budgetary subsidies, whereas the export

[9] *QER*, 1974/1.

of shoes received a 43 per cent, that of foodstuffs a 25 per cent, of steelwares a 35 per cent, of agricultural machines a 31 per cent, of rolling stock a 28 per cent government subsidy.[10]

In addition to budgetary subsidies export credits have gained considerably in significance. In the case of consumers' durables a 7–9 per cent interest one-year credit is granted with a maximum of 85 per cent of the value, whereas in the case of capital goods a maximum eight-year credit is made available. Besides the high rate of inflation and/or real interest level, the policy of export financing contains also some inherent subvention elements. Export credits are granted almost exclusively for stimulating Brazilian industrial exports directed to developing countries (first of all to Argentina). In 1970 6 per cent of the Brazilian industrial exports applied for credits amounting to the value of 18 million dollars (whereas export credits granted in 1968 totalled only 2 million dollars). In spite of the rather modest orders of magnitude the trend of increasing export credits is conspicuous.[11]

The simplified administration of export transactions has been one of the noteworthy incentives of the new economic policy. At the time of import substitution, the government attempted to restrict the flight of capital by demanding a detailed supply of data and specifying the endorsement of agreements by a great number of authorities before granting export licences. From 1964 on, 90 per cent of data requiring and endorsement prescriptions have been abolished, and the so-called bureaucratic barrier has vanished almost entirely.

Such a high-level system of incentives is naturally an inflationary factor in itself leading to imbalances under the circumstances of increased joining in the international division of labour. It should not be left out of consideration, however, that the percentage of the costs of protectionist-type development had been at least as high as that: in 1960 the Brazilian economy was isolated from the world economy by a 29 per cent average customs level and an intricate system of foreign exchange and administrative protectional measures. Within that, the effective protection of the industry—which was much more intensive than the average—amounted sometimes to several hundred per cent.[12] On the other hand, a considerable part of the subsidized industries have been capital intensive branches operating on high fixed costs, which reacted very sensitively to the cost of production. Thus, owing to additional exports, increased mass production resulted in a reduced specific cost level and in the reduction of the relative price level of products also on the domestic markets. The 36 per cent subsidy of export orientation therefore does not entail the social input of the same magnitude, and due to indirect effects, the concentration of incentives on exports demands less social costs. Incidentally, this is also borne out convincingly by the decreasing trend of the rate of inflation in the years of highest expansion.

Besides the transformation of the system of incentives, domestic accumulation, the directing function shouldered by the government as regards capital and technology imports, and the channelling of credits and direct capital investments to export-oriented industries have been the major factors in

---

[10] H. Hesse: "Export Promotion of Manufactured Articles." *Weltwirtschaftliches Archiv*, 1973/2. Kiel.
[11] *QER*, 1974/1.
[12] B. Kádár: *op. cit.*

recent years. Especially noteworthy is the activity aimed at the acceleration of the technological development of the industry as well as the creation of an independent technological basis. In the initial years of export orientation licenses purchased on the international technological market and/or agreements concluded with foreign capital were used chiefly for the updating of technologies. These technologies were applied in the course of processing export products, especially local raw materials and food. Brazil became one of the foremost technology importers of the world. According to estimates she spent an annual average of one billion dollars on patents and licenses as well as on the use of skilled labour by the beginning of the seventies.[13]

Owing to balance of payments considerations, the creation of an independent technical basis—which at the same time made it possible to import technologies —has been strongly emphasized in recent years. In 1973/74 the government spent 715 million dollars on the elaboration of specialized and up-to-date Brazilian technologies and on training Brazilian engineers, scientists and managers. In the pioneering industries of the private capital, such as space research, the utilization of nuclear energy, aeronautics, oil chemistry and electronics the technological development programmes were aided by budgetary credits and tax exemptions. According to plans, a considerable part of industrial exports will originate from the results of domestic technological and scientific basis by the end of the seventies.

The unfolding of public export developing activities has been of outstanding significance. State trading organizations intended to further the development of the export sector by technological aid, credits, vocational training facilities and diverse services. The support given by the state to further the penetration of external markets (fairs, exhibitions, organization of industrial fairs, etc.) as well as the participation of Brazilian diplomatic missions in export campaigns attracted considerable international attention. The settlement of relations with the Common Market, the diminishing of the disadvantages originating from Great Britain's joining the Common Market, the increasing initiation of Japan into the realization of the Brazilian development concepts, the fitting of the neighbouring Latin American states as well as of the countries of Africa and the Middle East into the global strategic framework of Brazilian raw material economy and market expansion—all these have become the tasks of Brazilian diplomacy. The Brazilian offensive aimed at the penetration of foreign markets is regarded as a useful pattern for other developing countries by a number of international organizations.

Due to political and economic reasons, the problems of regional tensions within the country became of outstanding importance in the system of economic equilibrium. During the first phase of industrialization, the backward northern and north-eastern regions of the country served almost as quasi-colonies and buyers' markets of the more advanced southern and central regions. However, in the phase of industrial maturity the underdevelopment of these regions has become a significant economic obstacle to expanding domestic markets, whereas in the international field it is highly detrimental to Brazilian efforts aimed at increasing national prestige. Backwardness is one of the main sources of the growing socio-political tension within the country. Thus, the acceleration of integration within the country became unavoidable both for the economy and for strengthening national unity. Besides, the exploration of territories

[13] *Világgazdaság*, September 20, 1973.

The Brazilian Way of Accelerating Growth

unexploited until that time held out promises of opening up new sources of growth.

In addition to the adoption of indirect means of economic policy (tax redemptions, credits), in handling private capital, the state made considerable investments in order to develop agriculture and industry in the backward regions. At the same time large-scale infrastructural investment projects have been realized with the aim to draw the backward and formerly unexplored Amazonas-basin into the economic circulation of the country. (Between 1964 and 1974 for instance, the mileage of surfaced roads increased from 20,000 kilometres to 62,000 kilometres.) Though the infrastructure development program regarded by many as a dubious prestige investment threw mainly political dividends, it undoubtedly had a part in the political consolidation of the military regime, and owing to the new, up-to-date highroads the direct physical presence of the Brazilian great power became clear to all neighbouring Latin American countries. At the same time, the extension of the economic frontiers played an economic role, too, in maintaining economic expansion and in the broadening of the raw material and energy basis.

3.2. THE ACHIEVEMENTS AND DRAWBACKS OF BRAZILIAN ECONOMIC GROWTH

It is a well-known fact that Brazil has emerged as one of the most rapidly growing economies in the past decade. Her rate of expansion can be compared only to that of Japan or to the dynamic development of some of the oil producing countries. Contrary to the development of the country in the period 1940–1960, economic growth made headway not only in the internal economy but also in the sphere of external economic relations. In addition to simple quantitative changes it brought about stepped-up development in an increasingly widening field of activities as well. The dinamics in the main fields of the economy is illustrated by Table 26.

The three-year periods clearly illustrate the gradual acceleration of the growth process. The rate of growth in the years 1971–1973 was three times

Table 26

The most important indices of the dynamics of the Brazilian economy (rate of annual increase, per cent)

	1962–1964	1965–1967	1968–1970	1971–1973
National product	3.1	4.2	8.8	10.8
Industrial production	1.6	3.3	13.2	13.5
Agricultural production	0.2	4.8	3.8	5.0
Exports	0.0	7.5	11.8	23.0
Imports	0.0	0.3	19.7	25.9
	(annual average, million dollars)			
Exports	1,351	1,664	2,310	4,298
Imports	1,407	1,419	2,413	4,841

Source: U. N. *Statistical Yearbook;* various issues of *FAO Production Yearbook; QER,* 1973. Annaul Supplement; *QER* 1974/1.

that of the initial three years, while the gross national product increased from an annual 23.4 billion dollars in 1964 to 36 billion in 1970, and to 50 billion in 1973. It is demonstrated by the data of the table that the main representatives of growth were industrialization and the joining in the international division of labour. Industry, especially the pioneering branches in technology were responsible for more than two-fifths of the gross national product. The joining in the international division of labour, the dependence on foreign trade and the intensity of the growth process are reflected by the fact that on the average of the period under review a unit increase of national product was followed by a two-unit-increase in exports even under the conditions of a high rate of economic expansion.

The increased efficiency of investments also reflects the measure of improved effectiveness in the national economy coupled with accelerated growth. While the increase per unit of the gross national product demanded a 4.2 per cent rise of investments on the average of 1960–1965, the national product increase per unit could be achieved by two units investments, despite concentration on branches with a higher capital intensity in the period 1970–1973. This means that, as a result of a more rational allocation of resources and the introduction of a new system of incentives, the effectiveness of investments has more than doubled.

The increased dynamics of the economy was accompanied by accelerated structural changes. The share of the manufacturing industry in the gross national product amounted to a mere 19 per cent in 1960, whereas in 1972 it made up 36 per cent. The industrial potential of Brazil took on highly significant dimensions in the wake of dynamic industrial development.

The 730,000 unit production of the car industry in 1973, the 13 million ton cement production, the 7 million ton steel production, and the 400,000 ton output of the shipbuilding industry are outstanding results even on an international scale.[14]

About 80 per cent of the rise in industrial production originated from increased productivity. Since the policy of industrialization gave preference to the development of technologically-based up-to-date branches of industry (metallurgy, electrical engineering, the vehicle producing industry and chemical engineering became the pioneering industries), the share of the heavy industries grew to 63 per cent within the manufacturing industry. The structural indices of the national economy, and primarily those of the industry, do not coincide with the assumptions relating to the correlations of the degree of economic growth and structural development, but reflect other factors rooted in the dimensions of the country and/or its specific development. The structural indices of Brazil, a country with a *per capita* gross national product of $500, did not resemble in the slightest degree the situation in the backward countries, in the peripheral economies of certain European capitalist economies or in the developed small capitalist countries, but increasingly show a similar picture to the conditions of the macro-economic structure of the leading capitalist nations. As a result of stepped-up development, Brazil has outgrown structurally the framework of underdevelopment and has become an *industrial-agrarian* country.

Taking the average of the period under review, the rate of agricultural development has lagged behind due to mainly external factors. The decrease in

[14] *U.N. Monthly Bulletin of Statistics*, November, 1973, NFA 1974.

coffee production between 1966 and 1972 was partly an after-effect of earlier measures taken on the basis of the economic policy to cut coffee production, and partly the consequence of unfavourable weather conditions and of the spread of the coffee-rot. The north-western parts of the country experienced an almost unprecedented drought at the beginning of the seventies. Under given conditions, even the rather modest growth rates of production ought to be regarded as relative achievements and as symbols of a diminishing dependence on the forces of nature.

The developments in the production structure coincide with the changes in the process of concentration. In the era of import-substituting industrialization the decisive part of production was supplied by small and medium enterprises, though owing to the scale of domestic markets the average size of enterprises (60 employees) exceeded the Latin American average. The penetration and gaining ground of the multinational corporations, and industrialization based on the cooperation of public capital, Brazilian big capital and the multinationals led to the gradual diminishing of the influence of the small and medium strata of the bourgeoisie and to an increase in the concentration of production. The rapid unfolding of this process could be observed first of all in the technologically most up-to-date industries. In the years of the preparation of the upswing, i.e. in the period of the qualitative development of the operating production basis, the advance of the concentration process had been rather modest, whereas accelerated concentration coincided with the establishing of new export capacities and new structures in recent years. It is a characteristic symptom that in 1970 already a handful of big enterprises concentrated the bulk of the capital in the individual branches of industry: the share of the biggest five enterprises in the textile industry amounted to 55 per cent, that of the biggest eight ones in the car industry to 90 per cent, of the biggest seven ones in engineering to 72 per cent, of the biggest seven ones in electrical engineering to 61 per cent, and of the biggest six ones in the food industry to 59 per cent.[15]

Dynamism and structural development, however, affected first and foremost the sphere of economy, while the regional imbalances and social inequalities within the country remained by and large unchanged. The fundamental objective of development strategy and industrialization was the acceleration of growth, which was coupled with the reduction of the backwardness in the international field. Thus, in almost all of the technologically advanced branches the overwhelming part of growth resources was devoted to the establishment of plants which on the whole could keep abreast with international technological development. Though the Brazilian objectives included the reduction of *regional* inequalities by means of creating new centres of growth and adopting various incentives, due to established preferential approaches in industrialization, the overwhelming majority of the technologically up-to-date industries settled in the southern part of the country which had a much more developed environment.

The share of the state of São Paulo in the industrial output of Brazil rose from 46 per cent in 1960 to 52 per cent in 1972, and in spite of the large-scale immigration of labour, *per capita* national product increased to $940.[16]

---

[15] *Nachrichten für Aussenhandel* May 23, 1973.
[16] *The Financial Times*, April 19, 1973.

Polarization in the *social* sphere exceeded even the extent of regional inequalities. The forced modernization programme, the concentration of economic activity on the industries and exports of the southern regions of the country have increased the incomes of groups living under more developed socio-economic conditions and enjoying, therefore, automatically higher incomes. On the other hand, the programme of compulsory accumulation, the further restrictions by monetary corrections and measures of price policy imposed upon the consumption of low income strata, which are unable to accumulate any savings, led to a process of income differentiation almost unparalleled in the international economic life.

Between 1967 and 1971 the share of the categories representing the highest wages, i.e. 16 per cent of the total labour force, increased from 33 to 39 per cent, whereas the gap between the lowest-category wages and the highest stratum of wage-earners (representing each 10 per cent of the total labour force) increased from 6 to 10 within four years.[17] By international standards, Brazil has one of the most differentiated wage systems.

Though the great number and political motivation of the unverifiable statistics on Brazilian income concentration warrant the highest possible caution by adopting necessary source-criticism, it might be stated that 5 per cent of the population belonging to the highest income category monopolized 27 per cent of the national income in 1960, and 37 per cent in 1970, whereas the share of 50 per cent of the population decreased from 18 to 14 per cent.

The intensity of the rate of increase of income concentration is unparalleled. The difference between the average *per capita* national income of those belonging to the highest and lowest income brackets respectively, increased from 3,200 to 6,300 per cent in the period under review. In 1970 *per capita* national product amounted to $6,500 in the case of almost one million Brazilian citizens, whereas in the case of nearly 50 million people it ran to a mere $100. Thus, the highest stratum, representing but one per cent of the population, maintained a living standard surpassing by far that of the US average; four per cent of the population reached the income level of West Europe, while the living standards of half of the country's population equalled those of the most backward African and Asian nations.

The overall income ratios, too, reflect the fact that the results of the rapid development are enjoyed by only a narrow stratum of the population, and at the same time the overwhelming majority of the population has to content itself with living on the peripheries of the economy. The opponents of the military regime, thus the famous Brazilian economist, C. Furtado[18] among others, point out that the Brazilian economic expansion is bound to fail because of basic social problems: it is impossible to enlarge the domestic market of consumers' articles owing to the demand-restricting effects of income concentration. Even General Medici, the former president of the country was obliged to admit: „In Brazil economy is on the upgrade, the living standards of the people on the downgrade".[19]

The most critical factor of Brazilian growth, which is accomplished under the circumstances of continuous disequilibrium, is undoubtedly the increasing inequality in the distribution of incomes. This process is regarded by the

---

[17] *The Economist*, December 15, 1973.
[18] C. Furtado: "Le Modèle Brésilien". *Revue Tiers-Monde* July–September, 1973.
[19] *IDB Economic and Social Progress in Latin America*, Annual Report, 1972.

economic ideologists of the Brazilian model as an inherent feature of a model optimalized to serve the acceleration of updating the economy. At the same time, the assertions concerning the effects of internal market restrictions cannot be substantiated sufficiently. On the one hand the 3.5 million cars on the roads, the 8 million television sets, the 7 million radio sets cannot be regarded as a magnitude reflecting a narrow market or the point of saturation, on the other hand the rate of increase in the output of products or product-groups consumed by lower income strata (e.g textiles, bicycles, cassava, sugar, bananas, rice) is not at all low according to international standards.

Following the succesful first phase of the export offensive, the rate of export increase, its effect on the overall business cycle in the second half of the seventies is going to slacken somewhat as compared to preceding years, whereas the growth-stimulating role of the expansion of the domestic market will increase in the second part of this decade. In 1972/73 there appeared the first modest signs that the purchasing power of wages were no more reduced by monetary corrections, the compulsory accumulation became less rigorous than before, and developmental burdens shifted upon the working class were eased to a certain degree. According to observers, increasing and differentiated domestic supply had a considerable part in the mobilization of the labour zeal of the Brazilian population. In the centres of growth it ceased to be a factor of enhancing political tension and became a means of gradually neutralizing the resistance against the political power channelling the energies of the masses into the sphere of economy.

Relying to a growing extent on *foreign capital* may entail even more serious consequences than social tensions originating from income concentration. Foreign capital had no essential part as a source of accelerated growth. However, in the wake of accelerated growth the rate of capital imports accelerated immensely in the early seventies: it amounted to $1.5 billion in 1971, $3.5 billion in 1972, and $4 billion in 1973. On the average of the seventies, annual capital imports made up 6 per cent of the national product and 30 per cent of gross investments[20]—a conspicuous order of magnitude even by international standards.

The significance of capital imports going beyond quantitative considerations manifested itself first of all in the taking possession of the key positions of industry. Thus, according to a survey in 1970, 8.5 per cent of all the capital functioning in the national economy, 31 per cent of industrial capital, 100 per cent of the motor car and rubber industries, 86 per cent of the pharmaceutical industry, 50 per cent of the aluminium and chemical industries, 35 per cent of the food industry, and about one-third of the export of industrial products[21] was controlled by foreign capital. This comprehensive control of strategically important branches represents the most imminent danger for Brazil. Any measure aiming at the extension of national economic sovereignty would result in the partial paralyzation of the key industries and thus of the development model as a whole.

Another factor of uncertainty is the high rate of indebtedness. Due to the stepped-up process of modernization, the import of equipment and technologies demanded the raising of substantial credits.

Foreign exchange indebtedness increased between the end of 1963 and

[20] *The Economist*, December 15, 1973.
[21] *Ibid.*

1973 from 3.5 billion dollars to 14 billion dollars. The level of foreign exchange reserves increased from 219 million[22] dollars to 6.4 billion dollars,[23] i.e. it grew by 6.2 billion dollars, thus diminishing the actual extent of indebtedness. Though running into debts is a deliberately planned process and, in addition to financing modernization, it is significantly linked to the process of expanding redemption and export capacities, nevertheless, the burden of debt services is still heavy. Public debt service burdens amounted to 31.5 per cent of export earnings on the average of 1960–1964, 27 per cent of 1965–1969, 17 per cent of 1970–1971, and 16 per cent of 1972–1973. It is true that parallel with increasing export capacities the level of debt service burdens reflects a downgrade trend, however, in case of some kind of depression or an international political and/or economic crisis of confidence in relation to Brazil, the yearly one billion debt service could become an extremely serious strain. The Brazilian policy of capital imports means an incessant race between the extension of export capacities on the one hand, and increased foreign exchange burdens on the other. In case of the slightest political and economic shock, the current successful policy might change over into a complete failure.

The most spectacular results of development manifest themselves in the field of foreign trade, more precisely in the sphere of export expansion. Within five years Brazil succeeded in trebling the volume of her exports and utilizing her joining in the international division of labour as a source of development— a unique achievement as compared to experiences of the world economy. In 1964 6 per cent, in 1973 12 per cent of the national product was realized by exports. Although the extent of the joining in the international division of labour is still extremely low as compared to the average of the developing countries, export orientation has furthered the less expensive enlargement of up-to-date industries requiring extensive markets.

In the period under review almost 30 per cent of the increment of Brazil's industrial production has been exported. Exports as well as import capacity developed in the wake of increased capital imports transformed Brazil into one of the most enticing buyers' markets among the developing countries. Her import potential ranking 16th on the international scale became an important means for furthering the expansive objects of Brazilian foreign trade policy, whereas by making use of the policy of import tie-up she skilfully exploited the market organizations of the developed countries.

The structural characteristics of Brazil's joining in the international division of labour are of no less interest. Related to the low effectiveness of protectionist industrialization, no significant exports of industrial goods could be realized despite the considerable industrial potential, and thus the foreign exchange *earning* sector was dominated by agriculture, mainly by coffee production. The developed industrial basis and the monoculture of coffee in exports represented one of the most characteristic contradictions of Brazil. Due to measures of economic policy aimed at the liquidation of this monoculture and to poor coffee harvests in succession, only by using up accumulated stocks could the Brazilian government fill up the export stocks. Though as a result of rising coffee prices the volume of coffee export has considerably increased in recent years, only 26 per cent of foreign currency earnings from exports were due to the coffee sector. At the same time, the reduction of stocks accumu-

---

[22] *U.N. Statistical Yearbook, QER*, 1973.
[23] *Survey of Current Business*, September 1967, September, 1973; *QER*, 1973. Annual Supplement.

lated in previous years set free significant monetary means, thus drying up one of the sources of inflation. Agriculture has proved to be by far not a lagging-behind sector in the field of exports, and the range of traditional export goods (coffee, cacao, wood and sugar) has been supplemented by products of a new, dynamic structure (soya beans, meat). As against the 86 per cent of 1964, the agrarian sector furnished 60 per cent of all export earnings.

The underlying cause of the relatively diminishing role of the agrarian sector as a foreign exchange earner has been the increasing export orientation of the extracting and first of all of the manufacturing industries. The raw material balance of Brazil used to show unambiguously a deficit, but owing to the fact that the raw material exports (iron ore, manganese) reached the level of 400 million dollars, the trade balance of the extracting branches took a turn for the better, and the export coverage increased from one-third to 50 per cent. Even more radical changes were experienced in the manufacturing industries. Following a three-year period of preparation, in 1967, i.e. on the threshold of the policy of export orientation, the volume of the export of industrial commodities had amounted to a mere 164 million dollars, whereas the share of industrial goods in total Brazilian exports had been but 10 per cent. Within six years, i.e. by 1974 industrial exports rocketed to the value of 3 billion dollars and made up one-third of total exports. Thus, the industrial sector has become the main driving force of exports supplying 50 per cent of the incremental exports.

The structure of industrial exports illustrates convincingly the vertical and horizontal advance of the process of modernization. Contrary to the example of other countries, Brazil's joining in the international industrial division of labour is founded only to a minor degree on comparative advantages originating from the wage-system.

Taking the average of the developing countries, nearly four-fifths of the industrial exports are made up of light-industrial goods, and the share of machinery in total exports is but 10 per cent. As a result of concentration on the industries of pivotal importance from the point of view of technology, the share of machinery in Brazilian industrial exports increased to 27 per cent as early as 1970. Thus, growing competitiveness makes itself felt not only in the light industries which have been pushed into the background in countries of high wage standards, but also in the branches where demand is more elastic in the international markets and in industries which are in the vanguard of technological progress. In addition to cars, electrical machines, machine-tools and chemicals, a whole range of new export goods have appeared which require highly developed technologies and organizational backgrounds as well as adequate financial resources, such as jet planes, oil and ore transporting vessels, up-to-date electronic equipment and motor vehicles.

The export of heavy industrial goods naturally reflects the consequences of the overall dependence on capital imports as Brazil has become a favourite production basis of the multinational corporations because of her broad buyer's market, low wage standards and political stability. Brazilian industrial exports therefore often take the part of industrial subsidiary plants of the parent companies or become covert means of penetration into other countries of unfavourable political and economic climate. It is impossible to mark off statistically the industrial exports of the multinationals, but according to estimates about one-third of the total exports of industrial goods is controlled by multinational corporations.

The structure of foreign trade indicates the fact that the most important trends of development policy are reflected by the sphere of external economy, too. As a result of structural development which has proved effective even related to international standards, the degree of industrial dependence has decreased, and the balance of the foreign trade relations of industry has shown a markedly improving tendency. While the ratio of imported industrial goods covered by industrial exports amounted to 12 per cent in 1964, this percentage improved as far as 45 per cent by 1973, thus approaching the average West European "late-comer" capitalist countries. It is by far not unfounded to assume that industry, too, will become a foreign exchange earning sector by the end of the seventies.

The analysis of the growth performance of the Brazilian economy in the past decade indicates the fact that up to this date the new military regime has realized the economic development targets which it has set for itself with a kind of operational precision adopted in well-trained armies. The country has become the scene of a spectacularly accelerated growth, modernization and structural development considerable even compared to international economic development, and thereby has highly bettered its world economic positions. The price for these growth achievements, however, has been intensified socio-economic inequality and increased dependence on foreign capital, first of all on the big multinational corporations.

3.3. THE IMPACT OF WORLD ECONOMIC CHANGES ON THE ECONOMIC GROWTH OF BRAZIL

The price-explosion of 1973, the recession which has unfolded in the developed capitalist countries in 1974/75, and/or the world economic changeover from one stage to another in the seventies—all these factors have naturally brought about new and difficult environmental conditions for all the countries depending heavily on the international division of labour, among them for Brazil, too. Being an oil importing country, the rise in oil prices alone has burdened Brazil's balance of payments with an additional two billion dollar foreign exchange expenditure. The recession taking shape in the industrialized capitalist countries put a break on the rate of growth of industrial exports and export-oriented industrialization respectively. In connection with the uncertain world economic situation and the growing long-term investment requirements of the world economy, there decreased the long-term capital supply financing structural transformation and modernization. Thus, Brazil's efforts for expanding her import capacity could rely less on external sources.

The unfavourable changes in the world economic environment have started a process of diminishing rate of growth in Brazil, too. The rate of growth of the gross national product was 11.3 per cent in 1973, 10 per cent in 1974, and 4.5 per cent in 1975. On the one hand, the global retardation of growth affected Brazil with a one-year's time-lag, on the other hand, the rate of growth was quite spectacular even in 1975, i.e. the rock-bottom of the world recession. The belated appearance of the recession was undoubtedly due to the fact that an extraordinarily fast growth rate of seven years is slackening at a much slower rate in a vast country with a varied economic potential than in smaller countries. Despite the rapid progress of the external economic sector only one-tenth of the national income was realized by exports and, therefore, the sphere of economy in direct contact with unfavourable world economic effects was from the very beginning much narrower. In spite of the relatively smaller weight of foreign trade within the national economy, the dynamics of growth in a so-called export-oriented growth model eventually depends on the trends of the world economy. Unfavourable world market tendencies therefore make themselves felt initially in the pioneering sectors of growth, but later on, are to be experienced in almost all the other sectors as well. Though a 4.5 per cent rate of growth should be considered undoubtedly an achievement— especially under the circumstances of a recession in the world economy—the dropping of the growth rate by 50 per cent necessarily has a serious shock-effect and may cause uncertainty and imbalances. This certainly reduces both domestic and foreign investments.

The dynamic development in 1974 was due mainly to the outstanding harvest, whereas the growth rate of the industry which is more susceptible to the world economic trends has dropped already from the earlier annual average of 14 per cent to 9 per cent. In the course of the further reduction of the growth rate in 1975 the rate of growth in the Brazilian industry dropped to 7 per cent which was mainly due to the sagging demand on the external markets. In the motor vehicle industry which has been the chief carrier of growth for two decades and reacts more sensitively to the market atmosphere, the volume of production diminished by 7 per cent, and steel production essentially came to a standstill. Expansion was achieved only by industries which made it possible to widen the structural bottle-necks, viz. power generation (38 per cent), the cement industry (11 per cent), the tractor industry (17 per cent), and oil refineries (12 per cent).

Naturally, the development of the external economic sector was the one affected to the greatest extent by the unfavourable changes in world economy. In 1974 the favourable sugar prices and rapidly expanding commodity stocks as well as the marketing of soya beans still increased foreign exchange incomes deriving from the selling of traditional export goods. At the same time, owing to the export-oriented industrialization, which had made great progress in the previous years, the export of industrial goods rose by more than 40 per cent, and the level of total exports exceeded that of the previous year by nearly 30 per cent, whereas in 1975 the influence of exports as the main driving force could not be felt anymore. The value of exports rose by 13 per cent to some 9 billion dollars, i.e. the increase in the volume of exports—taking into account the price inflation in world economy—has been negligible. For the first time in the past decade the rate of real growth of production surpassed that of exports, and thus the share of exports diminished.

The slackening of the growth of exports was accompanied by a sudden increase of imports. As a result of the realization of orders placed prior to March 1974, the rise in oil prices, and to no small amount due to the increase in the export of investment goods of the developed countries, Brazilian imports in 1974 more than doubled and rose to 14 billion dollars. Owing to the restrictions in Brazilian imports, the value of imported goods decreased somewhat in 1975, and amounted to 12.7 billion dollars. Thus, the deficit of the trade balance following the all-time high of 6 billion dollars in 1974, was again quite considerable in 1975, reaching the sum of 3.7 billion dollars. The unfavourable effects of the immense deficit of the trade balance both in volume and percentage was further aggravated by the adverse budget of services which amounted to 2.5 billion dollars in 1974, and 3.3 billion dollars in 1975. Thus the current items of the balance of payments showed a deficit of about 8.5 billion dollars in 1974, and 7.2 billion dollars in 1975.[18]

The task of balancing this immense deficit unprecedented not only in the history of Brazil but in the developing world as a whole fell to the share of capital imports. In the seventies the increasing joining in the international division of labour within the framework of the policy of "outward-looking" economic growth has been furthered by capital imports. Brazil has become one of the most important capital importers of the world. The import of loan capital totalled 6.5 billion dollars in 1974, and 4 billion dollars in 1975, whereas the so-called direct long-term capital import amounted to 900 and 830 million dollars, respectively. However, even this huge amount of capital imports did not suffice to balance the deficit of current items: the deficit of the balance of payments amounted to about 1.1 billion dollars in 1974, and to nearly 2.1 billion dollars in 1975.

In order to compensate the deficit of her balance of payments Brazil had to make use to a growing extent of her foreign exchange reserves accumulated in previous years. The value of the country's gold and foreign exchange reserves was 6.4 billion dollars at the end of 1973, 5.25 billion at the end of 1974, and according to estimates about 3.2 billion by the end of 1975. (Due to the deteriorating foreign exchange situation, the competent authorities have not published any information relating to foreign debts, debt services and foreign exchange reserves since April 1975.) The radical transformation of international payments positions is rather conspicuous: growth relies to an ever increasing extent on indebtedness. Foreign debts totalled 12.5 billion dollars at the end of 1973, 17.6 billion at the end of 1974, and about 22 billion at the end of 1975. Out of this nearly 60 per cent have been obligations payable to foreign private enterprises. As a result of the extremely high rate of indebtedness, average debt service burdens in 1974/75 came to about 2 billion dollars and consumed almost a quarter of the total export earnings. Owing to the international optimism concerning the long-term trends of Brazilian development, the increasing disequilibrium in 1975 has undoubtedly not resulted in a considerable reservedness on the international capital and credit markets, nevertheless the scope of foreign exchange manoeuvrability of Brazil in the field of external economy has narrowed down considerably in the past year. Brazilian economic growth founded to a considerable degree on capital imports has entered a critical stage during the past years, but until the end of 1973 the extent of foreign indebtedness hardly surpassed the order of magnitude of the foreign exchange reserves. Presently, debts amount to a sum seven times as high as foreign exchange reserves, and annual export capacities make up but 50 per cent of the debts. Thus, the situation of export capacities is much worse than the quantitative standards of international credit-worthiness. Foreign exchange reserves were sufficient only to cover the value of three-months' imports. This forces the government to adopt unambiguously a policy of reorientation as regards the country's external economic and economic growth objectives. The consequences of the price explosion of 1973 and the recession respectively, had to be paid for by the disequilibrium in the external economy, and subsequently by the reshaping of the development policy.

External economic imbalance and/or rising import prices have given a further impetus to the deepening of domestic disequilibrium. The economic growth of Brazil has traditionally taken the road of an inflationary economic policy. Taking the average of two decades, the rate of inflation has been around 35 per cent. Owing to the first results of the accelerated growth in the early seventies, major efforts were made in the economic policy to curb the process of inflation, and by 1973 the rise in the price index amounted only to an annual 15 per cent. As a consequence of new developments in the world economy, the rate of inflation again rose to 34 per cent in 1974, and it could be reduced to 26 per cent in 1975 only by means of stringent monetary restrictions and measures of credit policy. Though as a result of the rapid expansion of global inflation the scope of the imbalance of the Brazilian economy was not conspicuous as compared to international developments in the capitalist world economy, the realization of the developmental targets set at an earlier date by the Brazilian government has been hampered by the world economic environment.

Let us examine how the objectives and centres of interest of the Brazilian development plan have changed as a consequence of the recession.

Brazil achieved the fastest rate of growth of her economic history during the period of her 1st Development Plan (1970–1974). The 2nd Development Plan comprising the period 1975–1979 has been worked out at the time of accelerated growth. This plan aimed at the structural modernization of Brazil and the preparing of her regional as well as long-term role as a world power. Reckoning with a total investment of about 160 billion dollars, the plan envisaged an average annual 10 per cent increase in the gross national product, a 12 per cent rise in industrial production and a 7 per cent expansion of agricultural production. As a result of this, the gross national product of Brazil will have to rise from an annual 80 billion dollars in 1974 to 130 billion by the end of 1979, and the *per capita* gross national product from 700 to 1100 dollars (at 1974 prices). (See Table 27.)

The plan, extremely ambitious even if compared to other late-coming industrialized countries, was aimed at the diminishing of the structural disproportions of the economy. Thus, in industrial development the significance of the relatively slowly developing sector turning out investment goods, as well as the need to develop qualitatively the production of semi-finished goods was strongly emphasized. The steel, aluminium, petrochemical and pharmaceutical industries were also declared top-priority industries. The development targets of agriculture, which had been stepped-up considerably, too, were directed at the elimination of the sectoral imbalances between the industry and agriculture, the creation of an adequate export commodity fund as well as the ensuring of job opportunities. To curb the process of accelerated influx of population to the cities, a result of industrialization, the government emphasized the importance of creating new agricultural job opportunities.

The plan was launched at a time when the impact of the developments of 1973/74 reached the economic life of Brazil creating new external economic conditions for the growth process and the development plan. On the other hand, social tension grew as a consequence of these developments, and the constraining conditions made themselves felt in the political sphere, too. Thus certain shifts in the priorities of development strategy became necessary. Since the possibility of relying on the international division of labour as a source of growth has been radically reduced, the intensification of the internal division of labour was necessarily highlighted. Contrary to the experiences of the past ten years, the role of import substitution in economic development was stressed again as a result of foreign exchange restrictions. As much as 40 per cent of investments were channelled to import-substituting industries in 1975. The possibilities of import substitution, however, are significant not in agriculture or in the light industries but first and foremost in the spheres of so-called

Table 27

The main targets of the 1975–1979 development plan

Product	Unit of measure	1974	1979	Rate of growth per cent
Crop of grain	1,000 t	30,000	45,000	50
Oil seeds	1,000 t	10,000	16,000	60
Steel		7,500	43,500	580
Aluminium	1,000 t	120	190	58
Fertilizers	1,000 t	585	1,200	105
Iron ore	1,000 t	60,000	138,000	130
Generation of current (built-in capacity)	1,000 kW	17,600	28,000	59
Capacities of oil refineries	1,000 barrel/day	1,020	1,650	62
Tractors	piece	44,000	84,000	91
Ships	t	410,000	1,140,000	178
Rolling stock	t	122,000	214,000	75
Cement	1,000 t	17,100	26,190	53
Sulphuric acid	1,000 t	986	3,388	244
Wood-pulp	1,000 t	1,547	2,860	85
Exports	$ mill.	8,000	24,200	302
of this Industrial exports	$ mill.	3,200	14,800	462

Source: The Brazilian development plan for 1975–1979.

intermediate products and investment goods. The technological and organizational standards achieved by the Brazilian industry render it possible to extend the production basis in the category of investment goods as well, whereas the extensive scope of the domestic market is instrumental in establishing enterprise sizes which may promote the exploitation of advantages deriving from mass production. Thus, efforts aimed at the reduction of foreign exchange restrictions and structural imbalances of the economy fitted the policy of accelerated import-substituting industrialization.

The change in the priorities of sectoral development and/or the increased emphasis on import substitution, however, demanded some peculiar shifts in the income distribution and regional policies as well. One of the major requirements of export-oriented development policy was the maximum utilization of advantages originating from the extremely low domestic wage-level. The background to the expansion of Brazilian agrarian products turned out by cheap labour and to the achievements of the industry in the seventies was provided by the extreme inequalities of the Brazilian system of income distribution (conspicuous even within the capitalist world), the low wage standard, and finally by the forceful upholding of labour market stability. Under the circumstances of this unprecedented inequality in income distribution, it is natural that an economic policy presupposing the dynamic expansion of domestic markets cannot be imagined. The growing political tension, market strategy problems as well as the necessity to grant priority to the domestic market compelled the Brazilian government to pay more attention in its development policy to the improvement of income distribution. As a first step, minimum wages were raised by 40 per cent in 1975, and during the plan period the purchasing power of the lowest income strata was considerably increased so as to improve sales conditions on the domestic market.

The improvement of the conditions of income distribution necessitated certain modifications as regards the regional centres of gravity of development, too. The policy of export-orientation strengthened the relation with the world economy, and also established strong links among the southern and south-eastern regions of the country which had the most developed productive forces and industries. The light-industrial products and durable consumers' articles (e.g. cars, household appliances) which could be sold on external markets were turned out by the São Paulo–Rio de Janeiro–Belo Horizonte triangle, and at the same time it was this region which absorbed the main body of imported commodities, capital, and technology. The change in the priorities of development policy which has become necessary owing to external economic restrictions re-directs the greater part of market relations of the more advanced regions to the domestic market. This, however, requires the extension of the absorbing capacity of the more backward regions. The accelerated rate of agricultural development, the continuity of the extensive methods of agricultural development, the stepped-up rate of exploration of the country's interior—all these are integral elements of this process. The most backward north-eastern regions attain greater growth energies than ever by agricultural development projects on the one hand, and as recepients of newly founded basic industries on the other. Projects directly furthering the integration of the individual regions are to receive 15 per cent of the total investment allocations.

Besides the reshaping of the direct economic objectives, the increased reliance on the domestic market as well as the alleviation of socio-political tensions jeopardizing the security of the regime demanded an increase of social investments. Without the improvement of education, public health, housing, social insurance the shift in the economic priorities would prove by and large ineffective. The shifts in the priorities of development in 1975 have shown that within the last years of the present plan period the breakdown of investment expenditures will be as follows: development of economic infrastructure 25 per cent, so-called social integration 22 per cent, investments aiming at the development of human resources (education, public health, vocational training, improvement of nutrition, etc.) 15 per cent, regional integration 10 per cent, urban development 7 per cent, basic industries 15 per cent, and agriculture 6 per cent.

The high percentage within development expenditures of infrastructural, welfare and human investments not directly connected with the expansion of production may result in the reduction of the growth rate of production, and may make it impossible to fulfil the original 10 per cent growth rate target. On the other hand, a decrease in the growth rate of production could be a challenge to the realization of the planned increase of accumulation capacity, and may even jeopardize the objectives related to the improvement of the living standard.

The development of the sector of external economy too, raises some further problems. According to present-day Brazilian concepts, the orientation on the internal market, the diminishing comparative wage advantages put a break only on the increase of industrial exports the further expansion of which would have been restricted anyway by the limited capacity of the foreign buyers' markets. Thus, the bulk of Brazil's export commodity stocks would consist in the future too, of raw materials and agrarian products (soya beans, sugar, meat, etc.) and in these fields Brazilian competitiveness is founded on so-called natural advantages and not on any comparative wage advantages. The upsurge of demand on the world markets in 1976, the considerable shortage of foreign exchange, and especially the probable narrowing down of capital imports may again frustrate the hopes for the success of an import-substituting model, press for a policy of forced exports, and may hamper all efforts aimed at a modernization of income distribution. The shortage of capital prevents the government from taking strict measures of economic policy against the multinational corporations controlling 155 out of the 500 biggest industrial enterprises.

3.4. RECENT TRENDS IN THE EXTERNAL ECONOMIC AND FOREIGN POLICY ORIENTATION OF BRAZIL

At the time of accelerated growth, i.e. in the period 1968–1974, the foreign policy of Brazil served the aims of the "outward-looking" external economic policy and was aimed at the establishment or rather consolidation of the US-Brazilian axis. Cooperation with North American big business interests seemed the most convenient means of building up the organizational and external financial background needed for the structural transformation and technological development of the Brazilian economy. Brazil strived to consolidate her economic and political leading position in Latin America by a far-reaching identification with US policy in the sphere of international affairs. As an aspiring power sub-centre Brazil wanted to guarantee the "security" of the region for the US.

As a result of a new system of interrelated interests which came into being due to the rapid strengthening of Brazil's world economic positions and the emergence of up-to-date, competitive sectors in the Brazilian economy, and owing to the improved international bargaining position of the developing countries, the external and foreign policy of Brazil reached a crossroad and had to make a decision of both economic and political character. The further strengthening of the US-Brazilian axis would have gradually enabled Brazil to join "the Club of the Wealthy", i.e. the group of industrially highly developed countries, and also would have permitted her to improve her positions in the international division of labour as an "outworker" economy. However, certain powerful circles which were trying hard to increase the influence of Brazil as an international power realized that the maintenance of US dependence would jeopardize the consolidation of the great power status, first of all because of the resistance of the Latin American countries who might be joined by other developing nations as well. The reassessment of foreign policy was accelerated by the decline of US prestige in the international field, especially by the fiasco in Vietnam. In foreign trade, the new US trade act, as well as the adverse trade balance for Brazil, brought about a weakening in US-Brazilian relations. The pressure exerted by market problems of the recently established industries have brought new elements into the complex of external economic orientation. In addition to its efforts on the domestic markets which expanded somewhat slowly owing to the slackening rate of growth, the Brazilian heavy industry, which has gained strength in recent years, did its utmost to explore foreign buyers' markets. However, the buyers of the Brazilian products of the heavy industries, which have appeared but recently on the international markets are the neighbouring Latin American countries, African states from the opposite shore of the Atlantic, as well as other developing nations. Thus, the expansion of the Brazilian heavy industry on foreign markets is dependent on the strengthening of solidarity with the countries of the Third World. The pressure exerted by the export interests of the heavy industry is enhanced by the fact that the exporters of traditional Brazilian tropical products, too, hope for an improvement of their market positions as a result of the realization of the new world economic order.

The latest steps of Brazilian foreign policy and external economic policy reflect the dilemmas of external orientation resulting from the new interrelationship of interests. During his visit in 1976, Kissinger hailed Brazil as "a future great power" and made an unequivocal attempt to draw the country into the system of world political control as interpreted by the United States. From the point of view of Brazil's more and more independent foreign policy it is undoubtedly of great importance that the new international role of Brazil deriving from her increased power has been recognized by the United States. Of all the developing countries it was Brazil with whom the United States signed the first agreement on bilateral consultations; such agreements have been concluded up to that date only with leading industrial powers. At the same time, Brazil rejected the offer of taking responsibility for upholding the international balance of power, and emphasized her full scope to act. The loosening of the former special relationship with the United States was considerably promoted by the mitigation of the so-called geographical economic dependence. Taking the average of the sixties, more than 30 per cent of Brazilian foreign trade was realized with the United States. On the average of 1973/74, the share of the United States in Brazilian exports amounted to 20 per cent, in imports to 26 per cent. As regards foreign investments in Brazil, the share of the United States came to more than 40 per cent in the sixties, whereas on the average of 1973/74 it dropped to about 30 per cent.

Undoubtedly, the most spectacular and effective measure of the policy aimed at the diversification of external economic and foreign policy has been the concluding of a West German-Brazilian agreement on nuclear energy. In spite of the strong pressure exerted by the United States, Brazil is going to develop her nuclear industry (eight atomic power stations with a total capacity of 2.6 million kW) through placing orders with West German firms to the value of 4.5 billion dollars, and thus will enter the nuclear age in the eighties. The share of West Germany in foreign capital investments in Brazil amounted to 12 per cent already in 1974, and within that the share in the vehicle industry, the pioneer of economic growth, was 43 per cent, in metallurgy 26 per cent, and in general engineering 15 per cent. The international links of industries in the forefront of technological progress and industrialization are connecting Brazil increasingly with West Germany. Japan is similarly gaining rapidly in influence on the external economic connections of Brazil. The events of 1973/74 have enhanced Japanese efforts to develop a long-term raw material and food basis in Brazil with the help of the ethnic group of Japanese living in that country. Taking the average of 1973/74, the share of Japan in Brazilian foreign trade came to 7 per cent, whereas in foreign investments to 10 per cent. Increasing orientation on Europe is reflected by the agreements on consultation concluded with Italy, Great Britain and France as well as by President Geisel's itinerary for his visit to Europe in 1976.

The strenghthening of the relations with the developing countries is accomplished in three main spheres. Trade with other Latin American countries constitutes presently one-tenth of Brazilian foreign trade, its significance, however, in the export of industrial commodities and raw material import is much higher. The intensification of regional relations got a new impetus from recent measures of foreign trade policy. A close cooperation has been established with oil producing countries. A five-year agreement was signed with Saudi Arabia in May 1975 on cooperation in the fields of technology, economy and transport. The Brazilian sugar, meat and wood-pulp industries are to be developed by Saudi Arabian and Iranian investments, and in return Brazil will take an increasing part in the food supply of the oil producing countries. The Arab inhabitants of Brazil numbering about one million are going to take an active part in enticing oil capital: already in 1975 nearly one-fifth of direct capital imports came from Arab sources.

The stressing of the solidarity with the African countries has been mainly of foreign political significance. Brazilian foreign minister Silveira pointed out in his answer to Kissinger's statements: "The cooperation of Brazil with the United States is facilitated by the fact that both are belonging to the Western world and the American continent. However, the identity of the two nations is not confined to these common features. Brazil also regards herself as a member of the developing countries and especially as one of the community of the nations of Latin America. Moreover, Brazil has specific links with the nations of Africa, since Africa had been the cradle and breeding ground of a significant part of Brazilian culture. Brazil declares her solidarity with the demand of these countries to establish a more equitable international economic order, at the same time, however, she supports the view that such a new order should be established by mutual agreement and not by confrontation". (*Neue Zürcher Zeitung*, March 4, 1976.)

Brazil has been the first country among all the political allies of the United States which characteristically advocated an independent policy of cooperation as regards the problem of Angola. The improvement of the foreign relations with the African states provides favourable opportunities primarily to the expansion of the Brazilian industry on foreign markets.

The Brazilian policy labelled "responsible pragmatism" is also borne out by the efforts aimed at the improvement of foreign relations with the socialist countries. The first economic results of the newly established relations with the People's Republic of China made themselves felt already in 1976. The further development of trade contacts with the Soviet Union is facilitated first and foremost by the Brazilian leather and shoe industries which have lost considerable sections of the US market. The visit of Romanian president Ceauşescu (the first socialist head of state to visit Brazil) attracted great attention. A five-year trade protocol signed with Poland has stipulated an average annual mutual exchange of commodities to the value of 640 million dollars; this could considerably ease the burdens of Brazilian trade balance.

It is fairly unambiguously reflected by the above facts that the direction of the external economic and foreign policy orientation of Brazil has undergone a certain modification during the past two years. The policy aiming at the termination of the unilateral dependence on the United States seems to be an accomplished fact, but there are quite a few open questions deriving from this situation. Present-day Brazilian foreign policy settled on a compromise solution: on the one hand it regards the country as a member of the Western world in the broadest sense of the term, on the other hand, it emphasizes the strong links which bind Brazil to the developing countries.

3.5. THE CHARACTER OF THE ECONOMIC GROWTH PROCESS IN BRAZIL

It is an international truism to call Brazil "the new Japan". The growth processes of Japan and Brazil are frequently compared in international surveys. According to a favourite formula of Roberto Campos, former minister for planning in Brazil who is regarded the father of the Brazilian "economic miracle", the level of economic development and foreign trade potential of Brazil in 1972 corresponded to that of Japan in 1960; to make up for the twelve-years time-lag and to achieve the present Japanese standard, however, Brazil would not need twelve years. The general acceptance and fast spread of the views concerning the neo-Japanese nature of Brazilian development necessitate a brief comparison of the character of the growth process in these two countries.

The stressing of the neo-Japanese features is based on certain quantitative similarities of the growth process. The number of population of Japan in 1960, the value of the gross national product and, consequently, *per capita* national product, the low wage standard, the structural indices, the volumes of both exports and imports, the unheard-of dynamism of economic growth, the phantastic speed at which Japan succeeded in joining the international division of labour—all these facts undoubtedly show an almost absolute similarity to Brazil of the early seventies. In addition, apparent similarities are complemented by the rapid unfolding of direct Japanese-Brazilian relations, the accelerated pace of Japan's economic penetration of Brazil, the settlement of big Japanese monopolies in the country as well as the strong economic position of the Japanese minority numbering about one million.

Such far-reaching similarities seem sufficient for most of the observers to regard the character of the growth process in the two countries as identical. It should be added that the similarity of development of the two countries is corroborated by the dualistic nature of economic and technological development and development policies, termed "the policy of walking on both feet" as well as the forms of state intervention.

As a result of historical development, in both countries there are many symptoms of lasting coexistence of backwardness and development, despite the rapid growth. Japanese big industry conquering international markets has developed under the circumstances of close cooperation with peasant handicraft industry, whereas in Brazil similar phenomena were brought about by the development gap between the European-level southern and the northeastern regions which were backward even from the point of view of Latin American standards. In the case of "late-comers" such developmental dualism and pluralism are by far not rare phenomena: they resulted in a strong combination of extensive and intensive growth elements both in Japan and Brazil as regards regional, sectorial distribution, and can be observed even in the case of different pioneering sectors of the economy. Carrying through the policy of "walking on both feet" requires more sophisticated and complex economic mechanism the coordination of which presupposes consolidated political power. One of the characteristic features of present-day Brazilian development policy—recalling the situation of the one-time Japan—is the optimization of intensive growth in the more advanced regions and sectors, and at the same time, the development of the inland areas, the north-eastern regions by extensive methods.

Despite the similarities in growth and the methods of development policies, it would be an over-simplification to regard Brazilian economic growth as a neo-Japanese variant. Identification may be refuted by arguments of equal validity: Brazil is twenty times bigger than Japan and has natural resources by far outstripping those of Japan, thus guaranteeing adequate reserves for extensive development; as a consequence, she is rich in food and raw materials, has a broad potential energy basis, whereas viewpoints of environmental protection and costs of development are negligible. The historical background of the two countries is rather different, too. Japan avoided the stages of slave society and *laissez faire* capitalism, and right after feudalism rooted in the Asian mode of production state monopoly capitalism came into existence. The exogeneous growth factors are of less significance; their effects fit into the historically developed structures, and despite her important position in world economy, the symptoms of introversion characteristic of the Oriental societies are very accentuated until this very date. Self-sufficiency, protectionism, the negligible role of capital imports, a homogeneous scale of values originating from historical characteristics, attitudes and norms of behaviour, organizational forms, the higher level of discipline and technical skills of the labour force —all these are specific Japanese traits radically differing from the historical development of Brazil.

Due to the peculiarities of Portuguese colonialism, of all backward countries it is Brazil in the history of which the period of *laissez faire* enterprise is most protracted, where exogeneous factors have had a vital part for a long historical period. The development of the scarcely populated country has been started by the joining in the international division of labour and capital imports. In Brazil—disregarding the three decades of "inward-looking" development— world economic orientation has been more or less always present in the more than four and a half centuries of historical development. Market relations such as labour force, commodities and capital are of greater importance than in Japan or other Hispano-American countries. Slave import initiated with the aim to solve manpower shortage and later the large-scale immigration of Mediterranean, West, Central and Eastern European elements have resulted in very strong ethnical mixing as well as a heterogeneous scale of values, attitudes and cultural standards. The process of modernization in Japan was able to rely on the power of traditional factors—a rare phenomenon in world history. Development policy in Brazil, however, has had no such possibility, and could ease growth restrictions—due to the lower level of discipline, technical skill and heterogeneity of the population—only by extraordinarily strong economic incentives or by adopting means of coercion.

Thus, to talk about a neo-Japanese-type growth process in Brazil would seem exaggerated. Another assumption setting out from the basis of institutionalism would also sound attractive. According to such a theory, a "belated convergence" of modern structures founded on the models of Spanish and Portuguese colonialism is presently going on in Brazil, and this would be also part of the integration of the Latin American region. Brazil which had developed to a great extent within the framework of the *laissez faire* model at the time of colonial rule and had established her public administration but gradually, though unwaveringly, throughout her history, deemed it feasible in the seventies to adopt the elements of the Hispanic model which had been developing in the neighbouring countries for four centuries. Thus, according to this approach it would be not the Japanese model but the "demonstrative effect" deriving from the Latin American-Spanish growth model which determined the character of Brazilian economic growth.

There certainly does exist a process of institutional convergence between the Spanish and Portuguese regions of Latin America the sources of which, however, do not give any information on the the nature of the Brazilian course of development since under the conditions of a similar institutional system there evolved various types of economic growth in Latin America. Concerning Brazilian development, the use of direct international analogies will remain arbitrary. We should rather emphasize the heritage of history when examining the character of the present model. Brazil is the only developing country which, contrary to the general pattern of capitalist colonization, followed for a relatively long time the path of classical capitalist development, more exactly, that which had been shaped in North-Western Europe. As a consequence of the aggregated effects of colonization and semi-colonial subordination, the peculiar situation of a peripheral economy, and of a belated development, this course of development proved to be longer than for instance in the countries of Europe. The requirements and conditions of modernization and structural changes ripened only much later here. Contrary to the hesitant seeking of ways and means of the previous years, the military regime which came into power ten years ago has virtually but returned to the traditional road of capitalist industrialization.

Due to historic circumstances, the process of accelerated growth could not rely on moral incentives, and in addition, economic backwardness and the crisis of economic growth ten years ago imposed restrictions on the extensive adoption of economic incentives as well. Thus, coercive factors have played a major part in the mobilization of growth energies; compulsion has assumed dimensions almost unprecedented in Brazilian history and strongly reminiscent of the conditions of original capital accumulation. The connection between the character of power and the requirements of the growth process are by no means incidental.

The capitalist development of Western Europe received a major impetus from the plundering of the colonies and other nations. In the case of Brazil and other countries which have belatedly entered the era of modernization this factor was of no significance; the role of the colonies was taken by the backward northern and north-eastern regions of the country. The extent of regional polarization within the framework of the national economy surpasses by far Italian or even Yugoslav proportions. The backward regions, however, were unable to raise funds needed for the financing of accelerated growth. Compulsory accumulation covering the whole of the country and comprising socially the labour force and the small entrepreneurs has released new energies. Besides the factors mentioned, stepped-up capital imports and expansion on foreign markets are classical features of capitalist modernization.

Thus, large-scale international research into the ideologically and institutionally original elements in Brazil seem rather unnecessary. Brazil has essentially taken the classical course of belated capitalist development. As modifying elements the adoption of developmental characteristics and methods of the Japanese and Hispanic-American growth processes—different from those of the classical European modell—as well as the dependence of the development strategy on a "quasi imperialistic" power policy undoubtedly deserve certain consideration. As regards the methods of economic development, Brazil is a significant example, nevertheless, this should not be considered an independent model of the economic growth process. The Brazilian course of development proves unfeasible for smaller countries with a different historical background and lacking the abundant natural resources of Brazil.

CHAPTER 4

COLOMBIA: CASE STUDY OF A SWITCHOVER FROM IMPORT-SUBSTITUTING GROWTH TO EXPORT-ORIENTED ECONOMIC DEVELOPMENT

4.1. A SHORT REVIEW OF COLOMBIA'S DEVELOPMENT

Simón Bolivar, the hero of the independence movement of Latin America, an intuitive expert on comparing countries stated once: "Ecuador is a monastery, Venezuela a barracks, Colombia a university". For a better understanding of this definition illustrating strikingly the characteristic features of that era it seems worth-while to examine the following problems: which are the specific and unique features of the socio-economic development of Colombia; the criteria of her relations to the external world; to what an extent is it justified to speak about a Colombian model of development; whether this development reflects the common, general problems of all Latin American countries and to what an extent it represents features to be regarded as national characteristics.

Let us begin with the common traits. The main stages and directions of historical development, the heritage of colonial and semi-colonial past following the attainment of independence, the fundamental forms of backwardness and dependence, the linguistic and cultural factors show a considerable and essential similarity with all the other countries of Latin America. The population of about 25 million, the $400 *per capita* national product, the structure of the economy, the ethnic composition of the population (40 per cent mestizos, 30 per cent whites, 17 per cent mulattoes, 7 per cent Indians, 5 per cent blacks), the indices of production relations are all very close to the statistical averages of the Latin American countries, and therefore Colombia is often cited as a model country by authors of international comparative surveys. Besides the well-known similarities, however, it seems much more instructive to point out the historical and institutional differences which are generally disregarded. The most significant differences are the following:

a) The natural-geographical circumstances of the country are rather unfavourable. The country is intersected by the eastern and western ranges of the Cordilleras and thus is structured into isolated valleys. As a consequence of topographical conditions, transport and communications, the internal division of labour and the intellectual exchange among the regions of the country have remained underdeveloped until recent times and have constituted major obstacles to development. Though the country is potentially rich in natural resources, unfavourable *natural-geographical conditions* made it impossible to exploit these natural resources economically. Therefore, the growth of economy could not rely on the exploitation of certain abundant raw materials as in the majority of the Latin American countries. Thus, the development of Colombia may be rather instructive for other countries poor in natural resources.

b) Due to the unfavourable topographical conditions, several rival urban centres of growth have developed in the isolated valleys. Contrary to most of the Latin American countries, *economic life is more diversified in space*, more decentralized, and does not concentrate overwhelmingly in the capital. The capital was not able to subdue unequivocally the rival growth centres, its role as a decision-making centre proved to be weaker in the course of history, and economic and political centralism had not struck such deep roots as in the other Latin American countries. As opposed to traditional centralism, *localism* has proved to be a more decisive force right up to recent times. It is a thought-provoking though hardly traceable problem to establish the extent to which strong localism and/or geographically rather balanced economic power relations have historically contributed to the more decentralized character of political power and to the bloody and long-drawn nature of internal conflicts unparalleled in South America. (Such periods of internal strife had been the periods 1860–1880 and 1899–1903; in the most recent civil war between 1948 and 1957 more than 300,000 people lost their lives.)

c) As a result of historical development, *the extent of* clear-cut *dependence* in foreign relations *has been less* than in the neigbouring countries. Spanish colonial rule paid relatively little attention to the territory poor in precious metals and other natural resources, and regarded the colony as strategically important only from the point of view of the defence of the isthmus of Panama.

It was the very same considerations which rendered the country less attractive for foreign capital investors in the years following the attainment of independence. The by and large self-sufficient growth centres which developed in the river valleys covered their rather insignificant import requirements by exporting coffee (in the two decades preceding World War II, Colombian foreign trade amounted to a mere 20 per cent of the foreign trade of Argentina as against the 40 per cent of today), and since coffee was produced by domestic manpower and home-made implements, there was no demand for foreign capital. Up to the mid-twenties foreign capital had no part in production, and succeeded to penetrate only banking and foreign trade to a certain degree. Though the external economic relations of the country were founded on the monoculture of coffee, prior to the World Depression, during the heydays of the "outward-looking" period based on monoculture, export realized only one-sixth of the national product as against the 28 per cent average of Latin America.

It might seem paradoxical that the gradual gaining ground of foreign capital took place during the period of import-substituting growth. The capital investments of the United States, the power with the strongest positions in the region, amounted to 92 million dollars in 1950, 632 million in 1968, and 739 million in 1972.[1] Related, however, to the economic potential of the country, these positions are weaker than those of the USA in the other countries of Latin America. Due to violent localism and to a smaller dependence on foreign powers than that of most of the Latin American states, the broad masses were less affected by the so-called demonstration effect, i.e. the adoption of foreign, mainly Anglo-Saxon patterns of consumption, values and attitudes. "Criollismo", i.e. clinging to the Latin American way of life and traditions, has deeper roots here than in most of the Latin American countries. However, as a result of the activities of the mass media, certain changes make themselves felt, and Anglo-Saxon influence has become stronger.

d) Owing to the factors mentioned, *the economic role of the state is less significant* than in the other countries of Latin America. The liberal interpretation of the role of the state, the predominance of the protective, "night-watchman" functions over the direct developmental-political intervention was not shaken

---

[1] *Survey of Current Business*, October, 1970, September, 1972.

even by the stage of "inward-looking" economic development. The state has no entrenched positions in the sphere of production, and the state budget allocates only 9 per cent of the gross national product as against the 13 per cent Latin American average. Recently, the role of the state has involved tasks of "eliminating the blank areas of development", i.e. the development of fields neglected by private enterprise, and the creation of a coherent internal economic structure; this was attempted by means of primarily *indirect methods of economic policy*.

e) The diversification of political forces owing to the spatial decentralization of economic power, the terrible losses caused by the armed domestic conflict, as well as the lessons drawn from it, resulted in the establishment and mutual acceptance of certain rules of the power-game within the ruling classes in the past fifteen years. Besides Venezuela, Colombia is presently the only South American country where the ruling classes are able to operate *a political structure based on the two-party system*. The political positions of the two great parties, i.e. the Conservatives and the Liberals, taking turns every four years, are still strong enough; so these parties do not to resort to military dictatorship, though they are constantly threatened by growing popular discontent and the strengthening guerilla movements. The reformist conception of a cautious reduction of dependence on the United States and/or the international big monopolies, the easing of the self-imposed seclusion as regards socialist and other progressive countries, internal modernization and growth acceleration— all these are realized in rather an unprecedented way within the institutional framework of bourgeois political pluralism. The changeover from one stage of growth to another has not resulted in increased political coercive measures and institutional changes in power as in the cases of Brazil or Indonesia.

f) The Colombian course of development attracts additional interest by the present *changeover from one stage of growth to another*. True, in a less extreme form, the country has passed through the main stages of development of the Latin American nations, i. e. the period of export-oriented growth based on the semi-colonial division of labour which had ended with the World Depression, and the period of growth based on import substitution which had started in the thirties and showed the first signs of exhaustion in the mid-sixties. Though the total failure of the policy of import substitution and the unfolding of the growth crisis did not ensue, the negative aspects of growth brought about the re-evaluation of the crucial points of development strategy and an increased reliance on the international division of labour at the end of the sixties. Studying the means and limitations of these changes might be instructive for other countries as well. Before examining the changeover it would be useful to review briefly some important developments of the history of the Colombian economy.

Prior to the period of colonialism, the approximately one million Indians inhabiting the country had had an early type of state organization and society based on a certain degree of class-stratification (the states of Bacata, Tunja, etc.). The development of the Pre-Columbian society was not only impeded but also terminated by the Spanish conquerors. Owing to the lower level of organization of the Indian society, the colonialists virtually annihilated the Indian social organization (contrary to the cases of the neighbouring Ecuador and Peru), and thus, here the Hispanian colonial model could be put into effect in a relatively pure form.

The peopling of the colony began as a result of the gold rush, the legend of the fabulous treasures of El Dorado. However, after the richest precious metal reefs had been exhausted, the newly settled Spanish population generally remained in the country and switched over to agriculture. The territory, rather insignificant from the point of view of economy, was neglected by the colonial administration, control was lax, and therefore, the ties with "foreign lands" (mainly consisting of smuggling) were much stronger in the 18th century than in the case of other Spanish colonies in Latin America. It cannot be doubted that lively external relations had their part in the establishment of coastal trade centres, in the rapid spread of the ideas of Enlightenment and those of the American independence movement as well as in the development of cultural standards far ahead of the economic development. These circumstances explain the fact why it had been the Vice-royalty of New Granada, comprising the present territories of Colombia and Venezuela where discontent with the colonial rule had flared up at the earliest time and most vehemently.

In the first decades of newly won independence the extremely modest foreign relations of the country were based on the exports of sugar cane and coffee. Around the middle of the century, a veritable pioneer movement spread among the population of the formerly backward province of Antioquia in the western part of the country for planting coffee-shrubs on the western slopes of the Cordilleras. By the turn of the century an external economy founded on coffee monoculture had taken shape. Among the mostly smallholder coffee planters there developed a competitive spirit unprecedented on the continent, and this spirit of enterprise gave a strong impetus to industrialization as well. From the middle of the past century on enterprises in the provincial capital started to produce machines and implements for coffee-growing and mining. The inhabitants of Antioquia are regarded up to this date as the ideal type of entrepreneurs by Latin American sociologists.

Though the foreign trade relations of the country improved as a result of the development of the monoculture of coffee, the intensity of "outward-looking" proved to be weaker than in most of the Latin American countries; in the first third of the century only 10 to 16 per cent of the national product was realized by exports, and the productive apparatus was controlled almost entirely by domestic owners. The World Depression and the disastrous drop in coffee prices and coffee exports put an end to the stage of export-oriented growth, whereas the general decrease of the level of exports necessitated a radical increase in customs duties and restrictions imposed on imported goods. The radical deterioration of the terms of trade as regards industrial commodities on the one hand and raw materials (first of all coffee) on the other, as well as import restrictions by the state have radically changed the profitability conditions of the domestic industry, and thus the rate of industrialization has accelerated. In the thirties and forties this process was by and large a spontaneous one, i.e. a reaction to the crisis and to the emergencies of World War II. However, the economic role of the state was not extended, and neither was the structure of financial and economic management transformed. Protectionist policy was raised to the status of development strategy only in the post-war years and especially following the end of the Korean boom entailing a series of price declines. This strategy tried with more or less consistency to channel resources and funds from agriculture to industry, and from the exporting sector to branches of import substitution. Differentiated and overvalued exchange rates, quantitative restrictions and other administrative constraints such as customs, however, did not only set a limit to the level of imports, but also guaranteed considerable protection and profitability to the industry as well as tried to exploit this new possibility of growth.

During the first stage of protectionist economic policy, which may be termed the period of light industry, the rate of development increased, and this was furthered to a considerable degree by favourable world market conditions. The average annual rate of the national product amounted to 5.2 per cent in the period 1945–1957.[2] It decreased, however, and reached but 4.4 per cent between 1957 and 1967, and this—under circumstances of the high rate of the increase in population—meant nearly an actual stagnation of *per capita* production. Although growth achieved under circumstances of a stagnant volume of imports may be considered a good result, the reduced dependence on foreign trade is reflected by the fact that only 10 per cent of the national product was realized by imports in 1967.[3]

Ever more symptoms of deteriorating effectiveness and a decreasing rate of growth appeared in the sixties. Between 1957 and 1967 the growth rate of industry regarded as the main force of import-substituting growth dropped to 5.9 per cent, i.e. it expanded to a hardly greater extent than the national product, and therefore could not fulfil the functions of the main driving-force of progress. The investment rate of the national product amounting to an average of 20 per cent in the period 1950–1957, dropped to 17 per cent between 1957 and 1967.

As a consequence of the neglect of agriculture, the process of the influx of population to the cities accelerated (40 per cent of the population lived in cities and towns in 1950, whereas by 1970 this increased to 60 per cent). This inflated labour supply could not be absorbed by the rather modest industrialization: between 1960 and 1967 on the average industrial employment grew by 2 per cent annually, while the rate of overt unemployment increased from an annual 1.2 per cent in 1951 to 4 per cent in 1964, and in the capital city it reached 7 to 8 per cent.[4]

The increased troubles of the economic cycle may be regarded partly as a general symptom of the continued reliance on import-substituting growth, and partly as the consequence of Colombian peculiarities. It has been mentioned already that—even at the time of "outward-looking" economy based on coffee monoculture—the individual agricultural units pursued subsistence farming and exported only the surplus of products. Due to the characteristic topographical conditions, the urban and industrial centres, too, have essentially realized a self-sufficient way of life. Neither at the time of monocultural export-orientation, nor during the period of import-substituting growth could an internal economic integration be achieved. Compared to international standards, the national economic potential of Colombia and the internal markets of the individual, isolated regions are actually of midget proportions; this naturally did not make it possible to build up economical-size enterprises in the up-to-date industries. Thus, the changeover from the stage of light-industrial development into the one of heavy industrial development resulted in a loss of efficiency and the increase in the social costs of industrialization. By the mid-sixties the reserves of industrialization founded on the partial units

---

[2] C.F. Diaz Alejandro: *Tendencias y fases de la economia colombiana*. Fedesarrollo, Bogotá, 1972.

[3] *QER*, 1970, Annual Supplement.

[4] C.F. Diaz Alejandro: *op. cit.*

of the domestic market were gradually exhausted, and it became increasingly obvious that it is hardly conceivable to find the financial means for the necessary infrastructural investments at the given level of economic development without tapping new sources of growth.

Colombia poor in capital and natural resources has no other resource to rely upon but cheap labour. Average hourly wages in 1971 were 21 per cent, i.e. but a fraction of wages in developed capitalist countries, and amounted only to about 50 to 80 per cent of those of the developing Asian countries participating in the international industrial division of labour. Import-substituting industrialization made use of its growth energies in order to purchase technologies developed in the countries with high wage standards, but owing to the labour-saving character of these technologies, it was unable to exploit the international advantages inherent in cheap labour. A rational development strategy demanded the maximal utilization of comparative differences in wages based on cheap labour, i.e. it was made necessary to channel growth energies to the export-oriented production of labour-intensive commodities. The switchover from import-substituting growth to export-oriented production held out hopes for the exploration of new sources of growth, the improvement of employment conditions—an urgent requirement both socially and politically—as well as the elimination of imbalance in the external economy. Therefore, it is only natural that the development of the productive and export capacities of labour-intensive agricultural and industrial goods—which can be sold on the world market, too—has been priorized by the new development policy.

4.2. THE MEANS OF REALIZING AN EXPORT-ORIENTED ECONOMIC POLICY

Owing to the characteristic features of Colombian economic development, and also to the fact that both the policies of monocultural, export-oriented and import-substituting growth had been carried out without unnecessary extremes, and finally that the symptoms of exhaustion of import-substitution had been recognized in due time, the changeover from one stage of growth to another was effected without dramatic emergencies, economic or political crises. 1967 may be determined as the year of the decisive turn, although export incentives had been adopted at an earlier date, too. The system of export incentives basically served the interests of the policy of import substitution.

The political signs of the changeover had been the inauguration of President C. Llevas Restrepo, a liberal representing the reformist forces of the bourgeoisie, in August 1966, and concurrently the beginning of filling up public administration with experts. Following this, on March 22, 1967 the law-decree No. 444 was issued which organized foreign trade and foreign exchange policy on a new basis.[5] The outstanding feature of the law-decree was the fact that it linked the broadly founded system of export incentives with increased restrictions on the industry (contrary to the earlier 80 per cent, only 4 per cent of the turnover value remained unrestricted in 1967), as well as with the tightening of the conditions of foreign capital investments. Let us now examine the means the Colombian government made use of when switching over to an export-oriented economic policy.

[5] *QER*, 1972. Annual Supplement.

What made it necessary to couple the policy of opening up in exports with protectionism in imports? The protection of the two aspects of external economic relations by means of apparently contradictory policies is explained by the resistance of the industries accustomed to an atmosphere of protectionism. This protection was endorsed by the national bourgeoisie which has been afraid of the future competition of foreign capital and commodities in case of liberalization. After thirty years of protectionism there could not develop such export-oriented enterprises which exported the bulk or the total of their production. As a matter of fact, only 4 per cent of Colombia's industrial production was exported in 1967. At the end of the sixties the new economic policy could but hope that initially a modest, and later a greater percentage of the capacities of import-substituting enterprises, would be channelled to exports. Thus, as a starting point it was accepted by the economic policy that the profitability of the enterprises was connected basically with the internal market which was isolated from any competition because of protectionism, and the price level of which was extremely high. Import protectionism decreased only later on when the export-orientation of the industry increased and there appeared a number of overwhelmingly export-oriented enterprises.

Recurrent corrections of the rates of exchange and devaluations made it increasingly possible to rely on the defensive function of foreign exchange policy and to reduce the relative importance of customs and administrative restrictions. Import concessions and customs exemptions granted within the scope of the Vallejo Plan, the Latin American Free Trade Association and the Andean Group are covering an ever widening range of goods, and setting limits to the field of administrative protectionism. More than a third of total imports and half of the imports of investment goods was realized free of duty in 1970. Between 1960 and 1966 customs revenue made up on the average 16 per cent of actual imports, whereas for the period 1967–1970 the corresponding percentage dropped to 14 per cent. The monetary and trade political significance of preliminary deposits which had been of major importance at the time of import-substituting industrialization diminished, too. In the period 1960–1967 such deposits amounted on the average to 21 per cent of total imports, while between 1967 and 1970 they made up only 13 per cent.[6]

The system of export incentives, however, proved to be of a selective nature because of value and price distortions in the course of the protectionist policy. Development was aimed at the increasing of the export capacities of the non-traditional products, agricultural products, raw materials and industrial goods, and at the same time it was attempted to adopt such means of economic policy which might further clearsightedness in economic matters. The new system introduced in external economy in 1967 did away with the differentiated but fixed exchange rates which hindered acumen in economic matters, and switched over to uniform and occasionally corrected exchange rates wich favoured the consideration of the shifts in internal cost factors. This moving rate of exchange meant the systematic average annual 7 per cent devaluation of the Colombian peso in the period 1967–1972.

However, the international competitiveness of the new prices was not guaranteed by such occasional devaluations the extent of which was influenced by the foreign exchange position of the country highly dependent on coffee exports. Thus, it seemed necessary to introduce further fiscal incentives. In

---

[6] C.F. Diaz Alejandro: *Los mecanismos de control importaciones.* Fedesarrollo, Bogotá, 1973.

1967 the so-called CAT system (Certificado de Abono Tributario) was established which tried to promote the diversification in the case of exports of all products (with the exception of coffee, mineral oil, hides and skins) by issuing certificates to the value of 15 per cent of the exchange-value to be used fort the payment of taxes and also for business transactions. The so-called Vallejo Plan (PV) grants an additional incentive to the export of industrial goods: the import percentage of any exported industrial commodity is exempted from obligatory license applications, preliminary deposits and customs duties if an adequate agreement has been concluded between the beneficiary enterprise and the state foreign trade institution (ICOMEX). On the one band, this construction accelerates economic circulation by eliminating the lengthy process of import licensing, and on the other hand, by exemption of customs duties and putting an end to obligatory deposits it results in considerable cost savings. However, recently several critical voices have been raised objecting to this procedure saying that it stimulated the maximization of import percentages and diminished the inland cost ratio and net foreign exchange earnings. Such assumptions are not quite unfounded, nevertheless 65 million dollar worth of export and 26 million dollars worth of imports were realized in 1971 within the framework of this construction.[7] Actually, a 40 per cent import ratio is by far not unfavourable considering that of some import-substituting projects. This system of subsidies is complemented by a system of export credits, and the lower costs of pre-financing also represent some sort of subsidies.

According to some Colombian estimates[8], the above preferences resulted in a 20 per cent subsidy for the exports of non-traditional goods, and within that a 30 per cent subsidy to that of the exports of industrial commodities.

The different direct and indirect export subsidies granted amounted to a total of 543 million pesos in 1967, 1,261 million pesos in 1970, and 2,825 million pesos (estimate) in 1972.[9] The doubling of exports and/or the quadruplication of the exports of the non-traditional goods therefore called for the quadruplication of the budget subsidies. As it is indicated by the above ratio, the share of products turned out with a sub-optimal degree of efficiency is not going to increase in the course of the intensified export campaign.

As regards budgetary effects, the trend of growing strains has become manifest only in recent years. On the average of the period 1966—1967 12 per cent, in 1971 11 per cent, and in 1972 18 per cent, of the public revenue was appropriated to export incentives. Thus, the argument proved unfounded[10] according to which export incentives were a source of inflation and the direct involvement of the government deprived it of sources of investments, since the share of public revenue as compared to the national product increased from 8 per cent in 1960 to 9.6 per cent in 1971 as a consequence of the acceleration of economic development. The total of export subsidies made up one per cent of the national product on the average of the period 1960—1967, and this ratio did not increase on the average of the period 1968—1972 either. At the same time, the average annual growth rate of the gross national product was higher by 1.5 per cent. Besides the transformation of the system of incentives the shaping

[7] J.D. Teigero–R.A. Elson: *El crecimiento de las exportaciones menores.* Fedesarrollo, Bogotá, 1973.
[8] J.D. Teigero–R.A. Elson: *op. cit.*
[9] J.D. Teigero–R.A. Elson: *op. cit.*
[10] J.D. Teigero–R.A. Elson: *op. cit.*

of the economic policy and the institutional structures, too, has furthered the strategy of export orientation.

A state foreign trade development agency (PROEXPO) has been established for the promotion of foreign trade. This agency grants direct credits, technical aid and organizes market research, the training of managers, and the holding of fairs, ets. The organization of the public management of foreign trade was considerably simplified by the reform of 1967. Formerly, exporters had to contact 24 public authorities in order to get through a business; this number was reduced to 3 by the reform. The actually rather modest public investment programmes were made use of to an increasing extent in order to promote export orientation; this was reflected by the modernization of the four big sea ports and by the development of the road system leading to these ports. Declaring the ports of Baranquilla, San-Andrés and Cartagena free ports also gave an impetus to the expansion of export zones.

An important consequence of the system of incentives in external economic policy has been the joining in regional and sub-regional cooperation which ensures mutually favourable terms of trade as well as cooperation in development. Though the Latin American countries' share in Colombian foreign trade is still rather modest, they absorbed already 30 per cent of non-traditional Colombian export goods in 1971. It is worth-while to mention that following the joining of the Andean Pact the share of the four Andean countries in non-traditional Colombian exports rose from 10 per cent in 1968 to 22 per cent in 1971, i.e. the dollar value of these exports more than trebled within three years.[11] A new feature of Colombian trade policy is the extension of cooperation with the coffee producing countries, i.e. with Brazil and the Ivory Coast. This strengthened Colombia's positions on the international coffee market. Owing to economic and political reasons, protectionist import policy has not been wound up, nevertheless the shift in the priorities of development resulted in the gradual diminishing of the social costs of imports, the indirect relaxation of protectionism and the slow strengthening of import competition.

4.3. THE GROWTH ACHIEVED BY THE SWITCHOVER

Since unlike other Latin American countries, the Colombian state concentrated its development efforts on the external economy, changing and developing it to make it more competitive, we shall start our analysis of the growth achieved with a look at the external economic sector. The radical re-evaluation of economic policy took place in 1967, and thus the new stage can be illustrated by the developments which have taken place since that time.

Besides upholding a certain degree of parasitic import consumption, in the period 1953–1967 the role of foreign trade in economic growth was virtually restricted to the supply of imported machinery and raw materials covering the needs of import-substituting industrialization. The volume of imports amounted to 547 million dollars on the average of 1953/54, and to 576 million dollars on the average of 1966/67, i.e. it has practically stagnated during these 13 years and has ensured to an ever smaller extent the external conditions of extended reproduction. At the same time, exports fell from 596 million dollars

---

[11] J.D. Teigero–R.A. Elson: *op. cit.*

to 509 million dollars, and had no part in either enhancing import capacities, or improving the employment situation, or increasing economic efficiency.

The results of the change-over to export orientation manifested themselves already after the first year, and later on, the trend of increased turnover came to study during the whole period under review. Chain-indices prove the increase in exports: the rise amounted to 9 per cent in 1969, 21 per cent in 1970, 6 per cent in 1971, 8 per cent in 1972, and 32 per cent in 1973, whereas the corresponding rates of increased imports were as follows: 29, 8, 23, 7, 6 and 14 per cent.[12] Both the rate and evenness of development and its almost uninterrupted character are an unprecedented phenomena in the past two decades of Colombian economic history. The value of exports reached one billion dollars in 1973, i.e. it nearly doubled within six years. This is certainly an extraordinary achievement in a developing country poorly endowed with natural resources. While foreign trade turnover stagnated in the period of import-substituting economic growth, between 1963 and 1973 imports increased almost twice as fast as the national economy as a whole.

The dynamic character of the external economic sector is naturally but one aspect of development. The development of exports also contributed to the putting an end to the traditional monoculture and to the acceleration of structural development.

Though no artificial withholding of coffee exports was attempted, the monocultural character of coffee-growing has decreased to a considerable extent. While 80 per cent of export earnings originated from coffee exports in 1953/54, the corresponding percentage amounted to only 47 per cent in 1973, and to 45 per cent on the average of 1974/75. Thus, Colombia has ended the period of coffee monoculture. Since both the quantitative volume and the level of coffee exports have remained by and large stable, no economic shock effects have ensued in addition to the annual fluctuations. At the same time, dependence on the weather or world economic conditions of a single produce has diminished to a great extent, and this has manifested itself in the stabilization of both the general and the external economic development.

The increase in exports originated almost entirely from the rapid development of the production of non-traditional products. From the point of view of the diversification of the national economy as well as foreign trade the increase in the exports of non-traditional raw materials (such as cotton, sugar, tobacco, beef, fish, fodder, woodenware, etc.) is in itself a considerable achievement, since besides security and stabilization aspects it has meant the exploration of new sources of economic growth and the creating of additional export capacities. The accelerated rate of modernization is indicated by structural development, but first of all by the sharp increase in the export of industrial goods. At the time of economic growth founded on import-substituting industrialization the exports of industrial goods remained very low owing to worsening conditions of competitiveness and the lack of an adequate development policy; the share of industrial articles in overall exports increased from 2 to only 9 per cent. By 1973, however, industrial exports reached the level of 300 million dollars, and by 1974 511 million dollars; quantitative increase was sixfold within seven years, whereas the share in total exports grew to 30 per cent. It should be stressed that the exports of industrial commodities no longer consist entirely of textiles and other light-industrial goods, but in the seventies

---

[12] *U.N. Monthly Bulletin of Statistics*, May, 1974.

there have appeared machinery, metallurgical products and cement, too, on the export lists.

The statistical trends of development prove the fact that the more export-oriented development policy has achieved remarkable results within the sector of external economy itself. This development policy has reduced the monocultural nature of the economy and it has become the driving force of diversification and the accelerator of structural development. At the same time it has contributed to the reduction of traditional foreign trade disequilibrium. It happened for the first time in 1973 that the foreign trade balance of Colombia showed a considerable surplus. (According to Colombian statistics—contrary to UN data—Colombia had a favourable foreign trade balance already in 1972.)

Let us now examine the development of the other factors of external economic dependence characterizing developing countries. It is argued by those opposed to export-oriented development that a more open economic policy toward foreign markets leads to increased dependence on foreign capital and power centres, primarily to dependence on the United States, and also to reduced manoeuvrability. Since the assumption in itself is rather logical, let us now examine the validity of it in the case of Colombia.

Irrespective of the degree of development and the political system, the policy of joining in the international division of labour obviously involves not only the exchange of goods, but infers and requires closer cooperation in the sectors of capital and credit relations, technological development, and services as well. This in itself is by no means harmful and may prove even a precondition of progress if the extent of it does not jeopardize the decision-making autonomy of the national economy, and is not the sole or dominating source of development. Definining or even keeping track of such a sober measure, in addition to numerous other factors, is statistically almost impracticable, however, the indication of some parameters may be of a certain orientation value.

As regards the role of foreign capital, it has been pointed out already that one of the peculiarities of Colombian economic growth was the relatively modest part of foreign capital in the history of the country. Except for the capital imports of the twenties, capital influx on a larger scale was experienced only in the initial years of President Kennedy's "Alliance for Progress" programme when Colombia was considered a model state. The direct investments of US capital representing about 70 per cent of direct long-term foreign capital amounted to 632 million dollars in 1968, and to 739 million dollars in 1972, and within that the total investments in the manufacturing industry came to 195 million in 1968, and increased to 262 million by 1972.

Political impulses have a decisive part in the flow of US capital. The massive influx of North American capital ensued in the early sixties in connection with the objectives of US foreign policy, at a time when the level of economic activity was much lower, and its dynamics less intensive.

It is noteworthy fact that since the changeover from one stage of economic growth to another the volume of North American capital investments has increased by 107 million dollars, i.e. 17 per cent. The growth rate of capital investments has not surpassed that of the national product, and therefore, it would not seem justified to rank it among the sources of accelerated growth.

In addition to political aspects, capital investments in the industry were effected with the aim to circumvent and exploit internal protectionism. Since the changeover, industrial capital investments have increased by a total of

67 million dollars, i.e. 34 per cent between 1968 and 1972. In the period under review this increase, however, could not have been of vital importance in the 39 per cent growth of industrial production or in the 400 per cent (145 million dollars) increase of industrial exports. The comparison of growth indices proves that the relative part of foreign capital has diminished in the phase of export-oriented industrialization as compared to the period of import-substituting development, indicating the fact that the turn in economic policy and growth has been restricted primarily to the sectors controlled by the national bourgeoisie.

The above data might be complemented by some further facts regarding the role of direct long-term capital. Colombia, too, has joined the Andean Pact programme for restricting foreign capital, though only after protracted hesitation. Incidentally, for years already, Colombia has striven through her economic policy to involve foreign capital in the realization of her development targets not by means of ownership regulations but by contracts. Under the stipulations of these new-type contractual joint enterprises costs, risks and profits are shared equally; further obligations of the Colombian party are the management of the production process, training of the staff, quality control, whereas the foreign party undertakes to provide for development, supply of machinery and marketing of products. Though these cooperative forms are by far not universal, they eventually avoid bringing about capital-property relations.

Finally, let us examine available indices which are of considerable interest for evaluating present and future external manoeuvrability. In 1972 Colombia's foreign debts amounted to 900 million dollars, i.e. roughly the equivalent of her annual export earnings. This ratio is not unfavourable compared to international standards. The ratio of repayment obligations for previous credits as well as profit transfers and repatriations of the massive direct long-term capital came to 20 per cent in 1968, and 22 per cent in 1971.[13] Though the level of debt service burdens is higher than 17 to 20 per cent, i.e. the average considered acceptable by international standards, it must be taken into account that present burdens are but the subsequent payments for the industrialization policy of previous years, and have not brought about foreign exchange tensions in the new phase of economic policy. International payment positions are well demonstrated by the fact that the aggregate gold and foreign exchange reserves of the country amounted to 83 million dollars at the end of 1967, 543 million at the end of 1973, and 543 million at the end of 1974.[14]

Thus, the main indicators of the sector of external economy do not corroborate assumptions regarding increased dependence of the external economy; on the contrary, they reflect the global improvement of the external economic positions and, indirectly, the greater manoeuvrability in the field of foreign policy.

Another approach is to analyze foreign capital investments not from the point of view of dynamics or economic regulations but from the angle of absolute orders of magnitude. The share of foreign capital in the total industrial capital of Colombia is an estimated 10 per cent, a much favourable ratio than the average of the developing countries. According to the balance of payments data, altogether about 850 million dollars worth of direct and loan capital

---

[13] *QER*, 1973/3; Coyuntura Económica, Bogotá, April, 1973.
[14] *U.N. Monthly Bulletin of Statistics* May, 1974. Yearbook of International Trade, *QER*.

has been imported between 1968 and 1971, amounting to nearly 3 per cent of the average national product and 16 per cent of the total investments of the above four years. (According to the data of the Planning Office 6 per cent of savings in 1970 originated from foreign sources.) Taking the average of the developing countries, the above ratios have been 5 and 30 per cent respectively, moreover in the case of smaller developed capitalist countries capital imports used to equal 2 to 3 per cent of the national product. Thus, Colombia's dependence on capital imports is less than those of the similarly developed countries, and is by far not showing an upward trend.

The analysis of the geographical distribution of foreign trade, primarily that of exports, and the examination of dependence on the given external markets is of considerable interest. Colombian foreign trade was traditionally characterized by large-scale subordination to the United States, the dominating power in the region. Taking the average of 1960–1967, the United States took up 45 per cent of Colombian exports, whereas 48 per cent of Colombian imports came from the United States. If Colombia's export-oriented development strategy and industrial exports were to be traced back to the role of the North American multinational corporations or to the decisive activity of US capital, this would necessarily be reflected also by an increase in the trade turnovers of the parent companies. Such a trend, however, has not manifested itself, and US share in the foreign trade of Colombia was actually on the decline on the average of 1972–1974 (37 per cent) as compared to the previous decade, i.e. Colombia's dependence on her main buyers' market has decreased. The products of the new export-oriented industries have been sold to an increasing extent on new markets. The growing share of the small EFTA countries—which are indifferent from the political point of view of Colombia—, the Latin American countries as well as of the socialist states in Colombian foreign trade reflects that structural diversification is accompanied by the geographical diversification of foreign trade.

The achievements of the external sector of economy are naturally inseparable from the general survey of the overall economic development of the country. As a consequence of import-substituting growth, the role of the external economic sector was rather limited up to the end of the sixties, and the percentage of the gross national product realized by foreign trade was extremely small compared to the economic strength of the country (the share of exports in the national product amounted to a mere 9 per cent in 1967). The development trends of foreign trade and domestic economy were not at all parallel. In the course of the past six years, however, following the switchover from one stage of economic development to another, the growth of foreign trade has been twice as fast as that of the national product, and the share of exports in the gross national product has risen to 14 per cent. Though Colombia cannot be regarded yet as an open country from the point of view of world economy, recent trends of development have brought about increased dependence on foreign trade and a highly enhanced role of the external economy. At the same time, a certain synchronization could be detected between the overall development trends of the sector of external economy and the national economy.

The accelerated growth rate of the national economy has been the most outstanding feature of all internal economic developments. Whereas the average growth rate of the national product for 1960–1967 was 4.5 per cent, it was 6.2 per cent for 1968–1970, and 6.9 per cent for 1971–1973 (5.7 per cent for 1971, 7.8 per cent for 1972, 7.1 per cent for 1973). Even this growth rate proved insufficient to improve perceptibly the living standards of a population increasing by an annual 3.5 per cent, nevertheless it succeeded in ending the impasse of the *per capita* national product which had been virtually stagnating in the previous decade.

Within the framework of a modernized production structure the change-over to export orientation and the beginning of the exports of industrial goods have brought results primarily in industry. In the period 1960–1967, the growth rate of the manufacturing industry amounted to 5.8 per cent, whereas between 1968 and 1973 to 8.5 per cent. 40 per cent of the increased output originated from the rise in productivity, which was direct consequence of the stricter requirements of a more export-oriented industrialization as regards competitiveness. In the wake of the stepped-up rate of industrialization the share of industry in the national product grew from 16.7 per cent in 1960 to 19.6 per cent in 1967, and by 1972 it surpassed the 20 per cent level. One may get an idea of the importance of export orientation from the fact that the value of industrial production grew by about 500 million dollars and that of industrial exports by more than 250 million dollars between 1968 and 1973. i.e. about 50 per cent of the increment of industrial production was absorbed by external markets.

The changes in the structure of industry are demonstrated by the increased share in the total value of the production of industries producing investment goods: within one decade this has risen from 27 to 37 per cent. One of the chief consequences of accelerated industrialization in the domestic economy of the country has been the improvement in the employment situation. The number of employed labour force increased by 5.9 per cent in 1972, and according to estimates by 6.4 per cent in 1973. And since demand for manpower grew only by 3.8 per cent, demand was surpassed by employment, and a process of absorbing unemployment has started. It is a highly significant indicator that in the most export-oriented industries, i.e. textiles, ready-to-wear articles and leatherwares, employment has increased by 8 per cent.[15]

Efforts aiming at diversification accelerated the development of agriculture as well. Between 1960 and 1967 agricultural output grew by an annual average of about 2.5 per cent, i.e. much slower than the population increase and, therefore, *per capita* calory consumption decreased from yearly 2,260 units in 1961–1963 to 2,138 units on the average of the period 1967–1970. The accelerated growth rate came mainly as a result of diversification; the rate of increase of coffee crops was less than 2 per cent.

The expansion of the saving-investment capacity of the economy is also a significant positive aspect within the macro-economic structure. Regarding the average of the period 1960–1967, investments represented only 16 per cent of the national product. However, by 1970 this ratio has increased to 19.4 per cent. The growing effectiveness of investments, too, reflects the improvement in the net efficiency of macro-economic processes. Whereas a unit increment of national product required 3.5 units of investments on the average of 1960–1967, this was reduced to 3 units on the average of the period 1968–1973.

The development of the equilibrium of the economy has been less favourable. Some factors have been mentioned already: reduced external imbalances, the stabilization of the trade balance, and the spectacular increase in foreign

---

[15] *Coyuntura Economica*, Bogotá, April, 1973.

exchange reserves.[16] In a backward country grappling with the problems of basic disequilibrium any kind of industrialization policy, which strives to primarily eliminate external imbalances without trying to bring about any changes in the political power balance of the country, is due to encounter the problem of an increasing internal disequilibrium, i.e. the switchover from external to internal disequilibrium. The beginning of the exhaustion of import-substituting industrialization was indicated both by increased external and internal disequilibrium. In the period 1960–1967 the increase of the average annual internal disequilibrium, the cost of living index reflecting the rate of inflation amounted to 12 per cent, between 1968 and 1970 to 8 per cent, in 1972 to 11 per cent, and in 1973 to about 20 per cent. As it appears from these data, the diminishing of external and internal disequilibrium is a parallel process in the first three years following the economic changeover, while in the second part of the period the price of buoyant investment activity, rising employment, increasing exports and/or a highly improved foreign trade balance has been an even higher degree of internal disequilibrium than ever before. Although this is not an excessive price paid for reduced external disequilibrium and stepped-up economic growth, and should not be regarded as a substantial factor as compared to international standards, the switchover effected under the conditions of a political rotation system is by no means an indifferent factor from the point of view of political tensions.

Closely related to the problem of internal disequilibrium is a paradox of development: up to this date the acceleration of economic development has impeded and retarded the realization of a number of internal reforms, mainly that of the distribution relations. Due to the rapidly unfolding and successful policy of export orientation, those in charge of the economic policy dared not risk the acceleration of the land reform, being afraid of jeopardizing thereby the whole trend of the expansion of agrarian export capacities. On the other hand, conservative political circles, referring to the necessity of putting a stop to the deterioration of the employment situation wanted to make use of the policy of export orientation in order to further delay even the most urgent reforms. Thus, the stepped-up rate of economic growth has not been accompanied by major socio-political progress, and this may be harmful to long-range development.

The price explosion in 1973 and the world economic recession affected Colombia more seriously than most of the other countries. The presidential elections of 1974 proved to be an aggravating circumstance because the political leadership of the country concentrated all its efforts on the political problems, whereas the economy began to introduce the strategy of export-oriented growth just at that time. Production and/or the structure of supply reacted very sensitively to the fact that the country is relatively poor in raw materials. International demand for coffe and light industrial goods also fell, and thus, Colombian exports increased only by 15 per cent in 1974 (calculated at current prices), and export growth actually came to a standstill in 1975, while imports increased by nearly 50 per cent during these two years. In order to bring the growing disequilibrium to a close, new restrictive measures were introduced by the government, and thus, the government succeeded to limit the rise in the cost of living to 27 per cent in 1974, and 18 per cent in 1975. Owing to the slackening of external and internal demand, the growth rate of the gross

---

[16] *Principales aspectos de la evolucion económica colombiana.* Fedesarrollo, Bogotá, 1972.

domestic product fell from an annual 6 per cent in 1974 to 1.8 per cent in 1975. The construction industry and the manufacturing industries were especially hard hit by the effects of recession. Colombian economic policy regarded the economic problems as short-term ones, and therefore shifted the burdens of the recession by and large to the shoulders of the urban working population. At the same time—contrary to most of the developing countries—, it did not change that export-oriented character of the development policy. The upswing in the world economy in 1976 and the favourable situation in the coffee market therefore considerably alleviated the socio-political tension which has sharpened during the low-tide period of 1975, and thus, the continuity of development policy did not come to a stop.

To sum up, the conclusion may be drawn that Colombian development represents a peculiar way of economic growth shaped by the natural geographic and economic resources on the one hand, and by the country's historical experiences on the other hand. The present form of the model is by far not a clear-cut export-oriented type of development, it only reflects the adequate circumstances and consequences of the re-evaluation of the relationship between the international division of labour and economic development. Contrary to the world economic openings of other Latin American countries the characteristic features of the Colombian course of development are the lack of abundant natural resources, the relatively limited role of direct state intervention and/or the control of economic life by direct means and finally, a political structure characterized by a two-party system.

Protectionist development of the economy was rendered increasingly costly and slow by taking into consideration the spatial differentiation of the economy (created by geographical circumstances) and consequently, the limits to the balance of power. Effective development of the centres of growth by means of foreign trade was furthered by the policy of opening up to the world economy. The increased export orientation of the economy was based first and foremost on the cheap labour force available, and was directed towards the extension of the capacities of labour-intensive agrarian and industrial commodity exports. Due to scarce resources and the voluntarily set limits to cooperation with foreign capital, it became impossible to Colombian economic policy to develop parallelly export markets and domestic markets, and therefore it strove to accelerate the creating of export capacities both by enhanced austerity measures and by restricting consumption. It is a paradoxical phenomenon that politically liberal Colombia adopts stricter restrictions than the military dictatorship in Brazil.

The speed and efficiency of the changeover from one stage of economic development to another should be regarded as a considerable achievement. The new export capacities have been established within five years, proving the correct distribution and adoption of the means of development. Without this successful policy of opening, the Colombian national economy would have faced a serious crisis under the conditions of the present-day energy crisis.

The costs of establishing an export-oriented development policy are rather instructive, too. The costs of additional export capacity development per unit have not increased, the changeover and stepped-up growth have not demanded any additional costs, while, at the same time, the steadier character of accelerated development as well as the enhanced manoeuvrability of the external economy have been outstanding achievements. These successes are counterbalanced, however, by the further delay in carrying through the reform of distribution relations, the shifting of external disequilibrium to the sphere of domestic economy, the keeping of the living standards of the masses at a low level (necessary for the low wages level as a precondition of a successful switchover to export-orientation) as well as the increase in the socio-political tension. These negative features, however, were not so strong that a revolutionary situation or a domestic crisis should have come about, and it is likely that the further sharpening of tensions can be prevented by the strong reformist policies of the new president who is a representative of the left-wing of the Liberal Party.

In many respects the Colombian model may not be regarded as a typical development model for Latin America. It is a rather interesting fact that the bourgeois literature on "outward-looking" economies and the admirers of the "Brazilian miracle" regard just those features as the efficiency criteria of an export-oriented development strategy which are not very pronounced in the Colombian model. Such features are reliance on foreign capital, the presence of multinational corporations, participation in foreign market organizations, the adoption of the consumption patterns as well as the mental and behavioural attitudes and norms of the institutions of the developed countries, ect. Owing to the peculiarities of her history, Colombia remains an exciting field for research on comparative economic policy in the future, too.

CHAPTER 5
THE MEXICAN MODEL OF BALANCED ECONOMIC GROWTH

Not only in the literature on the Latin American countries but also in works dealing with the Third World, the peculiarities of the Mexican economic growth have attracted considerable attention. Due to its uninterrupted progress of three decades and its extraordinarily rapid development, the so-called Mexican growth model is often regarded as exemplary for the development of the Third World as a whole.

The experiences gained by the Mexican model have influenced the concepts of economic policy of the Andean integration, the so-called second Perón era, Panama and in certain periods those of other Latin American countries as well.

To put it in a concise form, the outstanding specific features of the Mexican growth process are as follows: in addition to a rapid and balanced rate of progress, an institutional stability and development unprecedented in the Third World; the diversification of the commodity structure of foreign trade and at the same time its almost unequalled one-sided and only slightly decreasing geographical dependence. Similarly to Brazil and Colombia, the growth strategy of Mexico has been characterized in recent years by an increasing reliance on international economic relations. However, while the strategy of economic growth is primarily influenced by efforts aiming at the establishment of a great power status in the case of Brazil, the elimination of an economy founded on a monoculture in Venezuela, the national control of decision-making centres and economic key-positions in Peru, and the parrying of the consequences of socio-economic disequilibrum as well as of diminishing regional positions in Argentina, in Mexico the maximization of growth in a technical sense goes hand in hand with some social reforms on the one hand, and acknowledgement of external dependence on her neighbour, the United States, on the other. Also the lessons to be drawn from her growth process are mainly related to the economic role of institutional factors, the components of industrial and agricultural development guaranteeing the dynamism and relatively balanced character of progress, as well as to the external economic phenomena of rapid progress in spite of external dependence.

5.1. THE INSTITUTIONAL CHARACTERISTICS OF GROWTH

The starting point of the analysis of the sources of Mexican economic growth should be the examination of the institutional factors. The liberalism imposed on Mexico in the second half of the past century started a process of unrivalled internal socio-economic polarization and external subordination. It was the pent-up tensions which led to the outbreak of the first anti-imperialist, anti-feudal bourgeois democratic revolution in the Third World (1910–1917). In the period of the bloody revolutionary wars and the increased revolutionary activity of the masses a coalition of interests was achieved among small capitalists, the peasantry and the working classes. As a result, the Constitution of February 1917, at that time considered outstandingly progressive, was promulgated.

The laws against the peasants issued during the past hundred years were all abolished by the Constitution of 1917. It also ordered the distribution of the latifundia on the basis of indemnification, and established a National Agrarian Committee to decide on the land claims of the peasants. The Constitution was the first to declare the public ownership of the subsurface natural resources thus serving as the legal basis of the expropriation of foreign interests at a later date. The establishing of the secular character of education and free elementary education were an outstanding achievement in raising the professional and cultural level of the population.

Traditional political structures and the power of the landowner-comprador stratum based on a system of an illusory parliamentary liberalism and the armed forces were crushed by the revolution. Mexico was the first country in the developing world where a power structure founded on a one-party system was established as early as 1929. The Institutional Revolutionary Party (PRI) which has been ruling ever since that time is not a mass party but the political organization of those in power, and is intertwined with the state administration. Leading positions can only be held by party members, and the presidents are elected every six year from candidates accepted by the narrow circle of party leaders. The centralized character of the political decision-making system is also reflected by the comprehensive sphere of authority of the President which is quite considerable as compared to international standards. Local organs, the federal states or the Congress are no counter-balances to the President, their only duty being to execute or support the government's policy. All major government decisions are made by the President whose powers are not limited by any institutional restrictions. The President in office plays a decisive role in the nomination of his successor; as a rule, one of his ministers follows him in office. This peculiar, rather unprecedented mechanism of succession resembling that of a constitutional monarchy is one of the determinating factors in guaranteeing strategical continuity. The main force and organized mass basis of the party includes three major organizations: the General Workers' Association of Mexico, the National Peasant Association and the National Association of Popular Organizations. Thus, the direct representation of the organized working people in the power mechanism has been stronger than in many other developing countries.

To consolidate its monopoly of power, the party put an end to the strong positions of the armed forces which had been considerable before and during the civil war. As a unique feature in Latin America, the army, numerically weak and poorly equipped, had no autonomous part in the political life of the country in the past fifty years.

The consolidated political mechanism rendered it possible to initiate structural reforms. Under the presidency of Lázaro Cárdenas, between 1934 and 1940, a radical land reform was carried into effect which covered 50 per cent of arable land. The nationalization of the oil industry in 1938 was the first large-scale nationalization and definite stand against foreign capital in the developing world, and this was followed by the nationalization of mineral resources and mining, and later on, of electric energy, the iron and steel industries as well as the railways in the early sixties. During its twenty-five-years development, the Mexican revolution has laid and gradually broadened the foundations of the mixed economy based on the co-existence of the public sector and the domestic as well as foreign private capital. This has ensured further favourable conditions to the increased indirect socio-economic role of the state. The most interesting characteristics of the Mexican growth model has been the fact *that for the first time in the history of the developing countries the economic role of the state has been permanently and on a large scale established*, and the interrelations of the public and private sectors, this "delicate equilibrium", have been moved into the focus of the growth process.

Our present aim is not to trace back the development of state intervention and presence. It will be sufficient to mention only some phenomena to demonstrate the national economic proportions and effects.

There are no exact data available on the scope of the direct economic activities of the state. According to statistics on the year 1970, the number of industrial and servicing enterprises owned by the state amounted to 136. Of these 18 operated in the sugar industry, 5 each in the textile and food industries, 4 in metallurgy, 3 in the chemical industry, 19 in mining, 2 in the car industry, 11 in commerce; all these enterprises partly supplied the poorer strata of the population with cheap commodities, and partly functioned as purchasing agencies.[1] Public enterprises are dominating or have a determinant role in transport and communication, mining, petroleum industry, petrochemistry, production of electric energy and metallurgy, i.e. in the infrastructure of production.

There are two distinct types of public enterprises. Part of them are independently founded public enterprises established by a Presidential decree, and part of them are such enterprises where the state has a majority share or at least has the right to enforce its influence. Various points of view have been taken into account in the establishment of the public sector. Since the end of the thirties Mexican development strategy has striven to nationalize or at least to ensure Mexican ownership in the basic sectors of the economy. Domestic private capital often refused to perform any import-substituting tasks and thus public enterprises had to shoulder responsibility for carrying into effect the official concept (e.g. in the paper industry). The structural delimitation of the public sector was often hindered by the fact that in order to guarantee employment, the state took over a number of bankrupt enterprises or firms. There were some cases when the extension of the public sector became necessary because—on the strength of social aspects—the state undertook to supply the population of some poorer districts with cheap commodities. Apart from political and power viewpoints or some exceptional cases when the requirements of the development strategy called for a different policy, public enterprises were induced to live on their own profits.

Public investment policy has a leading part in the implementation of the development strategy. Taking the average of the forties, more than 50 per cent of all investments were made out of the state budget and by the public sector. This ratio fell to 40 per cent in the past decade, nevertheless it is still great enough to determine the direction of development. Formerly, the part of the federal budget has been decisive as regards public investments,

---

[1] *Nachrichten für Aussenhandel*, January 9, 1971.

however, parallel with the extension and broadening of the public sector an ever growing percentage of investments was covered by the autonomous public enterprises. In the past ten years about 60 per cent of public investments have been made directly by the public sector, 36 per cent out of the federal budget, 4 per cent by the federal states and local authorities.[2] In order to enhance the effectiveness of the economic policy and to increase budgetary discipline, the federal budget and that of the public sector were merged in 1966, and thus the investments of the country are highly dependent on the political decision-making centre. The structure of public investments reflects the priorities of the various stages of development strategy. In the forties more than half of public investments had been assigned to the infrastructure, 18 per cent to agriculture, one-eighth each to industry and the development of social welfare. In the past decade, however, infrastructure has received only a quarter, agriculture one tenth, whereas industry a third and social projects one fifth of total investments.

Recently, the *federal budget* has been of much less importance in the mechanisms of economic management than in most of the Latin American countries. Taking the average of 1965–1970, 10.3 per cent of all expenditures were allocated by the federal budget. This rather modest percentage has come about as a result of various factors. The public sector disposes of an independent budget more significant than in other countries; neither its revenue nor its expenditures are parts of the federal budget. Similarly, the budget expenditures of the individual federal states are excluded from the central budget. (Incidentally, these sums are not significant at all.) Besides institutional elements the character of development strategy is of vital importance. Deficit financing, a common feature in Latin American countries, is incompatible with Mexico's balanced growth, moreover one of the important means of directing development are tax allowances, and therefore, the factors stimulating budgetary expansion are weaker. On the other hand, social aspects have been subordinated to efficiency by the distribution policy, and—contrary to some other Latin American countries, as for example Venezuela, Panama or Costa Rica—no measures financed by the budget have been and are taken to ease the socioeconomic conditions of the strata in the lower income brackets. In the last analysis the lower percentage reflects the view that incentive to the economy should be given mainly by granting preferences and not by subsidies.

However, the structure of the relatively smaller volume budget is more progressive than in other Latin American countries. About half of the budget revenue is made up of direct taxes as compared to the average ratio of one third of the Latin American countries. The ratio between direct and indirect taxes is a proof of the high standards of the Mexican fiscal system. The structure of expenditures, too, is rather favourable. Owing to the negligible numerical strength and political weight of the Army, the bulk of budget expenditures is allocated to development tasks. Thus, the development of public education receives generally one third, direct economic development projects 40 per cent of total expenditures, whereas the share of the armed forces does not exceed 5 per cent.

Examining the combined budget of the *federal authorities* and the *autonomous organization* of economic development (such as railways, petroleum industry, etc.) in 1972, a year of restrictive budgetary policy, public expenditures

---

[2] T. King: *Mexico–Industrialization and Trade Policies.* Oxford University Press, 1970.

represented 25 per cent of the national product, while in 1973 their share rose to 33 per cent (the share of the autonomous organizations in 1972 being 56 per cent).³ Thus state intervention could make use of a rather broadly founded financial basis. If the strength of national integration is gauged by the percentage of the national income distributed by the central power, Mexico undoubtedly ranges among the first as regards internal integration.

To sum up: the Mexican institutional structure is characterized by a high degree of stability, the significant role of the state and the highly developed centralization of the decision-making process. The President's supreme power of decision-making has necessitated the organization of a number of parallel decision-preparing and controlling committees whose spheres of authority overlap with those of the individual ministries, and which hinder quite frequently the efficient functioning of the organs of public administration. The control system is operating both by indirect and direct methods, and control mechanisms are relatively simple aiming at the regulation of relatively narrower spheres of activity regarded decisive at a certain time. The development of the system of economic decision-making is by far not exempt from contradictions, nevertheless neither the rate of growth, nor the speed of decision-making have been impeded by institutional restrictions up till now.

The interpretation of the contents of institutional frameworks and the aims of state intervention have been certainly not unambiguous and homogeneous in the course of the past fifty years. The conflicting interests and shifts in the power relations among the heterogeneous socio-political forces which have accomplished the Mexican revolution naturally manifested themselves even within the framework of the one-party system, and brought about political periods whose duration coincided with the terms of the presidents. Thus, during the presidency of Cárdenas it had been the left wing of the party which dominated, whereas under A. Camacho, Ruiz Cortinez, M. Alemán (1940–1958) and later on under Diaz Ordaz (1964–1970) the right wing gained the upperhand. During the presidencies of Lopez Mateos (1958–1964) and Echeverria (1970–1976) the centrist group became the most influential and pursued a pragmatic line both in domestic and foreign policy.

The shifts of emphasis in the political cycles reflect the changes in the ideology of the central power. The objectives of the revolution have manifested themselves in the ideological guise of "inward-looking" nationalism.

As a reaction to European liberalism and dependence on the external powers, revolutionary ideology has gone back to the historical-cultural national psychologic heritage of pre-colonial times. In trying to achieve moral and political identity as well as national unity, revolutionary Mexico wanted to rely on the revived value-system and mechanisms of Indian society and thereby to win the support of the Indian-mestizo majority comprising about four-fifths of the population. However, Indio nationalism which has dominated ever since 1910, gradually split up into two main phases. In the course of the first three decades, i.e. between 1910 and 1940, it was the social elements, the targets of building a new society which had a determining role, whereas in the second stage, i.e. since 1940, the problems of economic construction and growth-orientation have come into the fore.

The "inward-looking" tendency disappeared gradually. In spite of her past as a viceroyalty Mexico had refused to participate in any kind of inter-

³ *EIU* 1974/2.

national actions during the first decades after the revolution. In the early sixties, however, the joining into Latin American regional economic cooperation represented a step forward from Mexican isolationism towards a sort of Latin American identity, and the organization of the Olympic Games served already to a great extent the objective of calling international attention to Mexico. At the UNCTAD Conference held in Santiago in 1972, President Echeverria presented himself as a spokesman of the Third World, and since that time, the increasingly active presidential diplomacy—which has been motivated by the President's ambitions to be elected General Secretary of the United Nations—enhanced Mexico's role in international diplomatic life.

The slackening of the one-time revolutionary spirit, the forging ahead of the conservative forces in the long run was caused partly by the fact that the former leading revolutionary, political, military leaders have become a stratum of entrepreneurs or politicians. The growing political influence of the middle strata which have rapidly grown in numbers as a result of accelerated economic growth and education played also a great role in the above process. There is no other country in Latin America or in the whole Third World whose development process—due to the curbing of the power of the traditional ruling classes and the establishment of state presence—reflects to such a degree the increasing political and economic significance of the so-called middle strata. Though the revolutionary heritage is the *raison d'être* of the ruling circles, and time and again they are referring to "their revolution", the two processes mentioned have overshadowed the original revolutionary objectives and have increasingly emphasized a pragmatic attitude of cautious progress.

The unfavourable geopolitical position of the country, the external restrictions due to the proximity of a super-power, i.e. the United States, had naturally a major part in the emergence of this cautious strategy. The arch-reactionary Porfirio Díaz had stated that the main problem of Mexico "is the fact that it is situated too near to the United States and too far from God". And the restrictions deriving from the presence of an adjacent great power have delimited ever since the possibilities of socio-economic changes.

It is another aspect of the problem, rather in the range of national psychology, that Mexico has paid too high a price for carrying through the revolution. As a result of armed fights and wide-spread violence, more than one million people, i.e. at that time 7 per cent of the population had lost their lives. The socio-political disequilibrium in the wake of the armed struggles, the lack of experience of the new power and its internal contradictions have for a long time delayed the launching of economic construction work. Characteristically, *per capita* national product stagnated in the thirty-year period between 1910 and 1940, and the abject poverty of the broad masses became a permanent phenomenon and even increased. The laying of the foundations of the new revolutionary power required a historically rather long period, and the high price which had been paid scared off most of the social strata from accepting responsibility for decision-making and situations which might lead to violent conflicts.

Due to the above factors, the political cycle has not shown any major fluctuations in the past three decades; there occurred, at most, some shifts of emphasis the extent of which, however, has been negligible seen in the light of the historical events of the past fifty years. Mexico enjoys a high degree of institutional stability even by international standards, and this may prove a rather favourable precondition for her long-range development.

5.2. CHARACTERISTICS OF ECONOMIC GROWTH

The more than thirty years of uninterrupted, rapid and relatively balanced economic growth of Mexico may be regarded as a rare achievement even by international standards. In spite of one of the highest rates of population growth in the world, (3.5 per cent annually) *per capita* national product between 1940 and 1973 increased more than ten times, i.e. it rose from $75 to $820, and by the end of the seventies Mexico will have left the group of underdeveloped countries and practically will have entered that of the medium developed nations. Mexico has already overtaken Argentina, and she has become the second economic power of Latin America; she ranks third among the Third World countries and twelfth among all capitalist states. (See Table 28.)

Table 28

The main indices of the dynamics of the growth process
(percentage of average annual growth)

	1940–1964	1965–1970	1971–1973
Agriculture	4.6	2.8	3.4
Manufacturing industry	7.8	8.9	8.2
Exports	7.7	6.2	21.0
Imports	—	9.1	19.8
National product	6.3	7.1	7.2

Source: *Manual de Estadisticas Básicas, U. N. Statistical Yearbook, Monthly Bulletin of Statistics.*

The dynamism and relative evenness of growth are clearly reflected by the above data. The national product did not diminish all through these thirty years, and the annual growth rate was only twice lower than 4 per cent. The steady and even progress has not shown any signs of exhaustion but even accelerated in recent years. The favourable national economic effects of growth are indicated by the fact that the capital intensity was rather moderate according to international standards. The incremental capital-output ratio remained under 3 in the average of the period under review, i.e. the effectiveness of investments by far exceeds the average of the various groups of developing countries. Both the even and rapid rate of growth and its effectiveness prove indirectly the efficiency of Mexican economic policy.

Even more important than the rapid and even pace of growth has been its *balanced* character and/or the measure of structural development related to it. It is pointed out by the Hungarian economist J. Kornai[4] that in the case of smaller countries with narrower domestic markets it would be misleading to identify balanced growth with the parallel development of all productive sectors, since this would render the process of balanced growth the privilege of a handful of major powers. The diversified natural resources of Mexico, a country of almost continental proportions and considerable population, made is possible to unfold the economic growth process on a rather broad front.

The structural development and diversification of the one-time backward agrarian country demanded the accelerated development of the industry and

---

[4] J. Kornai: *Erőltetett vagy harmonikus növekedés* (Forced or Balanced Growth). Budapest, 1972.

within that of the heavy industry. In the course of the establishment of up-to-date branches, however, Mexicans did not kill the hen that lays the golden eggs, i.e. they did not absorb the growth energies of agriculture and mining to channel them into the industry as it happened in most of the Latin American countries pressing for an accelerated rate of industrialization; the main economic sectors were developed simultaneously, though at different rates. The ratio of the industrial and agricultural rates of development in Mexico has been 1.9:1 on the average of thirty years, whereas in most of the forcibly industrializing countries the extreme values have been 3 to 8:1. The production in the most important branch of the extractive industry, i.e. the oil-producing sector hardly lagged behind that of the manufacturing industry which constitutes a striking contrast to experiences in Venezuela, Peru, Chile, etc. The consequently though relatively but slightly differing growth rates in the different sectors did not result in the well-known problems of internal and external economy as well as political strategy so often encountered in other countries owing to the decapitalization and lagging behind of some branches. Moreover, one of the most interesting features of the Mexican model has been the fact that an original approach has been found for drawing agriculture into the process of growth.

Mexico's economic structure was radically changed by the structural development carried into effect unswervingly. Whereas in 1940 the share of agriculture in the national product had been 22 per cent, that of the manufacturing industry 17 per cent and of the oil-producing sector 3 per cent, according to data relating to 1972, these percentages changed to 11, 23 and 5 per cent respectively. As to the structural development within the industry, the share of the heavy industries in manufacturing production increased from 35 to 55 per cent between 1950 and 1970.[5]

The relatively parallel development of internal and external division of labour has been of a similar effect up to recent times. After the Korean boom, development relied to a major extent on the growth stimulating effect of import substitution; the degree of protectionism, however, was less conspicuous than in most of the developing countries; foreign trade turnover did not fall behind the overall expansion, and former imports were not substituted all of a sudden by more expensive domestic products of inferior quality.

Thus, accelerated growth and structural development took place concurrently without major sectoral disharmony and social sacrifices. The achievements of the growth process in Mexico refute the precepts of the so-called ,,turnpike" alternative of balanced growth expounded by Neuman and also by Dorfman, Samuelson and Solow,[6] according to which the turnpike of economic growth is identical with the parallel rate of development of the individual sectors as well as with constant input-output combinations and/or structures. Possibly owing to its pragmatic character, the Mexican strategy has succeeded in avoiding both the mathematical pitfalls of the American school of balanced growth and the political dogmatic drawbacks of the protectionist school of thought. However, it should be admitted that the favourable endowments of the country had a considerable part in this.

The balanced nature of growth is reflected by the development of *the internal*

---

[5] *EIU 1972.* Annual Supplement.
[6] Dorfman–Samuelson–Solow: *Linear Programing and Economic Analysis.* McGraw-Hill, New York–Toronto–London, 1958.

as well as of the external equilibrium. The economic policy highly impregnated with Keynesian elements has made use of deficit financing in Mexico, too, for promoting the expansion of the public sector, whereas the close external economic relations with the USA, a country maintaining a high cost-level, have strengthened the establishment of a surplus demand from outside. However, the inflationary process has unfolded to an extremely restricted extent, and, therefore, Mexico may be regarded in the long run as one of those Latin American countries which have the most stable internal economic balances. Likewise, no restrictions irreconcilable with harmonious growth have manifested themselves as far as the balance of payments were concerned. Although the trade balance of the country shows a chronic and increasing deficit, tourism, remittances of Mexicans working in the United States as well as capital imports which have considerably increased in recent years succeeded in ensuring the equilibrium of the balance of payments and rendered the introduction of a restrictive policy superfluous. The increase of the volume of foreign debts and/or debts service burdens kept pace with the expansion of export capacities and in the past ten years it has not surpassed 22 per cent,[7] i.e. it has not exceeded the limit of tolerance defined by international banking institutions.

Besides the quantifiable economic processes, balanced growth presupposes and requires the parallel development of the *qualitative elements*, i.e. the technological standards of production. The statistical measuring of qualitative progress is a problem still to be solved, nevertheless its indirect effects are clearly perceptible. In all the countries accomplishing accelerated industrialization the low efficiency, poor standards, and non-competitive character of the newly established branches of industry are rather general phenomena. Mexican development strategy, however, avoided big leaps, it took into account the requirement of stages in the development process, and initiated a more energetic rate of industrialization—first in the light industries and later on in the heavy industries—only after having realized the first steps of agricultural growth. The adherence to the norms of progressivity as well as the structural development on the basis of results achieved, had considerably contributed to the present progress of qualitative elements which can be measured by the international competitiveness and export capacity of the new industries. It is not by chance that of all Latin American countries it is Mexico where the share of industrial commodities in exports is the highest and where there appears the greatest interrelationship between production and the structure of foreign trade.

Examining the "human" factor of growth it appears that the stepped-up rates of consumption, education and public health, too, meet the requirements of balanced growth. The average of *per capita* personal consumption between 1954/56 and 1967/69 increased by 45 per cent which is the highest rate for all Latin America. According to the caloric norms established by FAO for the individual countries there are five Latin American countries (Uruguay, Argentina, Brazil, Paraguay and Mexico) whose *per capita* average daily caloric intake is higher than the actual needs.[8] Compared to the Latin American continent as a whole, the development of education shows a similarly dynamic picture. Mexico is the only country in Latin America (with the sole exception

---

[7] *IADB Annual Report*, 1972.
[8] *Economic Survey of Latin America*, 1970.

of tiny Costa Rica) where expenditure on public education is more than twice as high as the military budget. Despite the rapid increase in population the ratio of illiteracy diminished from 43 per cent in 1950 to 19 per cent in 1970, and at the same time the number of those participating in higher education rose from 65,000 to 260,000, i.e. it made up 0.6 per cent of the total population. Public education does not lag behind overall economic development. As regards the development of the public health system, it is not the present level but its dinamism which is rather conspicuous. The rate of infant mortality decreased from 30 to 12 per mille, whereas the average expectation of life increased from 48 to 60 years. The physiological and mental progress of the population was reflected by the growth achievements and probably also by the accelerated rate of growth of recent years: its significance is even greater as regards long-range development.

The above factors have demonstrated the outstandingly harmonious character as well as the parallel and relatively balanced exploitation of the national resources within the framework of the Mexican model. The pattern of growth, however, does not meet completely the requirements of balanced growth, since the distribution of incomes and the differing regional levels of development reflect the historical heritage of backwardness and the imbalances of the growth process. As a result of the revolution, the system of *income distribution*—as mentioned—is one of the less concentrated and polarized ones in Latin America. At the same time, it may be stated on the basis of the structure of income distribution that the achievements of accelerated growth have been monopolized mainly by the upper-middle and medium income brackets, while the popular masses, whose living standards are extremely low, have been excluded from the benefits of the growth process. The social principles of income distribution policy have made themselves felt only in the case of the highest strata partly including the former ruling classes, i.e. 5 per cent of the population. The share of the above strata within total incomes fell from 40 per cent in 1950 to 29 per cent in 1964, whereas that of the top 1 per cent diminished even more: from 23 to 12 per cent,[9] and represents the lowest ratio in present-day Latin America. The distribution system endeavoured to maximize the incentive factor and, therefore, channelled incomes towards the most efficient sectors disregarding thereby the aspects of social justice and equality. As a consequence, the share of the middle strata representing 25 per cent of the population rose from 28 to 42 per cent, whereas that of the 40 per cent of the population pursuing less efficient activities dropped from 14.5 to 11 per cent. Thus, the aspects of social justice and incentive manifested themselves in rather a contradictory way in the distribution policy of Mexico.

However, the inequalities and trends of income distribution reflect, in addition to social problems regional imbalances as well. In a huge country like Mexico, the development standards of the individual regions obviously represent considerable differences which have their roots in historical development. For a central power striving to bring about national unity and internal integration, which furthermore professes to be the heir of the revolution, it should be quite natural to pursue a development policy of balancing regional differences. During the first decades of the new power, there were made some efforts aimed at the levelling off the remote regions and the relatively developed central capital by means of developing agriculture. These efforts were com-

---

[9] *ECLA: Estudio sobre la distribucion del ingreso.* ONU, Santiago, 1967.

pletely thwarted, however, by the vigorous industrialization initiated in the fifties which at the time was founded on import substitution and consequently turned towards the domestic markets. As a result, the newly established industries were located entirely in the centres of consumption. The share in industrial production of the so-called federal district comprising the capital city and its environment amounted to 28 per cent in 1930 and—under the influence of the stepped-up rate of agrarian development—to 27 per cent in 1940, but rose to 40 per cent by 1960, and 43 per cent by 1970. Taking the national average as 100, the index of average *per capita* income in the federal district came to 261, in the state of Nuevo León to 186, in Lower California adjacent to the United States to 313, while in Guanajuto it amounted to a mere 49, in Tabasco to 40, and in Oaxaca to 27.[10] These enormous differences among the regions of the country have led to income polarization, and have considerably inhibited the socio-economic integration of the nation, while as a phenomenon of growth, they represent the concomitants not of a balanced but a forced policy of economic growth.

After having examined the characteristic features of the growth process, let us now briefly analyze the subject-matter of the manufacturing industry—the primary driving-force and leading sector of economic growth.

5.3. THE MAIN FACTORS OF MEXICAN INDUSTRIALIZATION

Tracking back the historical process of industrialization is highly instructive first of all from the point of view of the growth efficiency of the Mexican industry, unprecedented in the Third World. Contrary to the other countries of Latin America, Spanish colonial rule has bequeathed a more developed small industry in contemporary terms to the one-time centre of the Viceroyalty of New Spain which at that time also included the present-day US territories of Texas, New Mexico, Arizona and California. According to contemporary estimates, almost 30 per cent of the production of the viceroyalty had originated from industry,[11] mainly from the textile and metal manufacturing industries. Apart from a few episodes, the anarchy prevailing during the half century following the attainment of independence did not favour the process of industrialization, and even part of the existing small-scale industrial activity has been eliminated by the competition of imported goods. The export boom which had unfolded in the last thirty years of the past century had a stimulating effect on industrialization. In the three decades preceding the outbreak of the revolution industrial production increased by an annual average of 12 per cent.

A distinctive feature of the initial phase of Mexican industrial development distinguishing it from all other Latin American countries is the fact that it covered not only import-substituting branches producing for the domestic market but also the partial processing of exported mining products. Thus, the foundations of Mexican metallurgy have been laid as early as the beginning of the 20th century,[12] and this has left its mark on the character of

---

[10] A. Kuprianov: *Developing Countries: Regional Disproportions in Growing Economies.* Studies on Developing Countries. Budapest, 1973.

[11] T. King: *op. cit.*

[12] C. Furtado: *Economic Development of Latin America.* Cambridge University Press, 1970.

industrialization ever since. On the other hand, owing to the narrowness of the domestic market, the Mexican industry produced virtually for a stratum with high incomes, and relatively great quantities of high-quality commodities at that. In the three decades following the outbreak of the revolution, industrial production increased at an extremely low rate and on the average did not even reach 3 per cent. Employment fell from 874,000 in 1910 to 640,000 in 1940, whereas *per capita* productivity increased nearly threefold by an annual average of 3.7 per cent.[13]

Though the share of accelerated growth—initiated at the beginning of World War II—in the national economy was rather negligible, it could rely on a relatively developed industry as regards its structure, productivity and qualitative standards. The lack of competition, and later on the coming to an end of the wartime and Korean booms had a highly stimulating effect on import-substituting industrialization in Mexico, too. Conforming to the sequence of the complex nature of technological and organizational progress as well as to the problems of capital intensity, import substitution changed over from the light industrial stage in the forties and fifties to the *stage characterized by the heavy industry* in the sixties. The import of non-durable goods reached its peak value in 1952, and subsequently it began to decrease until on the average of 1951–1955 it covered only 6.5 per cent domestic consumption. The import of consumers' durables in turn experienced a peak value in 1955 when imports covered 11 per cent of domestic consumption. 80 to 90 per cent of investment goods came from imports prior to 1945, 70 per cent in the period between 1945–1950, 60 per cent between 1950–1955, 55 per cent between 1960–1965 and 50 per cent between 1965–1970.

Some Mexican economists look on the fifteen years of protectionist industrialization with a rather critical eye on account of its capital intensity and the general increase in the cost level,[14] nevertheless Mexico concentrated her import-substituting efforts, as shown above, primarily on the categories of consumers' goods and semi-finished products whose poorer quality or technological effectiveness, are less important. At the same time, the Mexican government gave proof of much more prudence and progressiveness as regards the category of investment goods which to a great extent influenced the speed, efficiency and quality factors of growth. Imports which are suitable for the transfer of up-to-date technology cover a considerably larger part of domestic demand (being proportionate with the development level of economy) than in Brazil, Argentina or Chile. Characteristically, the value of the net production in the manufacturing industry rose by 120 per cent between 1960 and 1970, within that, however, the rate of increase was 171 per cent in the industries turning out semi-finished products (chemicals, non-ferrous metals, steel and non-ferrous metallurgical products) and 85 per cent in branches manufacturing finished investment goods. Contrary to the conception of most of the countries, where following from the import-substituting ideology industrialization is founded on the manufacturing of finished products, Mexican industrialization forged ahead *vertically* and did not get stuck at the stage of assembling finished products.

A further characteristic feature of Mexican industrialization has been the relatively early emergence of *industrial exports*. In the ABC countries the share

[13] T. King: *op. cit.*
[14] P.G. Reynoso: "Veinticino años de politicana mexicana." *Comercio Exterior*, July, 1972.

of industrial goods in exports has reached 5 per cent in the second part of the sixties, i.e. in the third decade of import-substituting industrialization. In Mexico, however, the time lag is much shorter, and the five per cent threshold has been reached already in the forties. The structure of industrial exports is different, too. The industrial orientation has set out not only from the light industry as in the case of most of the developing countries, but heavy industry, too, has played a considerable role from the very outset. While in the fifties the driving force of exports has been the metallurgical products, their place has been taken by engineering products in the sixties. 10 per cent of industrial exports in 1955 consisted of textiles and about 80 per cent of metallurgical products. In the structure of industrial exports in 1973, however, the share of engineering amounted to 50 per cent, that of textiles to 14 and that of metallurgical products to 13 per cent.[15]

Watanabe[16] who sharply critized the structural development of industrial exports from the point of view of employment, stated that Mexican industrial exports are too much capital-intensive and half of a given order of magnitude creates the same number of job opportunities as for instance in South Korea. Owing to the high costs of protectionist industrialization, Mexican light industrial products are not competitive even on the markets of the neighbouring United States. The argument would be justified if opening up of new job opportunities were the exclusive function of exports. However, this is only one of the aims of industrial exports. Average hourly wages in the Mexican manufacturing industry came to 53 cents in 1970,[17] i.e. amounted to nearly the double of the wages level in the Southeast Asian countries. Due to the higher degree of organization of the working classes and the historically established industrial wages level, the indirect costs and social burdens of wages are higher, too. At the given wages cost standard, the international competitiveness of light industrial exports is undoubtedly going to diminish. In engineering, however, Mexican wages standards are still lower than those of the more developed countries, while competition on the part of the Southeast Asian countries as well as countries less developed than Mexico is almost negligible. Increasing specialization in exporting products of the heavy industry makes it possible to exploit the comparative advantages inherent in Mexico's natural resources and in her extractive industry, and accelerates the unfolding of vertical industrialization. The expediency of industrial specialization is shown also by the increasing dynamism connected with the rise in raw material prices.

By examining the role of industry from the aspect of *external economy*, it becomes clear that presently the chances of rational import substituting are rather limited. In the early seventies investment goods came to five-sixths of imports, and the majority of these capital goods cannot be substituted rationally. Thus, in addition to the expansion of internal trade, it is primarily export orientation which can give a new impetus to industrialization. While in 1960 about 5 per cent of the industrial production was exported, this ratio rose to 14 per cent by 1973. The significance of external markets is by far not determinant, nevertheless their importance is rapidly increasing.

However, the structural development of industry and the external economic

---

[15] *Yearbook of International Trade*, 1955. EIU, 1974/2.
[16] S. Watanabe: "Constraints on Labour-Intensive Export Industries in Mexico." *International Labour Review*, January, 1974.
[17] "Las industrias maquiladores de exportacion." *Comercio Exterior*, April, 1971.

relations are by no means exempt from the *negative aspects of growth*. Structural development was accelerated partly by the favourable natural-geographical endowments, and partly by the relatively strong bargaining positions of the trade unions. In the long run, the wages standards of organized labour have increased to a similar or even greater extent than the national income. In a number of industries the processes of labour force substitution and the supply of technical equipment were accelerated by the power of the trade unions. According to industrial surveys in 1970, taking the average of Mexican manufacturing industry, the turning out of a production value of one million dollars required only 117 workers, and within that this ratio was 235 workers in the less dynamic textile industry, 148 workers in the food industry, whereas only 90 workers in the dynamic electrical industry, 54 workers in the vehicles and chemical industries, and 34 workers in non-ferrous metallurgy. Thus, the productivity of industry, especially that of the heavy industry has been high even as compared to international standards, and has been characteristic not of the developing but of the category of the medium developed countries.

The number of employed in industry has been restricted by the increased capital intensity of industrialization the degree of which has exceeded the present level of development. In spite of the rapid industrialization in the period 1960–1970, industrial employment grew only by an annual average of 3.5 per cent, i.e. the rate of growth equalled that of population increase. Thus industry hardly had any part in absorbing unutilized surplus labour; it offered job opportunities to only 14 per cent of the employed labour fource in 1970.[18] The structural development and relative efficiency of industry resulted in limited *employment* possibilities. Although the number of fully and partially unemployed workers made up only 12 per cent of the total employed labour force, the ratio of hidden unemployment is much higher, because in order to accelerate the development of industry, Mexican economic strategy shifted the burdens of employment disproportionately on agriculture. In 1960 54 per cent of the labour force was employed in agriculture, and in 1970 this ratio still exceeded 50 per cent. There is no other country where, at the given level of general economic and structural development, the weight of agricultural population is even approximately as high as in Mexico.

Another serious drawback of industrialization is *the lack of an independent technological and scientific basis* and/or its insufficiency as compared to the development level of the country and the industry as a whole. Though Mexico rapidly catches up with more developed countries as regards *per capita* production, the gap is widening in the field of technological and scientific capacities not only between Mexico and the industrially highly developed countries but also in the relation of Brazil and Argentina. On the average of the past decade the number of technological and scientific R & D staff per 100,000 inhabitants has been only 6,[19] as compared to 260 in the United States, 250 in the Soviet Union, 150 in Japan, 110 each in Great Britain and the Federal Republic of Germany, 100 in France, and 40 in Italy. The neglect of scientific and technological development is reflected by the low level of research expenditures which originated almost exclusively from public sources: in Mexico these amounted to only 0.13 per cent of the gross domestic product as against

---

[18] S.T. Reyes: "Desempleo y subocupacion de Mexico." *Comercio Exterior*, May, 1971.
[19] M.S. Wionczek: "Las problemas de la transferencia de technologica." *Comercio Exterior*, September, 1971.

the 2–4 per cent average in the developed countries. The negative effect of technological and scientific backwardness can be measured by the fact that accelerated growth has turned the country into a mass importer of foreign technology. Such imported technology in turn has not adapted itself to the special circumstances of the domestic national economy, and at the same time, the lack of an adequate basis for controlling the process of adaption has demanded very high additional costs in foreign exchange. At the end of the past decade, Mexico has purchased foreign technology to the value of an annual 200 million dollars. It cannot be marked off statistically what percentage of payments on technological imports represents the actual price of such technology, and what percentage of it represents in a concealed form the repatriation of the profits of foreign companies. No doubt, however, a significant part of these costs is not connected with the technological development of the enterprises.

Incentives provided by economic policy had a major part in the achievements of Mexican industrialization. Despite the underdevelopment of the internal capital market, 40 to 50 per cent of the financial means of the banking enterprises are concentrated at the central bank and the Nacional Financiera development bank, and this guarantees a relatively favourable background to the realization of the industrialization targets set by the government. Contrary to other developing countries, protectionism has not become a determinant factor in economic development. *Protective tariffs* have been of secondary importance: even by international standards, Mexican customs are not high. Practically only food and textile products, easily to be substituted in domestic markets, have been guaranteed protective tariffs to the extent which has eliminated foreign competition.

The protection of new industries was realized mainly by means of an import licence system; decisions relating to this system were made by the Ministry of Industry and Commerce on the basis of recommendations by inter-departmental committees and often only after having consulted the President. Characteristically, vertical industrialization based on the utilization of domestic resources was facilitated mainly by export duties and licences. Export duties prevented the premature depletion of domestic raw material deposits on the one hand, and attempted to render local processing more economical on the other. Export duties were greatly expanded in the late forties reaching 12 per cent of the value of exports on the average of 1950–1960, and 7 per cent in the sixties.[20] Such duties have been for a long time the main means of the vertical development and export orientation of industry.

Government support for *export orientation* was realized by calling into being a funds for financing the export of finished products in 1963, and by establishing a system of credits for exports in 1964. These measures were followed by the starting of market research, public relations and the establishment of agencies of the Foreign Trade Bank in 1965/66, and finally all industries which settled near the US border and were producing for US markets received subsidies in the form of duty concessions and tax allowances. Towards the middle of the past decade, Mexico had a highly organized institutional structure furthering the export orientation of industry—an unparalleled phenomenon in the Third World.

Apart from the regulators mentioned, all other means of indirect guidance

---

[20] T. King: *op. cit.*

had but very little part in stepping up the process of industrialization. No system of price support was introduced, and owing to abundant capital imports and credit possibilities, the role of general tax allowances—with the exception of industries in the border regions—remained insignificant. According to estimates by B. Katz,[21] taking the average of the period 1956–1963, the total of all preferences granted to the economy amounted to 1.15 per cent of the national product in the United States, whereas to only 0.51 per cent in Mexico. Allowances in the form of raising depreciation rates were generally not accepted by the enterprises fearing control which such allowances implied, and no attempts were made to control industrial prices either. The main characteristic of industrial development has been the relatively simple operation of a regulating system concentrated on allocation and the key factors of external economic growth.

5.4. THE ROLE OF *EJIDOS* IN ECONOMIC GROWTH

The trends of Mexican agrarian development looking back on half a century's past provide an extremely interesting and instructive picture as regards the alternatives of agrarian development and the agrarian problem. It is noteworthy in itself that, as a result of the revolution, it has been Mexico which—as the first nation of all developing countries—has implemented a comprehensive land reform; the partial liquidation of the institutional limitations of agrarian development has markedly accelerated the development of agrarian productive forces. Mexican agriculture ranks among the first as regards its dynamism in the group of the developing countries and also as compared to international standards. The average annual rate of agrarian growth between 1940 and 1973 exceeded 4 per cent. Mexico, a country which formerly had been in need of importing food, became self-supporting in all the basic produce of national alimentation (wheat, corn, beans, rice). Dynamic agriculture created an adequate raw material basis for industrial expansion, and gradually even for exports. The organization of the distribution network also contributed to the fact that, contrary to most of the developing countries, the price level of agricultural produce remained moderate and did not represent a further inflationary factor.

The most interesting aspect of Mexican agrarian development, however, is not simply the land reform or the acceleration of the development of the productive forces. At the same time, the strategy of agrarian development represented the first and most outstanding experiment in the non-socialist world to solve the agrarian problem partly within the framework of collectivism, and to fit the traditional precapitalistic formations into the modern process of growth. The *ejidos*, i.e. the common land and community forms uniting the production relations of pre-Columbian Mexican village communities and medieval Spain, have for a long time attained a highly significant, though changing role in agrarian development.

Soon after their arrival, the Spanish conquistadors had reorganized the *calpullis*, i.e. the village communities of the disintegrating clan society, where in spite of the community of land, industrial and agricultural products as well as slaves had been in private ownership. The forms of land-ownership

---

[21] B. Katz: "Mexican Fiscal and Subsidy Incentives." *The American Journal of Economics and Sociology*, 1972/2.

established in Spain were partly adapted to local conditions, and soon, the two types merged and were transformed into a uniform village community, the so-called *ejido*. During the three centuries of colonial rule, an unceasing fight was waged between the *haciendas*, i.e. the emerging big estates, and the *ejidos*. The central power intervened by means of legislation in favour of the *ejidos* against the feudal oligarchs. Though the protection granted by the weakening central power succeeded in preserving the Indian village communities from feudal expropriation, it could not prevent their gradual losing ground in face of the unfolding system of big estates. At the beginning of the 19th century, in the last decade of colonial rule, only 40 per cent of the total population lived in *ejidos*. Agrarian capitalism unfolding after the attainment of independence, economic liberalism, and especially the economic and political oppression, expropriations and occasional violent liquidations under the dictatorship of Díaz completed the process of disintegration of the Indian community lands. By the turn of the century, 97 per cent of the land was concentrated in the hands of one per cent of landowners, and within that 10 per cent of the big estates was foreign owned.[22]

It was pointed out by the Constitution of 1917 that the pauperization of the agricultural population was connected with the liquidation of the *ejidos*, and therefore it made it possible for the former *ejidos* to regain their lands lost under or expropriated by the dictatorial regime of Díaz. If this did not prove feasible, the Constitution stipulated that the reorganization of the *ejidos* should be furthered by allotting them land from the state reserve plots. However, the new stratum which came to power in the wake of the revolution did not succeed in making it unambiguously clear, whether agriculture ought to be reorganized on the basis of collective Indian ownership or individual small holdings. As a result, the enforcement of the stipulations of the Constitution relating to agriculture made little progress and brought abroad serious contradictions. Whereas the Constitution, as well as certain measures in 1915, 1922, and 1925 strove to further the development of the *ejido* movement, the Colonization Act of August 2, 1923 emphasized the development of independent small-peasant farms. One tenth of the population worked in *ejidos* in 1923, and one fifth in 1934.

In the *ejidos* established during the first two decades following the outbreak of the revolution land was tilled collectively, 85 per cent of the yield was distributed among the members, whereas 10 per cent was allocated for investments, and 5 per cent went for taxes.[23] The initial system of cultivation and distribution had to be modified as early as 1934 on account of the extremely low incentive effects and poor yields. Ever since then, common arable lands have been allocated only in order to form investment funds or meet tax burdens in the majority of the *ejidos*, while the greater part of public lands was distributed among the members. In 1940 the so-called collective *ejidos* made up merely 6.6 per cent of the total *ejido* lands.[24] The *ejido* movement took on a mass character under the presidency of Cárdenas. According to Cárdenas' views the competition of the still existing big estates and/or private estates made it impossible to develop effectively the *ejido* movement, and therefore a considerable part of the big estates (about 17 million hectares) was

[22] M.G. Navarro: Mexico: *The Lop Sided Revolution*. Oxford University Press, 1970.
[23] M.H. Ortiz: *Reflexiones sobre una política agraria*. Investigación económica. ENE. 1961.
[24] S. Eckstein: *El ejido colectivo en México*. Fondo de Cultura Económica, Mexico, 1966.

distributed. Thus, nearly 50 per cent of arable land and 60 per cent of irrigated land was handed over to the *ejidos* by 1940, and more than 60 per cent of Mexico's rural population lived within the framework of the *ejidos*. However, owing to the high rate of population increase and the land-hunger of the peasants, the distribution of a great part of big estates did not furnish sufficient land for the establishment of viable peasant farms. Though the legal minimum size of the individual *ejido* plots was modified several times, the average size of an *ejido* plot was 6.5 hectares in 1960; within that half of the plots was smaller than 4 hectares, and only 15 per cent of them exceeded 10 hectares. In the overwhelming majority of the cases, farming on the basis of the *ejido* system practically did not differ from the activities of independent smallholders who were not members of any *ejidos*; the basic difference was the fact that *ejido* plots could not be sold, mortgaged or leased.

In the decades following the presidency of Cárdenas the economic strength of the *ejido* movement gradually decreased. (See Table 29.)

Table 29

The importance of the ejidos in the distribution of land in Mexico
(per cent)

	Economic units	Area		
		cultivated	irrigated	pasture
1940	50.2	47.4	57.4	22.5
1950	50.4	44.1	49.8	26.8
1960	53.3	43.4	41.6	26.3

Source: R. Stavenhagen: *Land Reform and Institutional Alternatives in Agriculture.* Vienna Institute for Development, 1973/9.

Data indicating the decreasing share of the *ejidos* since 1940 indicate two processes. The share of the *ejidos* has been reduced mainly on irrigated land which is cultivated by more up-to-date methods and the yield of which is higher, too. At the same time, the percentage of pastures and woods within the total of the *ejido* lands has increased. Thus, a certain relative increase in extensive farming can be seen in the development of the *ejidos*. As regards the sustenance of the agrarian population, the role of the *ejidos* is on the upgrade, whereas the trend of the so-called minifundization, i.e. the fragmentation of holdings also manifests itself most conspicuously in the *ejidos*.

The reasons for the losing ground of the *ejido* movement are the following: first of all the gradual shift to the right in the political leadership of Mexico since 1940; the changes in the political issues highlighted; the cessation of the privileged situation of the *ejidos*, and the granting of economic preferences to individual small and medium farmers as well. In addition to direct causes of political and economic-political character, there are, however, a number of other factors, too, originating from the nature of the Mexican growth process and from some objective technological processes.

Minifundization which has become a general phenomenon as a consequence of the spread of the so-called *ejidos* based on individual cultivation has certainly led to confusion in production. Neither the adoption of effective agrotechnics, nor any increase in productivity, incomes, saving or investment capacity could be achieved on the individually cultivated, almost dwarf-sized plots. Though the government has organized as early as 1926 a network of specialized banks in order to guarantee the *ejidos* adequate credit and organization facilities, the sources of credit and the scope of technical aid did not prove sufficient to revitalize the *ejido* sector of agriculture. The *ejidos* were unable to pay back 27 per cent of the credits granted between 1936 and 1961, and because of the great risks involved, their supply of credits gradually deteriorated.

The rigid institutional framework established for the protection of the one-time village communities also impeded the increase of economic effectiveness. Mexican economy as a whole, and within that the greater part of the agrarian sector has been highly competitive in the past two decades, whereas the sector of the *ejidos* is characterized by the elimination of restriction of the competitive elements. Dualistic economic environment in itself usually results in the losing ground of the weaker sector. While the prohibition of the sale of plots certainly hinders a concentration of landownership within the *ejidos*, it preserves the less efficient small holdings and does not offer any possibility to the expansion of the more efficient producers. On the other hand, though the statutes of the *ejidos* stipulate the election of a three-men committee for dealing with matters of common interest, this committee is headed by a president with extraordinarily great personal powers (land distribution, representation, management of the common propriety, etc.) who is responsible to the state authorities. The wide-ranging sphere of authority of the president offered immense possibilities to the handling down and transmittal of traditional Indian *caciquism*, since only the agrarian law of 1971 stipulated rather belatedly some measures in order to limit the personal powers of the president and the possibilities of abuse inherent in this sphere of authority.

In a homogeneous socio-economic environment, the lack of harmony between socio-political and economic optimums of the Mexican growth model would have not raised serious problems. However, the dynamism of the non-*ejido* sectors, and the restrictions serving the socio-political security within the *ejidos* have necessarily resulted in the steady secession of the most successful members of the *ejidos* and their migration to the private sector. As a consequence of the continuous loss in qualitative human resources, the gap between the competitiveness of the *ejidos* and the private sector of agriculture unceasingly widened, and the positions of the private sector grew ever stronger. According to Flores de la Pena,[25] Lira,[26] and Stavenhagen,[27] thirty years after the land reform which had crushed the power of the latifundia, there gradually emerged the modern, industrializing neo-latifundia organized on a commercial basis, as well as the highly specialized, commodity producing, independent small farms the productivity standards of which by far outstripped those of the average *ejidos*. At the end of the past decade 0.5 per cent of the estates turned out 32 per cent of the total agricultural production, whereas the share of 50 per cent of the estates came to a mere 4 per cent.

The dynamic development of the capitalist sector of agriculture has undermined the *ejido* movement in other respects, too. During the past twenty years the internal structure of the *ejidos* has undergone a considerable differentiation.

---

[25] H.F. de la Pena: *Crecimiento demográfico, desarrollo agricola, desarrollo economico.* Investigation Economica ENE, 1964.

[26] Lira: "Desarrollo agricola vs. desarrollo industrial." *Revista de Economia*, 1964, XXVII.

[27] R. Stavenhagen: *Social Aspects of Agrarian Structure in Mexico*, 1966. Cited by the *American Economic Review* June, 1971.

Besides the traditional self-supporting village communities, there came into being—especially in the central and northern regions of the country—up-to-date, mechanized, large-scale cooperative farms operating virtually like capitalist cooperatives. At the same time, the illegal leasing of the individual *ejido* plots has become a widespread practice, and hardly anything can be done against it. Stavenhagen cites the data of a survey in 1966, according to which in the cotton growing zone of Michoacán 55 per cent of the members have rented their plots to enterpreneurs who were not members of the *ejidos*, whereas in the vicinity of the capital the observed ratio was 8 to 30 per cent, and in the valley of the Yaqui illegal leasing reached 25 per cent.[28] The institution of the *ejidos* is weakened to a considerable extent partly by the *ejidos*' becoming capitalist enterprises, and partly by the fact that as a result of renting plots, there has emerged a strong stratum of agrarian entrepreneurs. Thus, the actual economic influence of the *ejidos* is certainly much less than indicated by the statistical averages.

The socio-political conception of the Mexican land reform no longer exerts such a great influence, and in the past decades the importance of the *ejido* movements also decreased. As a result, the political and economic circles of the right and the extreme right are tempted to a growing extent to reject the *ejido* movement as a total failure. It is often emphasized that the Mexican revolution has sacrificed agriculture and the rural districts, by giving priority to industrialization and the cities. However, it would be proper to speak about the failure of the *ejido* system only if the movement were to be judged solely on the basis of its direct economic performance or if it were to be regarded as a historic alternative to capitalist agrarian development and a means of socialist transformation in agriculture. Analysis taking into consideration only economic aspects would seem, however, rather a one-sided approach, whereas expectations of a socialistic development would seem unhistorical owing to the external limits originating from the proximity of the United States. The actual role of the *ejidos* should not be evaluated without taking into account the fifty-years' socio-economic development process of Mexico.

In the period under review, the *ejidos* did not represent a realistic alternative to the capitalist development of agriculture: their revitalization served mainly political and social objectives, and was not regarded as a direct means of agricultural development. During the initial period and the consolidation of the revolution, the revival of *ejido* movement proved to be an expedient device to mobilize the revolutionary energies of the peasantry still harbouring romantic nostalgies for the village community. The *ejido* movement undoubtedly restrained the disintegration of the peasant class which is a regular process inherent in accelerated growth. The land-hunger of Mexican peasantry, the large-scale agrarian over-population and the increase in population would not have made possible a simultaneous and dynamic development of agriculture as a whole. The so-called American course of capitalist development of agriculture has been excluded from the very outset from the feasible possibilities on account of the restricted amount of arable land available. The chances of the dwarf holdings to become large-scale, efficient, commodity producing agrarian enterprises were nil from the very outset. A capitalist development based on big estates could not be tolerated by the new revolutionary power

---

[28] R. Stavenhagen: *Land Reform and Institutional Alternatives in Agriculture*. Vienna Institute for Development, 1973/9.

for political reasons, whereas the unlikely American course offering results only in the distant future would have led to the rapid disintegration, pauperization of the peasantry, and eventually its influx to the cities. However, the capital-intensive, labour-saving character of Mexican industrial development (which is partly explained by the strong bargaining positions of the trade unions) has rendered the absorption of agricultural surplus population by the industry hopeless from the very beginning.

Under the prevailing conditions, the *ejidos*, relying on centuries old customs, traditional organizational and consciousness forms held out promises for the most efficient institutional framework and the lowest social price to be paid for the voluntary staying in the villages of the peasantry and for the postponement of the mass-scale influx to the cities. Thus, by means of keeping the population in the villages, the *ejidos* afforded a prolonged respite for the establishment of the modern sector and institutions. Without the *ejido* movement, Mexico would have hardly succeeded in reaching an 800 dollar *per capita* national product when the agrarian population amounted to more than 50 per cent of the total population.

On the other hand, the peasantry concentrated in *ejidos*, and thus within easy reach of state intervention, offered to the other sectors of economy extremely low-priced manpower reserves and simple kinds of agrarian products turned out at very low costs, and therefore became itself an indirect carrier of accelerated development. Thus, a peculiar type of functional dualism came into life in Mexican agriculture. The bulk of both export and domestic commodity funds is supplied by the private sector, mainly by the big estates. To the *ejidos* there falls the duty of reducing social and political imbalances and the upholding of the atmosphere of security sought for traditionally by the less dynamic strata of the population, as well as the absorption of manpower reserves.

Despite their decreasing economic weight and the process of becoming capitalistic enterprises, the *ejidos* still represent a significant and viable non-capitalistic formation (even compared to international standards) within a growth model which is developing virtually in a capitalistic way. Naturally, the *ejidos* cannot be exempt from the negative aspects of the socio-economic environment. As the world of the *ejidos* is alien to capitalism, the above aspects are even more harmful. However, without the *ejido* movement Mexico would have probably paid a much higher socio-political price for the accelerated rate of growth.

Finally, in spite of the present decline of the *ejidos*, the rather distant possibility should not be left out of consideration that in case of a radicalization of the socio-political development, the *ejidos* could regain their important role both in the political mobilization of the peasantry, and on a higher level of development of the productive forces, even in the socialist transformation of agriculture.

5.5. THE PROBLEM OF EXTERNAL ECONOMIC DEPENDENCE

External economic dependence, a characteristic feature of all underdeveloped countries manifests itself in a peculiar complex form in Mexico. It is not without interest to examine how and to what an extent the gaining momentum of the growth process has influenced external economic relations.

The Mexican Model of Balanced Economic Growth

The dimensions of the country as well as its historic development explain the fact that Mexico's joining in the international division of labour and also her dependence on world economy have not been so extensive as in other Latin American countries. In 1910, on the eve of the revolution, the share of imports in the gross national product amounted to only 8.3 per cent, i.e. it did not reach even half of the Latin American average.[29] During the fifty years since the outbreak of the revolution, the import-sensitivity of the economy has continued to decrease, and in 1965 it barely reached 6 per cent. Only as a result of the world economic orientation unfolding in recent years did it increase again to 7 per cent. Thus, similarly to other big countries, despite her relatively open character as regards world economy, *Mexico is dependent on international trade only to a small extent.*

For reasons enlisted above, the commodity structure of foreign trade as compared with the developing countries is highly diversified: Mexican economy has never been characterized by any kind of monoculture. Prior to the revolution, export earnings had been made up by precious metals, jute, caoutchouc, copper, lead, coffee and other products of lesser importance. In the period 1910–1940 there began the export of petroleum, whereas from 1940 on agricultural products, and beginning with the early sixties manufactured products have reflected the progress of *export diversification*. The export earnings of Mexico have never been highly dependent on the sales possibilities of a few products on the world markets, and owing to the higher degree of diversification and/or the smaller part of foreign trade in the national economy, the balance of the economy was less influenced by world economic trends than in other developing countries. One of the most interesting features of the Mexican growth process and the development of the external economic sector respectively, is the high structural development of exports. The share of industrial goods in exports was 6 per cent in 1955, 24 per cent in 1965, 40 per cent in 1970, and 52 per cent in 1973.[30] With the exception of the East Asian city-states, there has been no other developing country where the commodity structure of exports has developed so favourably, moreover the achievements of Mexico may be regarded as outstanding even compared to the group of belatedly industrialized countries. The structure of exports is thus characterized not only by diversification but also by the high ratio of industrial products, which are the pioneers of structural development and which are most in demand on international markets.

External economic dependence, a criterium of backwardness, is negligible both on the level of macro-economy and from the point of view of the commodity structure of foreign trade. Its most conspicuous manifestations are *the one-sided structure of geographical relations, capital relationships and technological and political dependence connected with these.*

The most vulnerable point of Mexican economy and especially that of the external sector is the high degree of dependence on the neighbouring great power, the United States. The enormous difference in the level of development and economic potential of the two countries indicates the fact that such dependence can only be unilateral. The Euro-centric foreign trade orientation of the 19th century has been gradually replaced during the first decades following the revolution and World War II by the predominance of the United

---

[29] C.W. Reynolds: *Changing Trade Patterns and Trade Policy in Mexico.* Stanford, 1967.
[30] BFA. October, 1973 Mexico; Wirtschaftsstruktur, EIU, 1974/2.

States. The share of the United States in Mexican exports amounted to 68 per cent in 1938, 77 per cent in 1948, 76 per cent in 1960, and 59 per cent in 1973, whereas the import ratios of the corresponding years came to 57, 86, 64 and 61 per cent.[31] This high degree of dependence which had inevitably developed during World War II and in the post-war years could not be diminished by the growing role of Western Europe and Japan in international trade or by the deepening of regional cooperation. Even the energetic measures by Mexico proved insufficient in cutting it back below the pre-war level which was dangerously high anyway.

The one-sided dependence in the field of capital relationships raises even more serious problems than the considerable dependence of foreign trade on certain geographical relations. The influx of foreign capital has started as early as in the chaotic years following the end of colonization, and the French intervention in the sixties of the 19th century also made the protection of foreign capital interests its pretext. The mass-scale influx of foreign capital, however, had taken place in the three decades preceding the revolution. For instance in the manufacturing industry there had been no noteworthy foreign investments before 1880, while 29 per cent of the total capital formation in the industry in the period 1886–1910 came from foreign sources. According to the calculations of Navarrete, in the first decade of the 20th century this ratio increased to two-thirds of total investments. At the outbreak of the revolution, the dependence of Mexico on foreign capital reached an all-time high compared to international standards: foreign capital investments were estimated 1.7 billion dollars, i.e. one and a half of Mexico's national product and nearly fifteen times the value of her exports at that time. One-third of foreign capital was invested in the railways, a quarter in mining, and the remainder went to the manufacturing industry, the petroleum sector, public works and trading enterprises. Within the framework of this liberal capital import policy, however, the Díaz regime strived to establish a certain balance between the positions of the great powers. Thus, the share of the United States in foreign capital investments was 38 per cent, that of Great Britain 29 per cent, and that of France 27 per cent.[32]

It is easy to understand that such an unprecedented degree of dependence on foreign capital inevitably led to the strengthening of the anti-imperialist elements in the ideology of the Mexican revolution. In practice, however, no definite stand has been taken against foreign capital for more than a quarter of a century. Moreover, the volume of foreign capital investments even increased to a certain extent until 1929, since the Mexican government attempted to reduce the risk of US intervention by upholding its liberal policy in regard to foreign capital. After a long delay, the first radical step against foreign capital was taken only in 1937/38, at a time when the retaliatory capacity of the leading capitalist countries was limited because of the imminent Second World War, and the internal political and institutional basis of the new Mexican power was consolidated. The nationalization of the railways and later on of the petroleum industry as well as the land reform directly affected the positions of foreign capital, and indirectly started a large-scale flight of capital. Foreign capital investments fell to less than 600 million dollars in 1939, equalling two-fifths of the national product and amounting to four

---

[31] *U.N. Yearbook on International Trade Statistics*. EIU, 1974/2.
[32] T. King: *op. cit.*

times the volume of Mexico's exports in that year. According to estimates,[33] following these nationalizations 10–16 per cent of Mexico's total capital stock, 52 per cent of transport and public works, 40 per cent of the extracting industry, and 6 per cent of the manufacturing industry remained in foreign hands. Thus, in spite of its heavy losses, foreign capital has succeeded in maintaining strong bridgeheads in Mexico.

From that time on, the dependence on foreign capital has been linked increasingly to the unfolding of general economic growth and development strategy respectively. The endeavour to eliminate dependence on foreign capital has lived on as a heritage of the revolution up to this date. Actual measures, however, are characterized by extreme caution, gradualness and the decisive role of economic aspects, as clearly indicated by the nationalizations and so-called Mexicanizations in 1960. As a consequence of the nationalization of the oil industry, Mexico lost a considerable part of her export earnings, whereas imports were greatly and continuously increased by the stepped-up growth process initiated in the forties, and the traditionally favourable foreign trade balance has shown an increasing deficit. Exports covered imports to the rate of 142 per cent on the average of 1935–1939, 98 per cent in 1940–1944, 73 per cent in 1950–1954, and only 60 per cent in 1970–1973.[34]

The enormous trade deficits and the burdens of debt services could not be balanced by the incomes from tourism and the remittances of the guest workers. Since the political leadership whose shifting towards the right began in the forties, did not want for political reasons to enforce radical restrictions, to introduce extensive control mechanisms or to take measures in order to shift the burdens of external disequilibrium on the domestic economy, the maintenance of the growth environment of a relative stability and balanced economic life required increasing capital imports and an economic policy furthering that aim. On the other hand, in addition to the points of view of the balance of payments, the technology and know-how needed for the technical updating of the economy often became possible only with the help of foreign direct capital. Thus, Mexican attitudes towards foreign capital have been characterized for decades by a peculiar dualism: *ideologically, Mexico proclaimed herself a bitter enemy of foreign capital, yet at the same time, the government accomplished a practical cooperation with foreign corporations*. The truism according to which Mexico's heart is on her left, while her wallet is on her right, is absolutely to the point.

The extent of dependence on foreign capital is very difficult to establish statistically since the data supply is rather inadequate. The nationalizations of 1937/38 affected mostly the positions of British and French capital, and because the shortage of capital during and after World War II further weakened the positions of European capital, the relatively equal dependence has become quite lopsided: more than four-fifths of direct capital investments have originated from the United States. While the sum total of U.S. direct capital investments amounted to only 414 million dollars in 1950, it increased to 980 million in 1960, 1182 million in 1965, and reached a round 2 billion dollars by 1972.[35]

The volume of loan-capital is indicated by current debts: in 1950 these

---

[33] E.V. Blanco: *La industria de transformación, 50 años de la revolución Mexico.* 1967.
[34] *U.N. Yearbook on International Trade Statistics.* EIU, 1974/2.
[35] Various issues of *Survey of Current Business.*

came to 650 million dollars, at the end of 1970 to 2.9 billion dollars, and by the end of 1972 to 4 billion dollars.

It is borne out by the available data that the capital imports of Mexico developed extremely dynamically, and cooperation with foreign capital was apparently by no means restricted by ideological differences. On the international capital and credit markets Mexico has been enjoying an extraordinarily high financial standing over the past two decades. Let us now examine the extent of dependence on capital imports under such circumstances.

The rate of increase in direct capital investments has been slower than that of the national product. While the annual import of direct capital had amounted to 17 per cent of all private capital investments on the average of 1940–1945, it came to a mere 10 per cent on the average of the period 1960–1965. In recent years both imported loan-capital and direct capital increased much faster than previously, the direction of such imports, however, did not change basically. Dependence on capital imports seems to be therefore to diminish in the long run. Direct long-term capital investments (about 3 billion dollars) and foreign debts made up only one-sixth of the national product and less than double of the annual commodity exports in 1972, i.e. the relative extent of dependence has shown a decreasing tendency since 1940 in spite of a more definite world economic orientation. The annual capital import amounted to 3 per cent of the national product in the early seventies and is proportionately much less than in most of the developing countries.

However, the slow though gradual decrease in the dependence of the national economy on capital imports is not at all a homogeneous process.

Foreign capital is gradually ousted from the productive branches and the infrastructure by the policy of Mexicanization, nevertheless this policy has to rely on imported new technologies and is therefore unable to take as resolute steps in the industrial sector as it would be necessary. Foreign capital compelled to relinquish its positions in certain sectors of the economy was channelled into the manufacturing industry which has become the main target of the penetrating new capital. According to estimates based on surveys, 6 per cent of the capital stocks of the Mexican economy and 20 per cent of all industrial capital investments are controlled by foreign capital whose positions are especially strong in the most up-to-date, pioneering branches, such as engineering and the vehicle industry (100 per cent), the chemical industry (78 per cent) and the metal industry (68 per cent).[36]

Thus, it is a regrettable fact that structural development reflects an increasing intertwinement of the Mexican economy with foreign, mainly North American capital. Quite a few Mexican economists have pointed out that the technological basis of the economy has become increasingly dependent on the technological standards of the United States.[37] Despite the diminishing quantitative elements of subordination, a complex dependence has developed in the most up-to-date sectors the effect of which makes itself felt to a lesser degree and in a more indirect way than before: nevertheless it has a stronger influence on the nature of the growth process. The problem is still more aggravated by the technological and structural dependence having become stronger on the neighbouring United States. This fact fundamentally sets a limit to all efforts aimed at increasing international manoeuvrability and permits

---

[36] EIU 1974/2.
[37] M.S. Wionczek: "La transmission de la technologia." *Comercio Exterior*, 1968/5.

only a very cautious progress. On the other hand, it may turn the modern sector, which is increasingly intertwined with foreign capital, against the social heritage of the revolution and the requirements of intensified internal integration and manoeuvrability in the field of external economy. This process has not reached a critical point yet, but may become a serious danger in the future.

As a consequence of the marked dependence on the economy of the United States, the economic progress of Mexico, which has accomplished an extensive opening up to world economy only a few years ago, has naturally been affected by the world economic recession. At the same time, the sudden rise in oil prices made the exploitation of oil fields which had been uneconomical on the basis of the previous price level appear attractive to Mexico, a traditional producer of oil, though she imported about one-tenth of her total oil requirements. Moreover, at an early stage of exploration works, huge oil fields were discovered in 1974, which are of considerable international significance as well. As a result of the rapid development of these fields, Mexico succeeded in earning a net 150 million dollars from oil exports already in 1975 as against the unfavourable balance of the previous year.

The beginning of the oil boom has released the balance of payments of Mexico at a critical time, partly by directly increasing foreign exchange export earnings, and partly by indirectly improving the international solvency of the country which rendered it possible to choose among various offers of loans in spite of the considerable degree of indebtedness. Finally, the dynamism of the Mexican economy which has developed without interruption for the last thirty years, also helped to retard the negative effects of world economic developments.

As a result of all the main factors mentioned, Mexico was not at all affected by the consequences of US recession in 1974, and despite a poor harvest, her gross national product still grew by 6 per cent. The unfavourable world economic effects reached Mexico only in 1975; at the time, however, the upward trend in the economy of the neighbouring United States contributed to warding off the disadvantages of retarded growth.

The lessons to be drawn from the Mexican course of economic growth are first and foremost that institutional reforms have to be made use of in an efficient way, that by taking into account historical traditions, as well as national endowments and international conditions, growth tends to accelerate. At the same time, attention is called also to the fact that a high price is to be paid for growth where efficiency is the first priority. Though the extent of external dependence generally diminishes, the presence of the United States is increasing in all the modern branches of the economy.

The new Capital Import Act of 1973, the increasing reliance on loan-capital imports, especially on the European and the Japanese money-markets, Mexico's growing activity within the group of non-aligned and developing countries, the trade agreements signed with the Soviet Union and China, as well as the organization of Latin American cooperation directed against the US—Brazilian axis are all signs indicating strengthening efforts to reduce dependence on the United States. However, owing to the dependence of the dynamic sectors on the United States, one can expect no great results along these lines in the foreseeable future.

INDEX

Agarwall, 79
Agreements with the CMEA countries, 186
Agricultural productivity, 62
Agricultural self-sufficiency, 28
Agricultural growth, 60
Ahmad, J., 130
American power politics, 144, 194
Andean Group, 79, 115, 132, 133, 142
Andean integration, 135, 141
Andean Pact, 229, 232
Arciniegas, G., 19
Arensberg, C. M., 20
Argentina, 28, 50, 54, 56, 60, 62, 65, 67, 68, 70, 71, 73, 77, 79, 83, 86, 87, 89, 92, 93, 94, 99, 104, 106, 109, 110, 119, 123,, 126, 127,131, 136, 137, 138, 141, 143, 172; agrarian protectionism 173; agricultural production 179, corn exports, 189; employment, 180; export, 191; external economic dependence, 231; external economic strategy, 168; foreign capital interests, 182; immigration, 175; importsubstituting industralization, 175; industrialization, 174, 178; investments, 174; Perónism, 176; Perónist development policy, 177; Perón's development concept, 191; protectionism, 177; public capital export, 94
Asian mode of production, 27
Australia, 62

Balassa, B., 118
Baudin, L. 15
Behrman, I. N., 120
Bolivar, S., 125
Bolivia, 78, 88, 126, 127, 131, 136, 143
Boxer, C. R., 41
Brandenburg, F., 52
Brazil, 40, 50, 56, 58, 61, 62, 67, 71, 73, 75, 77, 83, 86, 87, 88, 89, 92, 93, 95, 99, 108, 111, 118, 119, 121, 123, 126, 127, 131, 138, 140, 143, 144, 214, 215, 217, 229; accelerated growth 219; agrarian sector, 207; agricultural development 202, 213; budgetary subsities 199; capital imports, 205, 216, 219; coffee, 43, 203; coffee sector, 206; Development Plan, 211; development strategy, 192; economic development, 195; "economic miracle", 217; ecomonic policy, 104; economic role of the state, 196; export boom, 42; export financing, 199; export orientation, 198; external economic imbalance, 210; external economic policy, 215; foreign capital, 194; foreign trade, 206; immigration, 43; imports, 209; import technologies, 200; indebtedness, 210; industrial exports, 199, 207; industrialization, 219; inflationary economic policy, 210; investments, 195; 213, Japanese-Brazilian relations, 217; manufacturing industry, 202; mining, 58, 72, 90; ,,outward-looking" external economic policy, 214; protectionist-type development, 199; rise in industrial production, 202; Brazilian sugar, meat and pulp industries, 216; US-Brazilian axis, 214; sugar-plantation, 42

Calmon, P., 43
Cambon, P., 13
Campos, R., 193
Capital and technology imports, 35
Capital imports, 31, 89, 106
Capital investments, 90, 92, 231, 262
Cardenas, L., 37
Caribbean countries, 28, 77
Cartagena Agreement, 101, 132
Central American Common Market, 79, 115, 126, 130, 133, 137, 139
Centre vs. periphery relationship, 29
CEPAL (UN Economic Comission for Latin America) 110
Chile, 28, 33, 50, 54, 57, 58, 60, 62, 65, 70, 71, 78, 89, 94, 109, 118, 119, 126, 127, 136, 138, 141, 144
China, 75, 216
Cochrane, J. D., 131
Colmeiro, M., 21
Colombia, 28, 56, 71, 77, 83, 86, 92, 93, 108, 110, 126, 127, 141, 146; agriculture, 48; coffee, 224, 229; export orientation 230; external economic dependence 231; foreign capital, 222; import substituting growth 225; monoculture, 222, 230; role of foreign capital, 231
Colombian Federation, 125
Colonial heritage, 32
Common Market, 82, 86, 87, 127, 200
Costa Rica 54, 71, 78, 127, 130, 136
Cotler, J., 154
Crevea, R. A., 19
Cuba, 129

Delgado, C., 158
Dell, S., 114
Development strategies, 104, 173, 188
Diaz Alejandro, 225, 227
Dominica, 77, 86
Donell, G. A. 177
Dorfman, 245

Eckstein, S. 254
Ecuador, 28, 71, 72, 86, 109, 127, 131
EEC, 117, 187
EFTA, 87, 119, 127
El Salvador, 86, 126, 133
Elson, R. A., 228, 229
Encomiendas, 26
Energy crisis, 107
Engels, F. 152
Export of industrial products, 81, 230, 234
Export of raw materials, 146, 230
Export orientation, 31, 38, 81, 92, 97, 106
Export oriented industrialization, 66

Federal Republic of Germany, 251
Ferrer, A., 173
Foster, W. Z., 25, 30
France, 251
Frank, A. G., 19, 20
Free Trade Association, 132, 133, 135, 142
Freyre, G., 42
Furtado, C., 42, 61, 172

Garcia, Ponce, I. G., 153
GATT, 131
Germani, G., 175, 183
Gerol, E. H., 15, 17
Gerschenkron, A., 35
Glade, W. P., 32, 89
Godelier, M., 15
Great Britain, 200, 251
Guatemala, 127
Gunther, J., 175

Hacienda, 27
Haiti, 54, 71, 77, 86
Haring, C. H., 19, 24
Harrod, R., 114
Herrera, F., 140
Hispanian colonial model, 27
Honduras, 127, 130, 133, 136, 137

IMF, 132
Import-substituting growth, 44, 225
Import-substituting industrialization, 38, 59, 146
Import-substitution, 36, 64, 70
Inca mode of production, 17
Income distribution, 56, 123, 213
Inflation, 52, 78, 138, 196, 235

International division of labour, 39, 64, 75, 90, 100
„inward-looking" growth, 36, 63
Italy, 251
Ivory Coast, 229

Jamaica, 68, 116
Japan, 87, 110, 200, 215, 217, 251

Katz, F., 14
Kádár, B., 66, 117, 174
Kenya, 118
Kerekes, Gy., 60
King, T., 241, 249, 260
Kornai, J., 244
Kozma, F., 115

LAFTA, 79, 115, 118, 127, 129, 130, 139, 14g, 185
land reform, 62, 160
Latin America, 69, 75, 84, 88, 125
Latin American Economic System (SELA), 110
Lieuwen, E., 153
Lumbreras, L. G., 15

Macera, P., 27
Manchester, A. K., 40, 42
Mariategui, J. C., 15
Marx, K., 16, 18, 21, 22, 23, 152
Marxist model of development, 21
Métreaux, A., 15
Mexico, 34, 56, 60, 62, 67, 68, 71, 73, 75, 78, 86, 89, 92, 93, 95, 99, 106, 108, 109, 111, 118, 119, 121, 123, 127, 131, 136, 137, 141; agrarian development 253; agriculture, 245; budgetary policy, 241; capital controlling policy of Mexico, 102; chemical industry, 262; dependence on foreign capital, 261; distribution policy, 241; economic growth, 242, 244, 249; *ejidos*, 16, 253, 254, 255, 256, 257, 258; employment, 251; engineering and vehicle industry, 262; export diversification, 259; export orientation, 252; external dependence on the US, 238; external economic dependence, 258; mexican growth model, 238; 256; import substitution, 240; income distribution, 247; industrialization, 246; land reform, 35, 257; latifundia, 239; liberalism, 34; manufacturing industry, 245; metal industry, 262; multinational enterprises, 121; oil, 263, 261, 245; Mexican revolution, 30, 34, 239; trade balance, 246; multinational corporations, 99
Myrdal, G., 115

Navarro, M. G., 254
New hierarchy of American power politics, 144
Nicaragua, 127, 136

Index

Organization of the American States, 128
Ortega y Gasset, 20, 24
Ortiz, M. H., 254
Ots Capdequi, 19
Outward-looking, 31, 32, 34, 35, 222

Panama, 54, 68, 71, 78, 86, 109
Pan American Union, 128
Paraguay, 86, 126, 127, 131, 137, 143
Paz, P., 28
Pearson, H. W., 20
Pena, H. F., 256
Pena, I. P., 33
Perón, J. D., 37
Peru, 34, 57, 58, 72, 77, 78, 86, 88, 94, 109, 110, 126, 131, 136, 143, 146, 151; agricultural output, 163; capital import, 167; copper sector, 160; export orientation, 166; exports, 168; foreign trade dependence, 165; growth, 163, 164; income distribution, 156; industry, 161; investments, 164; land reform, 160; manufacturing industry, 163; oil, 167; services, 163; *ayllus*, 16; petroleum sector, 90
Poland, 216
Polányi, K., 20
Population increase, 48
Portugal, 41
Prado, J., 157
Prebisch, R., 132, 162
Price explosion, 208, 235
Protectionism, 89, 120

Quijano, A., 162

Raw material crisis, 78
Regional cooperation, 137, 140
Reynolds, C. W., 259
Reynoso, P. G., 249
Roel, P. V., 15, 29
Role of Latin America in world trade, 75
Rosenblatt, A., 18

Samuelson, P., 245
Service sector, 67
Shanahan, A. W., 19

Sloan, J. W., 131
Smithsonian agreement, 86
Socialist countries and Latin America, 88
Solow, R., 245
Soviet Union, 75, 216, 251
Spanish colonization, 19, 26
Spanish model, 24, 27
Stanislawski, A. K., 40
Stanley, J., 21
Stein, B. H., 21
Structural development, 49, 107, 121, 244
Sunkel, O., 28

Tariff protection, 118
Teigero, J. D., 228, 229
Tourism, 68
Tőkei, F., 18

UN Economic Comission for Latin America (CEPAL), 110
United States, 75, 82, 84, 89, 94, 95, 140, 151, 214, 243, 251; big corporations, 129; capital, 231; capital investments, 222; dependence on the US, 85, 110, 147; policy in Latin America, 84, 144, 194
Urquidi, V. L., 126
Uruguay, 54, 68, 70, 71, 77, 79, 86, 94, 126, 127, 136, 138

Vargas, G., 37
Veliz, C., 24, 25
Venezuela, 28, 56, 58, 60, 72, 75, 77, 86, 92, 99, 109, 111, 123, 127, 131, 136, 143, 146
Viner, J., 114

Watanabe, S., 250
Weber, M., 20
West European countries, 110
West Germany, 215
West Indies Federation, 116
Wionczek, M. S., 135
Wittman, T., 16, 21, 26
World War II, 249